Computation with
Recurrence Relations

Computation with Recurrence Relations

Jet Wimp
Drexel University, Philadelphia

Pitman Advanced Publishing Program
BOSTON · LONDON · MELBOURNE

PITMAN PUBLISHING LIMITED
128 Long Acre, London WC2E 9AN

PITMAN PUBLISHING INC
1020 Plain Street, Marshfield, Massachusetts 02050

Associated Companies
Pitman Publishing Pty Ltd, Melbourne
Pitman Publishing New Zealand Ltd, Wellington
Copp Clark Pitman, Toronto

First published 1984
© Jet Wimp 1984

AMS Subject Classifications: (main) 65D15
(subsidiary) 65D99

Library of Congress Cataloging in Publication Data
Wimp, Jet.
  Computation with recurrence relations.
  (Applicable mathematics series)
  Bibliography: p.
  Includes index.
  1. Functional differential equations. 2. Point
mappings (Mathematics). 3. Approximation theory.
I. Title. II. Series.
QA431.W64 1984   515.3'5   83-13226
ISBN 0-273-08508-5

British Library Cataloguing in Publication Data
Wimp, Jet
  Computation with recurrence relations.—
  (Applicable mathematics)
  1. Recursive functions
  I. Title   II. Series
  511.3   QAG.615
  ISBN 0-273-08508-5

Filmset and printed in Northern Ireland at The Universities Press (Belfast) Ltd, and bound
at the Pitman Press, Bath, Avon.

Dedication

Dedicated with profound gratitude to that loving and anonymous fellowship whose gifts are spirit, sanity, life itself.

Acknowledgement

I wish to thank Drexel University, which has been generous in its support of the writing of this book, and Alison Chandler, whose typing skills resulted in such a beautiful manuscript. Janet Heilman proofread both manuscript and page proofs, and her expertise saved me from much of the travail that usually attends the production of a book. Mark D. Cain has developed the graphics illustrating strange attractors.

Contents

Preface

The purpose of this book is to present applied mathematicians, numerical analysts, engineers, physicists and computer scientists with an in-depth study of that vast body of computational techniques based on the use of recurrence relations.

These methods can be traced back to the dawn of mathematics. The Babylonians used such a technique to compute the square root of a positive number, and the Greeks to approximate π. Much later, Lagrange (1789) used a computational scheme based on a two-dimensional non-linear recurrence to compute an elliptic integral.

Current interest in the subject, though, can be attributed to some almost offhand remarks made by J. C. P. Miller in an introduction to a table of Bessel functions he had compiled for the British Association for the Advancement of Science. This book, which was published in 1952, coincided with the vaulting growth of large-scale digital computers; although Miller's work was prefigured by some observations of Lord Rayleigh (1910) on the calculation of spherical Bessel functions, it was only with the advent of large-scale computers that the great promise of these algorithms could be brought to fruition.

What Miller perceived was that in a second-order linear recurrence which has solutions sufficiently differentiated asymptotically, there is a solution that may be uniquely characterized by one initial value and a knowledge of its growth. This led to an algorithm for computing certain solutions of the equation which required only a scant knowledge of their pointwise values.

Very rapidly Miller's technique was generalized to a whole host of computational problems. By now, virtually every important special function of mathematical physics may be computed this way. In addition, similar methods have been developed for computing the zeros of functions, eigenvalues of certain differential operators, and the coefficients for the expansion of functions in Taylor's series and in series of orthogonal polynomials. Further, the linearity of the underlying recurrences leads to very elegant statements about the growth of roundoff error and truncation error in the algorithm.

The theory of nonlinear recurrences, of which Lagrange's scheme furnishes an example, pursued a different path. Such algorithms are also quite useful, for instance in the evaluation of infinite products, solutions of functional equations, and a wide variety of definite integrals. In this situation error analyses are more difficult, but the convergence is usually extraordinarily rapid, much more so than for algorithms based on linear recurrences.

Roughly the first three-quarters of the book is devoted to linear algorithms, and the remainder to the nonlinear case. I have given throughout a very large number of examples for, as I see it, one of the primary purposes of this book is to put the reader interested in computer software in touch with the many specific successful applications of the algorithms. To this end, I have included an index of special functions whose computation is discussed.

I have also tried to make the development self-contained. Thus the book has substantial appendices incorporating the available material on the asymptotic theory of linear difference equations and the construction of recursion formulas for hypergeometric functions.

August 1983   Jet Wimp

List of symbols

Z	integers: $0, \pm 1, \pm 2, \ldots$
n, k	always integers
Z^0	integers: $0, 1, 2, \ldots$
$\mathcal{R}^p$	Euclidean p-space
$\mathcal{R}$	$\mathcal{R}^1$
$\mathcal{R}^+$	positive (>0) reals
$\mathcal{C}^p$	space of complex p-tuples
$\mathcal{C}$	$\mathcal{C}^1$
$\mathcal{R}_s$	space of real sequences
$\mathcal{C}_s$	space of complex sequences
$N_r(a)$	$\{z \in \mathcal{C} \mid |z-a| < r\}$ (open disk)
∂S	boundary of S, $S \subset \mathcal{C}$
ε_k	$1, k=0$; $2, k>0$
δ_{ij}	Kronecker δ
$x \wedge y$	minimum of x and y, $x, y \in \mathcal{R}$
$x \vee y$	maximum of x and y, $x, y \in \mathcal{R}$
∂p	degree of p, p a polynomial
$\partial_k p$	coefficient of t^k in p, p a polynomial
$f := g$	f as defined by g
$\mathbf{x}$	objects in a linear space
$\lin[\mathbf{x}_1, \mathbf{x}_2, \ldots, \mathbf{x}_\sigma]$	set spanned by $\mathbf{x}_1, \mathbf{x}_2, \ldots, \mathbf{x}_\sigma$
$Q(\rho; n)$	asymptotic form in Birkhoff series
$s(\rho; n)$	asymptotic form in Birkhoff series
$q(\rho; n)$	asymptotic form in Birkhoff series
$\sum'$	sum in which first term is to be halved
$\sum''$	finite sum in which first and last terms are to be halved

1 Introduction

Many of the most important techniques of numerical analysis developed since the beginning of this century can be subsumed under the theory of the vector system (or equation),

$$\mathbf{x}(n+1) = \mathbf{f}[\mathbf{x}(n), n], \qquad (1.1)$$

where $\mathbf{x}(n) \in \mathscr{C}^p$ and $\mathbf{f} : \mathscr{C}^p \times Z^0 \to \mathscr{C}^p$. The function $\mathbf{x}(n)$ is called the *solution* of the system and the graph of $\mathbf{x}(n)$, i.e., the set of points $\{\mathbf{x}(n)\}$, the *trajectory* of the system. Of course, to uniquely characterize the trajectory some auxiliary condition on $\mathbf{x}(n)$ is required. This may take the form of an initial condition, say $\mathbf{x}(0) := \mathbf{x}$, particularly for the case where $\mathbf{f}$ is nonlinear; another likely possibility is that information is known about the components of $\mathbf{x}(n)$ for different values of n. This constitutes, in analogy with the problem of differential equations, the boundary-value problem.

There are essentially two practical problems associated with equation (1). The first is, given a trajectory which is in itself of interest, how may an equation for it be derived? The second problem is, how may the equation be used, in a computationally efficient manner, to compute the trajectory?

For the case where $\mathbf{f}$ is nonlinear the derivation of the equation is often accomplished by the use of an integral invariant, a concept I will explain later, or by means of a functional equation. When $\mathbf{f}$ is linear the equation often arises as a special case of one of the large number of contiguous relationships for generalized hypergeometric functions or it may arise from considering the special properties of a related analytic scalar function $\phi(t)$: for instance, that $\phi(t)$ satisfies a differential equation. The components of $\mathbf{x}(n)$ may then, for example, represent the coefficients for the expansion of $\phi(t)$ in series of orthogonal functions.

How the equation itself is used typically depends on whether $\mathbf{f}$ is nonlinear or linear. When $\mathbf{f}$ is nonlinear one is generally interested not in the computation of the trajectory $\{\mathbf{x}(n)\}$ but on the limit

$$\lim_{n \to \infty} \mathbf{x}(n) = \mathbf{X}. \qquad (1.2)$$

In fact equation (1) is usually constructed in advance so that its limit is known to be $\mathbf{X}$. For linear equations this is never the case since the limit (2) always depends in a trivial algebraic manner on the initial or boundary conditions. When $\mathbf{f}$ is nonlinear, however, $\mathbf{X}$ may depend in a transcendental fashion on the initial condition even for the simplest equations. The equation then serves as an algorithm to compute $\mathbf{X}$. These kinds of algorithms have been known for many years, the most famous one bearing Gauss' name, though it is probably originally due to Lagrange. However, these algorithms have not, apparently, hitherto received a unified treatment in book form.

When $\mathbf{f}$ is nonlinear the equation (1) is usually constructed so that $\mathbf{X}$ is a root of an equation

$$F(\mathbf{X}) = c,$$

where $F: \mathscr{C}^p \to \mathscr{C}$ is an invariant for the algorithm, i.e., a function with the property

$$F[\mathbf{x}(n+1)] = F[\mathbf{x}(n)].$$

These invariants sometimes take the form of integrals or solutions of functional equations. The algorithms constructed this way are among the most rapidly convergent in all numerical analysis. In fact in Chapter 14 I construct an algorithm to compute a very general trigonometric integral and show that the convergence is quadratic in every case.

Cases where $\mathbf{f}$ is nonlinear but where attention focuses on the behavior of the trajectory rather than on the computation of $\mathbf{X}$ (which may not even exist) have come to the forefront recently in the study of discrete time dynamical systems, systems arising in multifarious disciplines: theory of turbulence, population growth, biosystems. One may ask whether $\mathbf{x}(n)$ has limit points or whether it is dense on some surface in $\mathscr{C}^p$ or how it changes as the initial point $\mathbf{x}$ is varied. The research going on in this area nowadays can only be described as frenzied and I have the space to do little more than offer the reader a tantalizing glimpse (in Chapter 12) of some of the ideas currently being pursued. However the mathematicians involved in this effort continue to be very productive and the reader after completing the chapter will find himself with a useful list of names for monitoring progress in the field.

In those cases where $\mathbf{f}$ is linear computing the trajectory itself is of primary interest. The reader may at first glance fail to understand how this could be a problem. Doesn't the equation offer an explicit format for computing $\mathbf{x}(n)$? The point is, of course, that $\mathbf{x}(0)$ is often not known, only asymptotic information about $\mathbf{x}(n)$, or perhaps the asymptotic behavior of $\mathbf{x}(n)$ compared to other solutions of the equation. It is an amazing fact, first articulated by J. C. P. Miller in 1952 (although the origins of the idea can be traced back at least to Lord Rayleigh), that $\mathbf{x}(n)$ can often be computed efficiently with no knowledge of $\mathbf{x}(0)$ whatsoever provided that $\|\mathbf{x}(n)\|/\|\mathbf{y}(n)\| \to 0$ as $n \to \infty$ for any solution $\mathbf{y}(n)$ not a constant multiple of $\mathbf{x}(n)$. This statement, strictly speaking, applies only to the homogeneous case, $\mathbf{f}[0, n] = \mathbf{0}$. However, powerful techniques (due primarily to F. W. J. Olver) based on the same principle have been devised to handle the nonhomogeneous case also.

The most exciting work on the linear case has taken place in the last 20 years. Among the mathematicians productively involved are F. W. J. Olver, C. W. Clenshaw, W. Gautschi, D. Lozier, J. R. Cash, J. Oliver and, I immodestly add, myself. By now a lexicon of techniques has been developed that can be used to do almost everything but peel apples, as Norbert Weiner once said of the Fourier transform. With these techniques we can now compute an array of important integrals and large classes of higher transcendental functions, solve eigenvalue problems, find zeros of functions, solve differential equations and determine the coefficients for the expansion of commonly occurring functions in series of orthogonal functions. It has been my purpose to give as many examples as I can of the use of these algorithms throughout this book.

My approach is to start with the very simplest linear cases, first $p = 1$, then $p = 2$. The $p = 1$ case, simple as it may be, nevertheless presents many interesting features. The $p = 2$ case is of course more significant and three full chapters are devoted to it. Although I generally disapprove of mathematical exposition which moves from the specific to the general there is, I think, in this situation a very good reason for it. The cases $p = 1$ and 2 are tractable and lend themselves to an elegant convergence and error analysis because then the solutions of the system adjoint to (1) can be written in a simple way in terms of the solutions to (1) itself. When $p > 2$ this breaks down, and much of the rich theory characterizing the simpler cases cannot be extended. Consequently the theory for $p > 2$, though equally important, is aesthetically less satisfying.

The nonlinear case is treated in the second half of the book.

The book contains three substantial appendices, all of which are quite intrinsic to developments in the book and they should probably be scanned before even Chapter 2 is attempted. Throughout the book referents preceded by an 'A', 'B' or 'C' refer to material in the appendices, for example, A-X, Theorem A.3, (A.12), etc. In Appendix A I have included a self-contained and virtually complete account of the theory of linear difference equations. Appendix B contains an exposition of the asymptotic theory of linear difference equations, first the classical results such as the theorems of Poincaré and Perron, next the very important research of Birkhoff and Trjitzinsky on the expansion of solutions in generalized asymptotic series and finally the recent work of Olver, which provides growth theorems allowing one to connect given solutions with their asymptotic forms. Appendix C contains a summary of the contiguous function relationships for generalized hypergeometric functions referred to earlier and which form the basis of many examples given in the text.

2 General results on the forward stability of recursion relations

2.1 Background

In this chapter I shall discuss computation based on the use of the recurrence

$$\mathscr{L}[y(n)] := \sum_{\nu=0}^{\sigma} A_\nu(n)y(n+\nu) = f(n), \qquad n \geqslant 0, \tag{2.1}$$

in the forward direction. I will assume that the functions A_ν satisfy the conditions given in Section A.1 so that the equation possesses a unique solution satisfying arbitrary initial conditions and such that the related homogeneous equation has a fundamental set. Generally speaking the problem is this: if $w(n)$ is a solution of (1) satisfying prescribed initial data, what error can one expect to encounter when w is computed from (1) for subsequent n? This problem is different in nature from the computational problems encountered in the more sophisticated Miller algorithm to which much of this book is devoted but it is important nonetheless. Many experienced numerical analysts are not fully aware of the hazards posed by such computations. Consider the following very simple first-order scalar equation provided by Gautschi (1972a tr. 1973).

Example 2.1 Compute the numbers

$$w(n) = n! \, [e^x - e_n(x)], \qquad n \geqslant 0,$$

where

$$e_n(x) = \sum_{k=0}^{n} \frac{x^k}{k!}, \tag{2.2}$$

and $x > 0$.

Note that $w(n)$ satisfies the recurrence

$$(n+1)y(n) - y(n+1) = x^{n+1}, \qquad n \geqslant 0. \tag{2.3}$$

We start with $w(0) = e^x - 1$. Taking $x = 1$ and rounding all computations to 4 SF produces Table 2.1. With the seventh entry one realizes something has gone awry, since, obviously, $w(n)$ must be positive. Things get

Table 2.1

n	$w(n)$ (approx.)
0	1.7180
1	0.7180
2	0.4360
3	0.3080
4	0.2320
5	0.1600
6	−0.0400
7	−1.2800

worse; in fact $w(10)$ is computed to be -1023. This disastrous accumulation of round-off error originated with the single rounding error in $w(0)$. ∎

The situation encountered in this example can be also found when the general difference equation $\mathcal{L}[y(n)] = f(n)$ is used in the forward direction. Whether or not small arithmetical errors introduced at one stage in the computation will tend to grow with succeeding computation until they overwhelm (in the sense of relative error) the true solution depends, as we shall see, on the relative growth properties of the desired solution versus the growth properties of other solutions of the original as well as the related homogeneous equation $\mathcal{L}[y(n)] = 0$.

These are among the matters to be investigated in this chapter.

As is well known, the nonhomogeneous equation $\mathcal{L}[y(n)] = f(n)$ can be written as a system

$$\mathbf{y}(n) + A(n)\mathbf{y}(n+1) = \mathbf{f}(n),$$

where $A(n)$ is the $\sigma \times \sigma$ matrix

$$A(n) = \begin{bmatrix} \dfrac{A_1(n)}{A_0(n)} & \dfrac{A_2(n)}{A_0(n)} & \cdots & \dfrac{A_{\sigma-1}(n)}{A_0(n)} & \dfrac{A_\sigma(n)}{A_0(n)} \\ -1 & 0 & \cdots & 0 & 0 \\ 0 & -1 & \cdots & 0 & 0 \\ \vdots & \vdots & & \vdots & \vdots \\ 0 & 0 & \cdots & -1 & 0 \end{bmatrix}, \tag{2.4}$$

$$\mathbf{y}(n) = [y(n), y(n+1), \ldots, y(n+\sigma-1)]^{\mathrm{T}},$$
$$\mathbf{f}(n) = [f(n)/A_0(n), 0, 0, \ldots, 0]^{\mathrm{T}}. \tag{2.5}$$

This matrix formulation can be quite useful for analyzing the stability properties of the scalar equation. Also we will have occasion to consider the matrix equation in itself, not necessarily arising from some given scalar equation. $A(n)$ then will be some given $\sigma \times \sigma$ matrix, nonsingular for $n \geq 0$.

When $A(n)$ *is* singular for certain values of n problems arise in the existence and uniqueness of solutions which must often be resolved on an *ad hoc* basis. Existence and uniqueness questions for difference equations are discussed in great detail in the standard works, Milne-Thomson (1960), Meschkowski (1959), and K. S. Miller (1968), the latter being particularly good on matrix systems.

2.2 Homogeneous systems

The analysis in this and subsequent sections draws heavily on the work of Gautschi (1967, 1972a tr. 1973).

We start with the system

$$\mathbf{y}(n) + A(n)\mathbf{y}(n+1) = 0, \tag{2.6}$$

where $A(n)$ is a $\sigma \times \sigma$ matrix, defined for $n \geq 0$, Det $A(n) \neq 0$, and $\mathbf{y}$ is a σ-vector.

Let $\mathbf{w}(n)$ be the desired solution of (6). I assume that $\mathbf{w}(n) \neq \mathbf{0}$, $n \geq 0$. Let $\bar{\mathbf{w}}(k)$ for a fixed $k \geq 0$ be any vector approximating $\mathbf{w}(k)$ with relative error $\varepsilon > 0$, i.e.,

$$\bar{\mathbf{w}}(k) := \mathbf{w}(k) + \|\mathbf{w}(k)\| \, \mathbf{e}, \qquad \|\mathbf{e}\| = \varepsilon. \tag{2.7}$$

Here $\| \cdot \|$ denotes any convenient vector norm. Also for any matrix B, $\|B\|$ will be the matrix norm induced by $\| \cdot \|$.

Let $\mathscr{F}_\varepsilon(k)$ denote the set of all such vectors. We have, trivially,

$$\sup_{\bar{\mathbf{w}}(k) \in \mathscr{F}_\varepsilon(k)} \frac{\|\bar{\mathbf{w}}(k) - \mathbf{w}(k)\|}{\|\mathbf{w}(k)\|} = \varepsilon. \tag{2.8}$$

Now consider (7) to be an initial condition. Then each vector $\bar{\mathbf{w}}(k) \in \mathscr{F}_\varepsilon(k)$ generates a 'perturbed' solution $\bar{\mathbf{w}}(n)$ of (6). By analogy with (8), I wish to investigate the quantity

$$\sup_{\bar{\mathbf{w}}(k) \in \mathscr{F}_\varepsilon(k)} \frac{\|\bar{\mathbf{w}}(t) - \mathbf{w}(t)\|}{\|\mathbf{w}(t)\|}, \qquad t \text{ fixed}, \quad t \neq k, \quad t \geq 0.$$

Let

$$Y(n) := [\mathbf{y}^{(1)}(n), \mathbf{y}^{(2)}(n), \dots, \mathbf{y}^{(\sigma)}(n)],$$

be a fundamental matrix of solutions of (6); see Miller (1968).

For every solution of (6), in particular for $\bar{\mathbf{w}}(n)$, one can find a constant vector **c** so that

$$\bar{\mathbf{w}}(n) = Y(n)\mathbf{c}.$$

From the condition (7) we find

$$\begin{cases} \bar{\mathbf{w}}(n) = Y(n)Y^{-1}(k)(\mathbf{w}(k) + \|\mathbf{w}(k)\|\,\mathbf{e}) \\ \qquad = \mathbf{w}(n) + \|\mathbf{w}(k)\|\,Y(n)Y^{-1}(k)\mathbf{e}, \end{cases} \tag{2.9}$$

so

$$\sup_{\bar{\mathbf{w}}(k)\in\mathscr{F}_\varepsilon(k)} \frac{\|\bar{\mathbf{w}}(n) - \mathbf{w}(n)\|}{\|\mathbf{w}(n)\|} = \frac{\|\mathbf{w}(k)\|}{\|\mathbf{w}(n)\|} \sup_{\|\mathbf{e}\|=\varepsilon} \|Y(n)Y(k)^{-1}\mathbf{e}\|.$$

The last term on the right can be computed in terms of the matrix norm of $Y(n)Y(k)^{-1}$, so we get

$$\sup_{\bar{\mathbf{w}}(k)\in\mathscr{F}_\varepsilon(k)} \frac{\|\bar{\mathbf{w}}(n) - \mathbf{w}(n)\|}{\|\mathbf{w}(n)\|} = \frac{\|\mathbf{w}(k)\|}{\|\mathbf{w}(n)\|} \|Y(n)Y^{-1}(k)\|\,\varepsilon. \tag{2.10}$$

This provides an indication of how the relative error at n compares with the initial relative error at k.
Let

$$\alpha(k, n) := \frac{\|\mathbf{w}(k)\|}{\|\mathbf{w}(n)\|} \|Y(n)Y^{-1}(k)\|. \tag{2.11}$$

Let $Z := \{\mathbf{z}^{(1)}, \mathbf{z}^{(2)}, \ldots, \mathbf{z}^{(\sigma)}\}$ be another fundamental matrix for (6). Then it is readily verified that for some constant matrix K

$$Z = Y \times K,$$

so

$$Z(n)Z^{-1}(k) = Y(n)Y^{-1}(k).$$

Thus the matrix on the right does not depend on the choice of the solutions $\{\mathbf{y}^{(h)}(n)\}$, only on the system.

It is thus natural to adopt the following as a measure of stability for the computation of $\mathbf{w}(n)$ by forward recursion from (6).

Definition 2.1 *The* index of stability *for the forward computation of* $\mathbf{w}(n)$ *at k is*

$$\alpha(k) := \sup_{n > k} \alpha(k, n).$$

If $\alpha(k) < \infty$ (6) *is said to be* stable *for* $\mathbf{w}(n)$ *at the point k.* ∎

This property can be shown to be equivalent to one sometimes called *weak stability* (see also Section 7.3) defined as follows:

Definition 2.2 *If for each $K > 0$ there exists a constant $C \equiv C(K) > 0$ such that*

$$\sup_{\substack{n > k \\ k \leqslant K}} \alpha(k, n) = C < \infty,$$

then the recurrence (6) is said to be weakly stable for $\mathbf{w}(n)$. ∎

To establish the equivalence, 'stability at some $k \Leftrightarrow$ weak stability' we proceed as follows:

$$\alpha(k + 1, n) = \frac{\|A^{-1}(k)\mathbf{w}(k)\|}{\|\mathbf{w}(n)\|} \|Y(n) Y^{-1}(k+1)\|$$

$$\leqslant \|A^{-1}(k)\| \frac{\|\mathbf{w}(k)\|}{\|\mathbf{w}(n)\|} \|Y(n) Y^{-1}(k) A(k)\|$$

$$\leqslant \|A^{-1}(k)\| \|A(k)\| \alpha(k, n) = \text{Cond } A(k) \alpha(k, n).$$

(Cond $B := \|B\| \|B^{-1}\|$ is called the *condition number* of B; see Isaacson and Keller (1966, p. 37).)

Similarly,

$$\alpha(k, n) = \frac{\|\mathbf{w}(k)\|}{\|\mathbf{w}(n)\|} \|Y(n) Y^{-1}(k+1) A^{-1}(k)\|,$$

so

$$\alpha(k + 1, n) \geqslant \frac{\|\mathbf{w}(k+1)\|}{\|\mathbf{w}(k)\|} \|A^{-1}(k)\|^{-1} \alpha(k, n).$$

But

$$\|\mathbf{w}(k)\| \leqslant \|A(k)\| \|\mathbf{w}(k+1)\|,$$

so

$$\frac{\|\mathbf{w}(k+1)\|}{\|\mathbf{w}(k)\|} \geqslant \|A(k)\|^{-1}.$$

Combining the above results gives

$$[\text{Cond } A(k)]^{-1} \alpha(k, n) \leqslant \alpha(k+1, n) \leqslant [\text{Cond } A(k)] \alpha(k, n),$$

and taking a sup over n gives

$$\alpha(k + 1) \leqslant [\text{Cond } A(k)] \alpha(k),$$

which shows the equivalence (and much more).

An important global property of stability is formulated in

Definition 2.3 *If*

$$\sup_{\substack{n,k \geqslant 0 \\ n>k}} \alpha(k, n) = C < \infty,$$

then the recurrence is said to be stable *for* **w**(n). ■

Obviously stable $\Rightarrow$ weakly stable.

Definition 2.4 *When neither Definition 1 nor 3 is satisfied, the term* stable *is to be replaced by the term* unstable *in either situation.* ■

By the equivalence of vector norms in finite-dimensional spaces the properties of stability and instability are norm-independent.

The definition of stability in Definition 2 is rather stringent. However, even if a system is theoretically stable, if C is very large the desired solution may be difficult to compute accurately. Conversely though the system may be unstable for **w**(n) the function can usually be computed accurately on small subsets of Z^0.

Example 2.2 Second-order systems will be discussed in more detail later on but it seems a good idea to offer an example to clarify these definitions. The system

$$\mathbf{y}(n) + \begin{bmatrix} -2 & 1 \\ -1 & 0 \end{bmatrix} \mathbf{y}(n+1) = \begin{bmatrix} 0 \\ 0 \end{bmatrix}$$

has a fundamental matrix

$$Y(n) = \begin{bmatrix} n & 1 \\ n+1 & 1 \end{bmatrix} = [\mathbf{y}^{(1)}(n), \mathbf{y}^{(2)}(n)].$$

We have $\|\mathbf{y}^{(1)}(n)\|_1 = 2n+1$, $\|\mathbf{y}^{(2)}(n)\|_1 = 2$. A straightforward computation shows that for $w(n) = y^{(1)}(n)$

$$\alpha(k, n) = \frac{(2k+1)(2n-2k+1)}{(2n+1)}, \qquad n>k.$$

Thus

$$\alpha(k) = 2k+1, \qquad k \geqslant 0.$$

The recursion is stable for the computation of $y^{(1)}(n)$ at any point k and also weakly stable, but not stable. ■

It is easily shown that the system with fundamental matrix

$$\begin{bmatrix} \lambda_1^n & \lambda_2^n \\ \lambda_1^{n+1} & \lambda_2^{n+1} \end{bmatrix},$$

is stable for $y^{(1)}(n)$ (and hence weakly stable and stable at any point) provided $|\lambda_1| > |\lambda_2|$.

The following provides a criterion for instability.

Theorem 2.1 (Gautschi) *Let $\sigma \geq 2$ and let the recursion relation (6) have a solution $\mathbf{y}^*(n)$ such that*

$$\lim_{n \to \infty} \|\mathbf{w}(n)\| / \|\mathbf{y}^*(n)\| = 0. \tag{2.12}$$

Then the recursion is unstable for $\mathbf{w}(n)$ at every point.

Conversely if (6) is unstable for $\mathbf{w}(n)$ in the following manner:

$$\overline{\lim_{n \to \infty}} \, \alpha(k, n) = \infty \qquad \text{for some fixed } k \geq 0, \tag{2.13}$$

then a solution $\mathbf{y}^*(n)$ of (6) exists having the property (12).

Proof Note that for any nonsingular constant matrix B and any matrix $W(n)$

$$\overline{\lim_{n \to \infty}} \, \|W(n)B\| = \infty \Leftrightarrow \overline{\lim_{n \to \infty}} \, \|W(n)\| = \infty, \tag{2.14}$$

because

$$\|W(n)B\| \leq \|W(n)\| \, \|B\| \qquad \text{and} \qquad \|W(n)\| = \|W(n)BB^{-1}\|$$
$$\leq \|W(n)B\| \, \|B^{-1}\|.$$

Now assume (13) holds. The solutions $\mathbf{w}(n)$ and $\mathbf{y}^*(n)$ are linearly independent. Therefore they may be completed to form a fundamental matrix

$$Y(n) := [\mathbf{w}(n), \mathbf{y}^*(n), \mathbf{y}^{(3)}(n), \ldots, \mathbf{y}^{(\sigma)}(n)].$$

Let's assume (without loss of generality) that we are working with the l^1 norm. We have (see Isaacson and Keller (1966, p. 9))

$$\|Y(n)/\|\mathbf{w}(n)\| \| \geq \|\mathbf{y}^*(n)\| / \|\mathbf{w}(n)\|.$$

Now applying the result (14) to the matrices

$$W(n) := Y(n)/\|\mathbf{w}(n)\|, \qquad B := \|\mathbf{w}(k)\| \, Y(k)^{-1}, \tag{2.15}$$

shows $\overline{\lim}_{n \to \infty} \, \alpha(k, n) = \infty$ for every $k \geq 0$. Thus the recursion is unstable for $\mathbf{w}(n)$.

Conversely assume (13) holds. Let

$$Y(n) := [\mathbf{y}^{(1)}(n), \mathbf{y}^{(2)}(n), \ldots, \mathbf{y}^{(\sigma)}(n)],$$

be a fundamental matrix for (6). The relation (14), again applied to the matrices (15), yields

$$\varlimsup_{n \to \infty} \| Y(n)/\|\mathbf{w}(n)\| \| = \infty.$$

Let $Y(n) := [\mathbf{y}^{(h)}(n)]$ and define r_n so that

$$\|Y(n)\| = \|\mathbf{y}^{(r_n)}(n)\|, \qquad 1 \leq r_n \leq \sigma.$$

Obviously there exists an integer r, $1 \leq r \leq \sigma$, such that $r_n = r$ for infinitely many n. Letting $n \to \infty$ over this subsequence of Z^0 shows that

$$\varlimsup_{n \to \infty} \| \mathbf{y}^{(r)}(n)\|/\|\mathbf{w}(n)\| = \infty,$$

and taking $\mathbf{y}^*(n) := \mathbf{y}^{(r)}(n)$ concludes the proof of the theorem. ■

Note the recursion is always stable for any nontrivial solution in the scalar case $(\sigma = 1)$ for then $Y(n) = \mathbf{y}(n)$ is the scalar solution and $\alpha(k, n) \equiv 1$.

Theorem 2.2 *Let, for some fundamental set of solutions* $\{\mathbf{y}^{(h)}(n)\}$, *the quantities*

$$\|\mathbf{y}^{(h)}(n)\|/\|\mathbf{w}(n)\|, \qquad n > n_0, \quad 1 \leq h \leq \sigma,$$

be bounded. Then the recurrence is weakly stable for $\mathbf{w}(n)$.

Proof We have

$$\alpha(k, n) \leq \frac{\|\mathbf{w}(k)\|}{\|\mathbf{w}(n)\|} \|Y(n)\| \, \|Y^{-1}(k)\|.$$

But

$$\frac{\|Y(n)\|}{\|\mathbf{w}(n)\|} = \frac{\max_h \|y^{(h)}(n)\|}{\|\mathbf{w}(n)\|},$$

and the theorem follows. ■

2.3 Nonhomogeneous systems

We now consider

$$\mathbf{y}(n) + A(n)\mathbf{y}(n+1) = \mathbf{f}(n), \qquad n \geq 0. \tag{2.16}$$

I will assume once again that Det $A(n) \neq 0$, and that $\mathbf{f}(n) \not\equiv 0$.

Let

$$K := \{n \in Z^0 \mid \mathbf{w}(n) \neq \mathbf{0}\},$$

where $\mathbf{w}$ is the solution to be computed. I will assume K is infinite.

The general solution of (16) is

$$\mathbf{y}(n) = Y(n)\mathbf{c} + \mathbf{w}(n),$$

where $Y(n)$ is a fundamental matrix of solutions of the homogeneous equation. As before, if $\bar{\mathbf{w}}(n)$ is a solution of (16) whose initial vector $\bar{\mathbf{w}}(k)$ approximates $\mathbf{w}(k)$ to within ε, in accordance with (7), one finds that the equation (9),

$$\bar{\mathbf{w}}(n) = \mathbf{w}(n) + \|\mathbf{w}(k)\| \, Y(n) Y(k)^{-1} \mathbf{e},$$

holds also for the nonhomogeneous system. Thus equation (10) holds and Definition 1 of the previous section still makes sense with one minor change, that is, $n, k \in K$.

Definition 2.5 *The recurrence (16) is said to be* stable *for* $\mathbf{w}(n)$ *if there exists a constant* $C > 0$ *such that*

$$\sup_{\substack{n > k \\ n,k \in K}} \alpha(k, n) = C < \infty,$$

$\alpha(k, n)$ *as in (11). Otherwise (16) is said to be* unstable *for* $\mathbf{w}(n)$. ∎

The following result follows immediately from Theorem 1.

Theorem 2.3 *Let* $\sigma \geqslant 2$ *and let the associated homogeneous recursion (16) have a solution* $\mathbf{y}^*(n)$ *such that*

$$\lim_{n \to \infty} \|\mathbf{w}(n)\| / \|\mathbf{y}^*(n)\| = 0.$$

Then the nonhomogeneous system (16) is unstable *for* $\mathbf{w}(n)$ *at any point.* ∎

2.4 The first-order scalar case: forward vs. backward recursion

For $\sigma = 1$ the difference equation (6) is

$$y(n) + a(n)y(n+1) = f(n), \qquad n \geqslant 0.$$

We find

$$\alpha(k, n) = \left| \frac{w(k)}{w(n)} \frac{y^*(n)}{y^*(k)} \right| = \frac{\rho(n)}{\rho(k)},$$

$$\rho(n) := |w(0)y^*(n)/w(n)|,$$

(2.17)

where

$$y^*(n) := \prod_{j=0}^{n-1} [-a(j)]^{-1}, \qquad n \geq 0,$$

is a solution of the homogeneous equation. (Empty products are to be interpreted as 1.) Going from k to n we find the error is increased if $\rho(n) \geq \rho(k)$ and decreased if $\rho(n) < \rho(k)$. If $\rho(n)$ is monotone decreasing, the recursion relation may be used safely in the *forward direction* to compute $w(n)$. On the other hand, if

$$\lim_{n \to \infty} \rho(n) = \infty, \tag{2.18}$$

then by Theorem 3 the recurrence is unstable for $w(n)$. Note, however, that if $\rho(n)$ is monotone increasing there exists the possibility that the recurrence may be used safely in the backward direction starting with some (possibly very large) initial value of $n = k$. This is the philosophy behind the Miller algorithm, which I will discuss in future chapters. When $\overline{\lim} \, \rho(n) = \infty$ the error in computing $w(n)$, being amplified by a factor $\rho(n)/\rho(k)$, can be made arbitrarily small by taking k large enough.

Example 2.3   (The partial sums of the Taylor series for e^x)   Let

$$w(n) := n! \, [e^x - e_n(x)],$$

$e_n(x)$ as in (2). The relevant equation is (3). A solution of the associated homogeneous equation is $y^*(n) = n!$ so

$$\rho(n) = \frac{|e^x - 1| \, n!}{|w(n)|} = \frac{|e^x - 1| \, n!}{|e^x - e_n(x)|} \to \infty, \qquad n \to \infty,$$

for x fixed. Thus the equation is unstable for $w(n)$. ∎

Example 2.4   (The incomplete Gamma function)   Let

$$\phi(n) := \frac{1}{\Gamma(n+\alpha)} \int_0^x e^{-t} t^{n+\alpha-1} \, dt, \qquad \alpha > 0. \tag{2.19}$$

Integration by parts shows $\phi(n)$ satisfies

$$y(n) - y(n+1) = \frac{x^{n+\alpha} e^{-x}}{\Gamma(n+\alpha+1)}. \tag{2.20}$$

Since a solution of the homogeneous equation is $y(n) \equiv 1$, $\rho(n) = 1$ and $\alpha(k, n) \equiv 1$. Thus (20) is stable for the computation of $\phi(n)$. ∎

Example 2.5   (Gautschi (1961a))   Suppose we wish to compute the exponential integrals

$$E_n(x) := \int_1^\infty t^{-n} e^{-xt} \, dt, \qquad x > 0, \quad n \geq 0. \tag{2.21}$$

$w(n) = E_{n+1}$ satisfies the equation

$$\frac{x}{(n+1)} y(n) + y(n+1) = \frac{e^{-x}}{(n+1)}.$$ (2.22)

$\rho(n)$ is found to be

$$\rho(n) = \frac{x^n E_1(x)}{n!\, E_{n+1}(x)}.$$

It can be shown that if $x \leqslant 0.61006 \ldots$, then $\rho(n)$ decreases monotonically from 1 to 0 as n increases so that (22) is stable for E_n. If $x > 0.61006 \ldots$, then $\rho(n)$ first increases to a maximum at $n = n_0$ and then tends monotonically to 0 so again (22) is stable for E_n. If the error amplification $\rho(n_0)$ is acceptable (22) may be safely applied to the computation of $E_n(x)$ for all n. ∎

2.5 The computation of successive derivatives

This topic provides a very interesting application of the material in the preceding sections. Sources for this material are Gautschi (1966, 1972a) and Gautschi and Klein (1970).

Let $f(z)$ be analytic on some set A; it is desired to compute the quantities

$$\phi(n) = \frac{d^n}{dz^n} \left[\frac{f(z)}{z} \right], \qquad z \in A - \{0\}.$$ (2.23)

From the identity

$$\frac{d^n}{dz^n} f = \frac{d^n}{dz^n} \left(\frac{f}{z} \cdot z \right) = \frac{z \cdot d^n}{dz^n} \left(\frac{f}{z} \right) + n \frac{d^{n-1}}{dz^{n-1}} \left(\frac{f}{z} \right),$$

we see that $\phi(n)$ satisfies

$$y(n) + \frac{zy(n+1)}{(n+1)} = \frac{(d^{n+1}/dz^{n+1})f(z)}{(n+1)}.$$ (2.24)

It is surprising that the stability of the computation of $\phi(n)$ depends only on the value of f at 0 provided f is analytic in a large enough disk containing 0 and z.

Theorem 2.4 *Let $f(\zeta)$ be analytic in the disk*

$$N_r(z) := \{\zeta \mid |\zeta - z| \leqslant r\}, \qquad r > |z|,$$

and assume $f(\zeta)/\zeta$ is not a polynomial. Then (24) is stable for $\phi(n) \Leftrightarrow f(0) \neq 0$.

Proof   We need a preliminary result. Let $\Delta \subset N_r$ be a disk centered at 0 which excludes z. By Cauchy's integral formula

$$\phi(n) = \frac{n!}{2\pi i} \int_{\partial N_r} \frac{f(\zeta)\,d\zeta}{(\zeta-z)^{n+1}\zeta} - \frac{n!}{2\pi i} \int_{\partial \Delta} \frac{f(\zeta)\,d\zeta}{(\zeta-z)^{n+1}\zeta}.$$

A nontrivial solution of the homogeneous equation is

$$y^*(n) = (-1)^n n!/z^n.$$

Letting the radius of $\Delta \to 0$ gives

$$z\frac{\phi(n)}{y^*(n)} = f(0) + \frac{(-1)^n}{2\pi i} \int_{\partial N_r} \left(\frac{z}{\zeta-z}\right)^{n+1} \frac{f(\zeta)}{\zeta}\,d\zeta.$$

On $\partial N_r \; |\zeta - z| = r > |z|$ so

$$w := \left|\frac{z}{\zeta-z}\right| < 1, \qquad \zeta \in \partial N_r.$$

Thus

$$\lim_{n\to\infty} \frac{z\phi(n)}{y^*(n)} = f(0). \tag{2.25}$$

Also

$$\rho(n) = |f(z)y^*(n)/z\phi(n)|. \tag{2.26}$$

$\Leftarrow$:   Since $f(\zeta)/\zeta$ is not a polynomial the set $K = \{n \mid \phi(n) \neq 0\}$ must be infinite. Suppose $f(0) \neq 0$. Because of (25) there are constants C, C' and $n_0 > 0$ such that

$$0 < C \leqslant \left|\frac{z\phi(n)}{y^*(n)}\right| \leqslant C', \qquad n > n_0.$$

Thus

$$\alpha(k, n) = \frac{\rho(n)}{\rho(k)} = \left|\frac{z\phi(k)}{y^*(k)} \middle/ \frac{z\phi(n)}{y^*(n)}\right| \leqslant \frac{C'}{C},$$

for k, n sufficiently large, and this shows stability.

$\Rightarrow$:   By contradiction. Assume $f(0) = 0$. Then $\underline{\lim}_{n\to\infty} \rho(n) = \infty$, so by (18) the equation is unstable for $w(n)$.   ■

Example 2.6   (Gautschi and Klein)   For fixed $x \neq 0$ compute

$$\phi(n) := \frac{d^n}{dx^n}\left[\frac{\sin x}{x}\right], \qquad n \geqslant 0. \tag{2.27}$$

The recurrence relation is

$$y(n) + \frac{xy(n+1)}{(n+1)} = \frac{\text{Im}\,(i^{n+1}e^{ix})}{(n+1)}.$$

Using the method described in A-IX this equation may be solved and we find

$$\phi(n) = \frac{(-1)^n n!}{x^{n+1}} \,\text{Im}\,[e^{ix}e_n(-ix)],$$

e_n as in equation (2). Thus

$$\rho(n) = \left| \frac{\sin x}{\text{Im}\,[e^{ix}e_n(-ix)]} \right|.$$

Since $e^{ix}e_n(-ix) \to 1$ as $n \to \infty$ this confirms the instability asserted by the theorem. It might be thought that $\phi(n)$ could be computed by backward recursion starting, say, with $\phi(N) = 0$ for some large N, but this procedure, to be more thoroughly analyzed later, is not appropriate unless x is rather small compared to N.

Using the elementary properties of the confluent hypergeometric function one finds

$$e^w e_n(-w) = 1 + \frac{(-1)^n e^w w^{n+1}}{(n+1)!} \,\Phi(1; n+1; -w)$$

$$= 1 + \frac{(-1)^n e^w w^{n+1}}{(n+1)!} [1 + O(n^{-1})].$$

Consequently for x not an integral multiple of $\pi/2$ there will be constants C_1, C_2 and an $n_0 > 0$ such that

$$0 < C_1 \frac{(n+1)!}{|x|^{n+1}} < |\rho(n)| < C_2 \frac{(n+1)!}{|x|^{n+1}}, \qquad n > n_0.$$

Although these estimates are not uniform in x they provide a good indication of what actually happens. For a given x the curves of $\rho(n)$ start out decreasing, attaining a minimum value near $n = x$, and then begin to grow essentially monotonically. Although the recursion is unstable for $\phi(n)$, it is stable in the region $0 \le n \le |x|$. Clearly, however, the recursion cannot be used in an unrestricted fashion in either the forward or backward direction. ∎

The derivatives $\phi(n)$ of the previous example arise in the theory of cardinal interpolation. Let $f: R \to C$ and a real number $h > 0$ be given. The series

$$F_h(z) = \frac{h}{\pi} \sum_{m=-\infty}^{\infty} \frac{f(mh)}{z - mh} \sin \frac{\pi}{h}(z - mh), \qquad z \in C,$$

is called the *cardinal series of f* (with respect to h). Obviously, $F_h(mh) = f(mh)$, $m \in Z$. Further, it is known that if f is an analytic function bounded in the strip $R \times [-ia, ia]$ and

$$\int_{-\infty}^{\infty} |f(\tau \pm ia)|^2 \, d\tau < \infty,$$

then the series converges uniformly on compact subsets of the strip and, further, $\lim_{h \to 0} F_h(z) = f(z)$; see Kress (1972). For some functions the representation is *exact*, e.g., when f is entire with

$$|f(z)| \leq c e^{\rho |\mathrm{Im}\, z|},$$

for some $c \geq 0$, $0 \leq \rho < \pi/h$, then $F_h(z) \equiv f(z)$ with uniform convergence on compact subsets of C. Often the derivatives of $F_h(z)$ are required, which means one needs the quantities

$$\frac{d^n}{dz^n} \sin \frac{\frac{\pi}{h}(z - mh)}{(z - mh)} = \frac{\pi^{n+1}}{h^{n+1}} \frac{d^n}{dx^n} \frac{\sin x}{x}, \qquad x = \frac{\pi}{h}(z - mh).$$

Gautschi (1972a) has observed that this series arises in communication theory and, for $h = 1$, has to do with the reconstruction of a function with a limited frequency band when its sample values at equidistant points are known.

2.6 Scalar equations of higher order; minimal and dominant solutions

The stability of the nonhomogeneous scalar equation $\mathcal{L}[y(n)] = f(n)$ is rather easy to appraise. Basically this is due to the fact that in the scalar case the fundamental matrix for the related homogeneous system assumes a simple form, i.e.,

$$Y(n) = \begin{bmatrix} y^{(1)}(n) & y^{(2)}(n) & \cdots & y^{(\sigma)}(n) \\ \vdots & \vdots & & \vdots \\ y^{(1)}(n+\sigma-1) & y^{(2)}(n+\sigma-1) & \cdots & y^{(\sigma)}(n+\sigma-1) \end{bmatrix},$$

where $\{y^{(1)}(n), \ldots, y^{(\sigma)}(n)\}$ is a fundamental set for the scalar homogeneous equation $\mathcal{L}[y(n)] = 0$.

Let $w(n)$ be the desired solution of $\mathcal{L}[y(n)] = f(n)$ and define

$$\mathbf{w}(n) := [w(n), w(n+1), \ldots, w(n+\sigma-1)]^T.$$

Theorem 2.5   *Let the recursion relation $\mathcal{L}[y(n)] = 0$ have a solution $y^*(n)$ with the property*

$$\lim_{n \to \infty} w(n)/y^*(n) = 0,$$

where $w(n)$ is the desired solution of the nonhomogeneous equation (1). Then $\mathcal{L}[y(n)] = f(n)$ is unstable for $w(n)$ at every point.

Proof Let $\mathbf{y}^*(n) := [y^*(n), y^*(n+1), \ldots, y^*(n+\sigma-1)]^T$. We have, with $\|\cdot\|$ the l^1 norm,

$$\frac{\|\mathbf{w}(n)\|}{\|\mathbf{y}^*(n)\|} = \frac{\sum_{k=0}^{\sigma-1}|w(n+k)|}{\sum_{k=0}^{\sigma-1}|y^*(n+k)|} = \frac{\sum_{k=0}^{\sigma-1}\left|\dfrac{w(n+k)}{y^*(n+k)}\right|\dfrac{1}{\pi_k}}{\sum_{k=0}^{\sigma-1}\dfrac{1}{\pi_k}},$$

where

$$\pi_k := \prod_{\substack{j=0\\j\neq k}}^{\sigma-1}|y^*(n+j)|.$$

Thus we have

$$\frac{\|\mathbf{w}(n)\|}{\|\mathbf{y}^*(n)\|} \leqslant \max_{0\leqslant k\leqslant\sigma-1}\left|\frac{w(n+k)}{y^*(n+k)}\right|.$$

Consequently $\lim_{n\to\infty}\|\mathbf{w}(n)\|/\|\mathbf{y}^*(n)\| = 0$, and by equivalence of vector norms this holds for any other norm. The result now follows by an application of Theorem 1. ∎

In general it is the rate of growth of the desired solution of a homogeneous difference equation with respect to the other solutions (i.e., the other solutions in a completed basis) that determines whether or not a computational scheme based on the equation will be successful. This applies to schemes based not only on forward recursion previously discussed but also techniques based on backward recursion, to be discussed in later chapters.

The appropriate concepts will now be introduced.

Definition 2.6 *Let $\phi(n)$, $\psi(n)$ be (complex-valued) functions defined for $n \in Z^0$ and let*

$$\phi(n)/\psi(n) = o(1), \qquad n \to \infty.$$

Then I say ψ dominates ϕ (or ϕ is dominated by ψ). ∎

Definition 2.7 *A nontrivial solution $\mu(n)$ of the homogeneous difference equation $\mathcal{L}[y(n)] = 0$ is called a* minimal solution *if $\mu(n)$ is dominated by each of the other solutions in some fundamental set of solutions containing $\mu(n)$.* ∎

(This means, of course, that $\lim_{n\to\infty} \mu(n)/y(n) = 0$ for any other solution $y(n)$ linearly independent of $\mu(n)$.)

Definition 2.8 *A nontrivial solution $\sigma(n)$ of the homogeneous difference equation $\mathscr{L}[y(n)] = 0$ is called a* dominant solution *if $\sigma(n)$ dominates each of the other solutions in some fundamental set of solutions containing $\sigma(n)$.* ■

If a minimal solution exists it is, up to a constant multiple, unique; for if $\mu(n)$, $\mu'(n)$ were two minimal solutions then simultaneously $\mu(n)/\mu'(n) \to 0$ and $\mu'(n)/\mu(n) \to 0$, which cannot be.

Dominant solutions are not unique (if $\sigma > 1$).

A corollary to the previous theorem can now be stated in terms of these definitions.

Corollary *Let $\mu(n)$ be a minimal solution of $\mathscr{L}[y(n)] = 0$. Then this recurrence is unstable for $\mu(n)$.* ■

Example 2.7 Consider the difference equation of Example B.3:

$$xg(n) + (n+2)g(n+1) - 2g(n+3) = 0, \qquad x > 0.$$

As shown there, the equation has a fundamental set $\{y^{(h)}(n)\}$ with the property

$$y^{(1)}(n) = n^{n/2}(2e)^{-n/2}\left(1 + \frac{c_1}{n^{1/2}} + \frac{c_2}{n} + \cdots\right),$$

$$y^{(2)}(n) = n^{n/2}(-1)^n(2e)^{-n/2}\left(1 - \frac{c_1}{n^{1/2}} + \frac{c_2}{n} + \cdots\right),$$

$$y^{(3)}(n) = n^{-n}(-xe)^n n^{-3/2}\left(1 + \frac{d_1}{n} + \frac{d_2}{n^2} + \cdots\right).$$

Thus the equation possesses a minimal solution $y^{(3)}(n)$. ■

Example 2.8 The equation

$$(n+b)(n+c)y(n) - (n+1)[(2n+b+c+1)+z]y(n+1)$$
$$+ (n+1)(n+2)y(n+2) = 0, \qquad -\pi < \arg z \leqslant \pi,$$

satisfied, for instance, by $w(n) := (b)_n(c)_n n!^{-1}\Psi(n+b, b+1-c, z)$ (see Example B.2) possesses a minimal solution provided $z \neq e^{\pi i}$ and none otherwise. In the former case $w(n)$ is the minimal solution. ■

Example 2.9 The equation

$$y(n) - y(n+2) = 0$$

possesses no minimal solution, for a fundamental set is $y^{(1)}(n) = 1$, $y^{(2)}(n) = (-1)^n$. ∎

There seem to be no simple necessary and sufficient conditions which will guarantee $\mathcal{L}[y(n)] = 0$ has a minimal solution. However, in many cases the asymptotic theory of linear difference equations developed in Appendix B can be used to determine the existence of such solutions. Of particular importance are the Theorems B.1 and B.2.

Demonstrating stability is generally more difficult than demonstrating instability and usually requires more asymptotic information than furnished by either of the preceding theorems. The construction of the matrix $Y(t) Y^{-1}(k)$ and the estimation of it can be a formidable problem, even in the scalar case. I give an example to show that in some cases the computations may actually be performed.

Example 2.10 (Repeated integrals of the error function) The functions

$$\phi^{(1)}(n) := i^n \operatorname{erfc} x := \frac{2}{\sqrt{\pi}} \int_x^\infty \frac{(t-x)^n}{n!} e^{-t^2} \, dt, \qquad x > 0, \tag{2.28}$$

$$\phi^{(2)}(n) := \frac{2}{\sqrt{\pi}} \int_{-\infty}^x \frac{(t-x)^n}{n!} e^{-t^2} \, dt, \qquad x > 0, \tag{2.29}$$

satisfy the difference equation

$$y(n) - 2xy(n+1) - 2(n+2)y(n+2) = 0. \tag{2.30}$$

Neither of the preceding theorems suffices to indicate whether this equation has a minimal solution. However, it is known that

$$\phi^{(1)}(n) = \frac{e^{-x^2}}{\sqrt{\pi} \, 2^n} \Psi\left(\frac{n+1}{2}, \tfrac{1}{2}; x^2\right).$$

The result (5.8) shows

$$\phi^{(1)}(n) \sim K \lambda^n n^{-(n+1)/2} e^{-x\sqrt{(2n)}} \left[1 + \frac{c_1}{n^{1/2}} + \frac{c_2}{n} + \cdots\right],$$

$$K = \frac{e^{-x^2/2}}{\sqrt{\pi}}, \qquad \lambda = \sqrt{\left(\frac{e}{2}\right)}.$$

Gautschi also gives the lead term of this result (1961b, p. 230). By the discussion in Section B.2 it follows that there is a second solution of the equation $y^*(n)$ with the property

$$y^*(n) \sim K \lambda^n (-1)^n n^{-(n+1)/2} e^{x\sqrt{(2n)}} \left[1 - \frac{c_1}{n^{1/2}} + \frac{c_2}{n} + \cdots\right].$$

$\phi^{(2)}(n)$ may be identified as a constant (independent of n, that is) multiple of $y^*(n)$ since $(-1)^n\phi^{(2)}(n; -x) = \phi^{(1)}(n; x)$. Thus for some $K, K' > 0$ we have

$$\phi^{(1)}(n) \sim K\lambda^n n^{-(n+1)/2} e^{-x\sqrt{(2n)}}[1 + O(n^{-1/2})],$$
$$\phi^{(2)}(n) \sim K'\lambda^n (-1)^n n^{-(n+1)/2} e^{x\sqrt{(2n)}}[1 + O(n^{-1/2})].$$

A direct computation shows

$$Y(n) = \phi^{(2)}(n) \begin{bmatrix} r(n) & 1 \\ (2n)^{-1/2}r(n) & -(2n)^{-1/2} \end{bmatrix} (1 + O(n^{-1/2})),$$

where

$$r(n) := \frac{\phi^{(1)}(n)}{\phi^{(2)}(n)} = O((-1)^n e^{-2x\sqrt{(2n)}}). \tag{2.31}$$

Let us first examine stability of the recurrence for the computation of $\phi^{(2)}(n)$. A little algebra yields

$$\alpha(k, n) = \sqrt{\left(\frac{k}{2}\right)} \frac{|r(n) - r(k)|}{|r(k)|} (1 + O(n^{-1/2}))(1 + O(k^{-1/2})),$$

in the 1-norm. Thus by (31) the recurrence (30) is stable for $\phi^{(2)}(n)$ at any point, i.e., weakly stable. But since

$$\alpha(k) \approx Mk^{1/2}, \qquad k \to \infty,$$

the recurrence is not stable for $\phi^{(2)}(n)$.

Since $\phi^{(1)}(n)$ is a minimal solution, the equation is unstable at any point for $\phi^{(1)}(n)$; cf. Theorem 5.

Gautschi (1961b, 1977) has conducted the definitive study of the computation of these and other closely related integrals by both forward and backward recursion. See also Amos (1973) and Bardo and Rueden-berg (1971). For some numerics on the computation of these integrals see Kaye (1955). ■

For the second-order equation stability is assured by easily verifiable conditions.

Theorem 2.6   *Let $\{y^{(h)}(n)\}$ be a fundamental set of the equation*

$$y(n) + a(n)y(n+1) + b(n)y(n+2) = 0,$$

with $y^{(1)}(n) \neq 0$.

(i)   *If*

$$y^{(1)}(n)/y^{(1)}(n+1) = O(1), \qquad y^{(2)}(n)/y^{(1)}(n) = O(1),$$

then the equation is weakly stable for the computation of $y^{(1)}(n)$ and in the

$\| \cdot \|_1$ *norm*

$$\alpha(k) \leqslant \frac{C}{|D(k)|} (|y^{(1)}(k)| + |y^{(1)}(k+1)|) \sup_{j=k,k+1} (|y^{(1)}(j)| + |y^{(2)}(j)|),$$

for some $C > 0$.

  (ii) *If*

$$\overline{\lim} \left| \frac{y^{(1)}(n)}{y^{(1)}(n+1)} \right| < 1, \quad \overline{\lim} \left| \frac{y^{(2)}(n+1)}{y^{(2)}(n)} \right| < 1, \tag{2.32}$$

then the equation is stable for the computation of $y^{(1)}(n)$.

Proof (i) We have

$$\alpha(k, n) \leqslant \left(\sup_{n > k} \frac{\|Y(n)\|}{\|\mathbf{y}^{(1)}(n)\|} \right) \| Y(k)^{-1} \| \|\mathbf{y}^{(1)}(k)\|,$$

and the result follows easily. (Note $D(k) \neq 0$.)

  (ii) Choose N so that

$$\left| \frac{y^{(1)}(n)}{y^{(1)}(n+1)} \right| < A < 1, \quad \left| \frac{y^{(2)}(n+1)}{y^{(2)}(n)} \right| < B < 1, \qquad n > N.$$

Then

$$|D(k)| := |y^{(1)}(k)y^{(2)}(k+1) - y^{(2)}(k)y^{(1)}(k+1)|$$

$$= |y^{(1)}(k+1)y^{(2)}(k)| \left| 1 - \frac{y^{(1)}(k)y^{(2)}(k+1)}{y^{(1)}(k+1)y^{(2)}(k)} \right|$$

$$> |y^{(1)}(k+1)y^{(2)}(k)| (1 - AB), \qquad k > N.$$

Thus we have

$$\alpha(k, n) < \frac{C}{|y^{(1)}(n+1)y^{(2)}(k)|} \max_{j=k,k+1} [|y^{(1)}(n)y^{(2)}(j) - y^{(2)}(n)y^{(1)}(j)|$$

$$+ |y^{(1)}(n+1)y^{(2)}(j) - y^{(2)}(n+1)y^{(1)}(j)|]$$

$$< \frac{2C}{|y^{(1)}(n+1)y^{(2)}(k)|} \max_{j=k,k+1} [|y^{(1)}(n+1)y^{(2)}(j)| + |y^{(1)}(j)y^{(2)}(n)|]$$

$$< 2C \left(1 + \frac{|y^{(1)}(k+1)| \, |y^{(2)}(n)|}{|y^{(1)}(n+1)| \, |y^{(2)}(k)|} \right)$$

$$< 4C, \qquad n > k > N.$$

Since $D(k) \neq 0$, $\alpha(k, n)$ is finite on the grid $0 \leqslant k < n \leqslant N$. Thus α is bounded. ∎

Additional material on the stability of scalar equations of arbitrary order can be found in Oliver (1967).

3 First-order equations used in the backward direction: the Miller algorithm

3.1 Introduction: the algorithm

I shall begin this chapter with an

Example 3.1 Consider the computation of

$$w(n) := \int_{-1}^{1} t^n e^{-xt} \, dt, \qquad n \geq 0, \quad x \neq 0, \tag{3.1}$$

by use of the difference equation

$$y(n) - \frac{x}{(n+1)} y(n+1) = \frac{(-1)^n e^x + e^{-x}}{(n+1)}, \tag{3.2}$$

with the initial condition $y(0) := x^{-1}(e^x - e^{-x})$. Application of the theory of stability for forward recursion found in Chapter 2 reveals that (2) is not stable for the computation of $w(n)$. In fact an integration by parts shows

$$w(n) = \frac{e^{-x} + (-1)^n e^x}{(n+1)} + O(n^{-2}), \tag{3.3}$$

while a nontrivial solution of the homogeneous equation is given by

$$y^*(n) = \frac{n!}{x^n}. \tag{3.4}$$

The result (3) used in equation (2.17) gives

$$\rho(n) \sim \frac{(n+1)!}{|x|^{n+1}} \times \begin{cases} 1, & n \text{ odd}; \\ |\tanh x|, & n \text{ even}. \end{cases}$$

Theorem 2.3 shows the recurrence is unstable for $w(n)$ at every point.

But consider the following stratagem. Let N be a 'large' integer and consider the sequence $y_N(N), y_N(N-1), \ldots, y_N(0)$, where $y_N(N) = 0$ and $y_N(n)$ is generated from the difference equation (2) used in the backward direction, i.e.,

$$y_N(n) - \frac{x}{(n+1)} y_N(n+1) = \frac{(-1)^n e^x + e^{-x}}{(n+1)}, \qquad n = N-1, N-2, \ldots, 0.$$

Table 3.1

n \ N	4	8	12	16
		$y_N(n)$		
0	2.327 386	2.350 395	2.350 402	2.350 402
1	−0.758 774	−0.735 767	−0.735 759	−0.735 759
2	0.832 854	0.878 869	0.878 885	0.878 885
3	−0.587 601	−0.449 555	−0.449 507	−0.449 507
4	0	0.552 182	0.552 373	0.552 373

To fix ideas, let $x = 1$ and take $N = 4, 18, 12, 16$. We obtain Table 3.1 for the values of $y_N(n)$, $0 \leqslant n \leqslant 4$. ∎

Obviously some kind of convergence is occurring in the example. In fact we can show easily that $\lim_{N \to \infty} y_N(n) = w(n)$. By the theory in Appendix A I can write

$$y_N(n) = cy^*(n) + w(n).$$

Using the initial value $y_N(N) = 0$ and solving for c yields

$$y_N(n) = w(n) - \frac{w(N)y^*(n)}{y^*(N)}.$$

Since

$$|w(n)| \leqslant \frac{2e^x}{(n+1)}, \qquad x > 0, \tag{3.5}$$

I have

$$|y_N(n) - w(n)| \leqslant \frac{2e^n n! \, x^{N-n}}{(N+1)!},$$

and the assertion follows. Of course the accuracy deteriorates with increasing n but, nevertheless, for $n \in$ any *fixed* subset of Z^0 and for any fixed x, $w(n)$ may be computed as accurately as one wishes by this method, at least theoretically. There is as yet no assurance that some potentially destructive build-up of round-off error caused by finiteness of the computer representation of numbers will not occur. I will return to this topic later.

This kind of an algorithm—by now actually a class of algorithms applicable to many different kinds of difference equations under many different conditions—is called a *Miller algorithm* after J. C. P. Miller who used a variant of the algorithm in 1952 to construct tables of the Bessel functions $I_n(x)$. Apparently, though, the idea is much older. It can be traced back at least as far as a 1910 paper of Lord Rayleigh.

The algorithm of the previous example displays two features which characterize the general class of Miller algorithms:

(i)   the difference equation is used in the *backward* direction;
(ii) no initial (i.e., for $n = 0$) values were required.

The second feature, in the general Miller algorithm, takes the form that only global information, if any, is required to single out the desired solution of the equation. Such information may be very general; for instance, it may only involve knowing the sum of some series of the form $\sum c(k)w(k)$. This is one of the most attractive features of the algorithm, since in almost any practical situation the necessary information is available.

The following notation will be useful.

$$E_N(n) := w_N(n) - w(n), \tag{3.6}$$

denotes the error of the algorithm and

$$E_N^*(n) := E_N(n)/w(n), \qquad w(n) \neq 0, \tag{3.7}$$

the relative error.

3.2   Convergence and error analysis

Consider the general first-order nonhomogeneous equation

$$y(n) + a(n)y(n+1) = b(n), \qquad n \geq 0. \tag{3.8}$$

We assume as before that $a(n) \neq 0$. Suppose one wishes to compute a solution $w(n)$ of (8) by the use of Miller's algorithm, i.e., one sets $y_N(N) = 0$ and computes $y_N(n)$ for $n = N-1, N-2, \ldots, 0$ from

$$y_N(n) = b(n) - a(n)y_N(n+1). \tag{3.9}$$

There are, of course, two apparently distinct problems associated with this computational scheme. First, does the algorithm, done in infinite precision at each step, converge? And, second, how does error in the algorithm accumulate assuming that at each step random errors are introduced into the computations?

I shall show that for the simple first-order equation both of these problems can be treated at the same time. The analysis is essentially that of Gautschi (1961a).

Let $\{\varepsilon_N(n)\}$ be an infinite matrix of errors. What one actually computes when one applies the Miller algorithm to (8) is a quantity which satisfies

$$z_N(n) = [b(n) - a(n)z_N(n+1)][1 + \varepsilon_N(n)], \qquad n = N-1, N-2, \ldots,$$
$$\tag{3.10}$$

with $z_N(N) = \varepsilon_N(N)$. We rewrite this as

$$z_N(n) + a(n)(1 + \varepsilon_N(n))z_N(n+1) = b(n)(1 + \varepsilon_N(n)),$$

and (8) as

$$w(n) + a(n)(1 + \varepsilon_N(n))w(n+1) = b(n) + a(n)\varepsilon_N(n)w(n+1).$$

Subtracting these two equations and letting

$$E_N(n) := z_N(n) - w(n)$$

gives

$$E_N(n) + a(n)(1 + \varepsilon_N(n))E_N(n+1) = \varepsilon_N(n)w(n). \tag{3.11}$$

Solving this equation by the procedure in A-IX gives

$$E_N(n) = y_N^*(n) \left\{ \frac{\varepsilon_N(N) - w(N)}{y_N^*(N)} + \sum_{j=n}^{N-1} \frac{\varepsilon_N(j)w(j)}{y_N^*(j)} \right\}, \tag{3.12}$$

where $y_N^*(n)$ is any nontrivial solution of the homogeneous equation associated with (11). Further, (A.24) shows $y_N^*(n)$ may be written

$$y_N^*(n) = y^*(n)\varepsilon_N^*(n),$$

where $y^*(n)$ is any nontrivial solution of the homogeneous equation associated with (8) and $\varepsilon_N^*(n)$ is an appropriately chosen nontrivial solution of

$$y(n) - (1 + \varepsilon_N(n))y(n+1) = 0. \tag{3.13}$$

We thus obtain

$$E_N(n) = y^*(n)\varepsilon_N^*(n) \left\{ \frac{\varepsilon_N(N) - w(N)}{y^*(N)\varepsilon_N^*(N)} + \sum_{j=n}^{N-1} \frac{\varepsilon_N(j)w(j)}{y^*(j)\varepsilon_N^*(j)} \right\}. \tag{3.14}$$

Equation (14) with $\varepsilon_N(n) \equiv 0$ produces the following convergence criterion:

Theorem 3.1 *Let $y_N(n)$ be computed, as described, from (9) in exact arithmetic. Then*

$$\lim_{N \to \infty} y_N(n) = y(n) \Leftrightarrow \frac{w(N)}{y^*(N)} = o(1),$$

for every nontrivial solution $y^*(n)$ of the homogeneous equation

$$y(n) + a(n)y(n+1) = 0.$$

We then have

$$|y_N(n) - w(n)| = |y^*(n)| \left| \frac{w(N)}{y^*(N)} \right| = O\left(\left| \frac{w(N)}{y^*(N)} \right| \right), \qquad N \to \infty. \quad \blacksquare$$

Suppose a substitution $Y(n) = \lambda(n)y(n)$ is made in the original difference equation. Let $Y(n)$ be computed from this new equation by an application of the Miller algorithm and then $y(n)$ computed from $Y(n)$. Can $\lambda(n)$ be chosen to substantially reduce the error $[Y_N(n)/\lambda(n)] - y(n)$? Gautschi (*ibid*) has observed that the computations cannot be improved in such a facile way. Indeed, this is obvious from the representation

$$w(n) = y^*(n) \left\{ C - \sum_{k=0}^{n-1} \frac{b(k)}{y^*(k)} \right\}.$$

A description of the error accumulation follows from (14).

Theorem 3.2 *Let* $w(n) \neq 0$. *Then*

$$\left| \frac{z_N(n) - w(n)}{w(n)} \right| \leq \left| \frac{y^*(n)}{w(n)} \right| \left\{ (1 + \varepsilon)^{N-n} \left[\frac{\varepsilon + |w(N)|}{|y^*(N)|} \right] \right.$$

$$\left. + \varepsilon \sum_{j=n}^{N-1} (1 + \varepsilon)^{j-n} \left| \frac{w(j)}{y^*(j)} \right| \right\}, \qquad 0 \leq n \leq N, \qquad (3.15)$$

where

$$\varepsilon = \sup_{0 \leq n \leq N} |\varepsilon_N(n)|.$$

Proof The theorem results by taking as a solution of equation (13)

$$\varepsilon_N^*(n) := \prod_{k=0}^{n-1} (1 + \varepsilon_N(k))^{-1}. \quad \blacksquare$$

Example 3.2 Consider Example 1. Using the value given by (4) for $y^*(n)$ and the inequality (5) yields

$$\left| \frac{z_N(n) - w(n)}{w(n)} \right| \leq \left| \frac{n!}{x^n w(n)} \right| \left\{ \frac{x^N (1 + \varepsilon)^{N-n}}{N!} \left[\varepsilon + \frac{2e^x}{N+1} \right] \right.$$

$$\left. + 2\varepsilon e^x x^n \sum_{j=0}^{\infty} \frac{(1 + \varepsilon)^j x^j}{(j+n+1)!} \right\}.$$

Using the fact that $(j+n+1)! \geq j!\,(n+1)!$ produces

$$\left| \frac{z_N(n) - w(n)}{w(n)} \right| \leq \frac{n!\,[x(1+\varepsilon)]^{N-n}}{N!\,w(n)} \left(\varepsilon + \frac{2e^x}{N+1} \right) + \frac{2\varepsilon x^n e^{(2+\varepsilon)x}}{(n+1)!}.$$

The latter term represents the contribution to the error produced purely by accumulated round-off error. Since it is independent of N it cannot be diminished by taking N large. ■

Example 3.3 (General moment integrals) Let

$$\phi(n) := \int_a^b t^n \psi(t)\, \mathrm{d}t, \qquad -\infty < a < b < \infty, \quad n \in Z^0,$$

where $\psi \in C^1[a, b]$ and satisfies the linear differential equation

$$\psi'(t) + d\psi(t) = g(t), \qquad d = \text{const.}$$

A direct computation shows that $\phi(n)$ satisfies

$$y(n) - \frac{d}{(n+1)} y(n+1) = c(n), \qquad n \geq 0, \tag{3.16}$$

with

$$c(n) := \frac{t^{n+1}\psi(t)}{(n+1)}\bigg|_a^b + \frac{1}{(n+1)} \int_a^b t^{n+1} g(t)\, \mathrm{d}t.$$

We have

$$y^*(n) = \frac{n!}{d^n}, \qquad |\phi(n)| \leq M\lambda^n,$$

$\lambda = \max(|a|, |b|)$, for some M. Thus Miller's algorithm for the computation of $\phi(n)$ based on (16) converges. The contribution of pure round-off error, i.e., the last term in (15), is bounded by

$$\frac{\varepsilon M(\lambda |d|)^n}{n!} e^{\lambda |d|(1+\varepsilon)}.$$

Thus if the absolute relative errors are bounded, round-off error cannot accumulate with N. The Miller algorithm is obviously the method of choice for the computation of such integrals. Consulting the treatment of forward recursion given in Chapter 2 we find

$$\rho(n) := \left| \frac{\phi(0) y^*(n)}{\phi(n)} \right| \geq \frac{|\phi(0)|\, n!}{M(|\lambda|\, d)^n} \to \infty,$$

and so the recursion is unstable in the forward direction for the computation of $\phi(n)$. Moreover, a value for $\phi(0)$ is required. ■

4 Second-order homogeneous equations: the Miller algorithm

4.1 The algorithm

I will write the difference equation in the form

$$y(n) + a(n)y(n+1) + b(n)y(n+2) = 0, \qquad n \geq 0, \quad b(n) \neq 0. \qquad \textbf{(4.1)}$$

(In some situations $b(n)$ *does* vanish for certain values of n but it is often possible in these cases to make a change of variable $Y(n) = \lambda(n)y(n)$ which will produce an equation of the desired type.)

As usual let $w(n)$ be the desired (nontrivial) solution of the equation. I assume that a convergent *normalizing series* involving $w(n)$ is known:

$$S := \sum_{k=0}^{\infty} c(k)w(k) \neq 0. \qquad \textbf{(4.2)}$$

Let N be an integer ≥ 0 (actually N will be large) and define $y_N(n)$, $0 \leq n \leq N+1$, by

$$y_N(n) := \begin{cases} 0, & n = N+1; \\ 1, & n = N; \end{cases}$$

$$y_N(n) + a(n)y_N(n+1) + b(n)y_N(n+2) = 0,$$
$$n = N-1, N-2, \ldots, 0.$$

$$\textbf{(4.3)}$$

We take as an approximation to $w(n)$

$$w_N(n) := S y_N(n)/S_N, \qquad \textbf{(4.4)}$$

$$S_N := \sum_{k=0}^{N} c(k)y_N(k). \qquad \textbf{(4.5)}$$

Note that $\sum_{k=0}^{N} c(k)w_N(k) = S$ and $w_N(n)$ satisfies the equation (1).
We say that the algorithm converges if

$$\lim_{N \to \infty} w_N(n) = w(n). \qquad \textbf{(4.6)}$$

Sometimes an initial value $w(0) \neq 0$ is known. Then one can replace (4)

Table 4.1 Computation of $I_n(1)$.

n	$N=4$	$w_N(n)$ $N=6$	$N=8$
0	1.265 927 978	1.266 066 460	1.266 065 876
1	0.565 096 950	0.565 159 364	0.565 159 103
2	0.135 734 072	0.135 747 732	0.135 747 669
3	0.022 160 665	0.022 168 435	0.022 168 425
4	0.002 770 083	0.002 737 123	0.002 737 120

by

$$w_N(n) = \frac{w(0) y_N(n)}{y_N(0)}, \qquad y_N(0) \neq 0. \tag{4.7}$$

The algorithm so defined is called the *simplified Miller algorithm*.

Example 4.1 Calculate $I_n(x)$ from the recurrence

$$y(n) - \frac{2(n+1)}{x} y(n+1) - y(n+2) = 0, \tag{4.8}$$

and the normalization

$$1 = \sum_{k=0}^{\infty} \varepsilon_k (-1)^k I_{2k}(x). \tag{4.9}$$

(Here $\varepsilon_k = 2 - \delta_{k0}$; see *List of Symbols*.)

Some sample computations are given in Table 4.1 for $x = 1$. The entries in the last column are accurate to the figures given. This is essentially the algorithm given in Miller (1952). Another possible normalization relation for this example is

$$e^x = \sum_{k=0}^{\infty} \varepsilon_k I_{2k}(x). \quad \blacksquare \tag{4.10}$$

As one might suspect, the success of the method depends on the rate of growth of the desired solution $w(n)$ compared with other solutions of the equation. Recall that for any homogeneous linear difference equation the set $\mathcal{M}$ of minimal solutions is a one-dimensional subspace of the solution set $\mathcal{S}$ of the equation. (Any minimal solution $y^{(1)}(n)$ may be completed to a basis $\{y^{(1)}(n), y^{(2)}(n), \ldots, y^{(\sigma)}(n)\}$ of $\mathcal{S}$. For the case $\sigma = 2$ the actual construction of $y^{(2)}(n)$ may be carried out by the formula A-X.) Further, no matter what the order of the equation is a minimal solution is uniquely determined by *one* initial value. These observations suggest that minimal solutions are ideal candidates for the application of the Miller algorithm.

If the equation has a minimal solution but $w(n)$ is not a constant multiple of that solution then forward recurrence (which requires two initial values) should be used to compute $w(n)$. When the equation possesses no minimal solution either backward or forward recurrence may be unstable. However, there are special techniques which depend on averaging solutions which can then be used (see Section 4.7).

The reason the algorithm works so well on I_n is that

$$I_n(x) \sim \left(\frac{x}{2}\right)^n \Big/ n!, \qquad n \to \infty, \tag{4.11}$$

while the behavior of a linearly independent solution is

$$(-1)^n K_n(x) \sim \frac{1}{2}\left(\frac{-2}{x}\right)^n (n-1)!, \qquad n \to \infty. \tag{4.12}$$

Thus I_n is strongly minimal. A good question is, how may K_n be computed (without the use of initial values)? This question is addressed in Sections 5.2 (Example 5.2), 5.7, 7.3.

If only one value of $\{I_n\}$ is required, the Miller algorithm loses some of its attractiveness. Henrici (1979) has shown that in this case it may be preferable to use the fast Fourier transform on an appropriate generating function. At any rate, often many of the values $\{I_n\}$ *are* required, for instance, when it is necessary to evaluate an infinite series involving the functions.

Before I discuss the problems of convergence and error propagation I will give a formalization of the Miller algorithm due to Shintani (1965) which is very useful computationally since it allows one to increase N in $y_N(r)$, r fixed, without recomputing $y_N(N+1)$, $y_N(N)$, ..., $y_N(0)$.

Theorem 4.1   (Duality theorem) $y_N(n)$ *satisfies*

$$b(N)y_N(n) + a(N+1)y_{N+1}(n) + y_{N+2}(n) = 0,$$
$$0 \leqslant N = n, n+1, n+2, \ldots, \tag{4.13}$$

with the initial conditions

$$y_n(n) = 1, \qquad y_{n+1}(n) = -a(n), \tag{4.14}$$

and

$$S_N(n) := \sum_{k=n}^{N} c(k)y_N(k), \tag{4.15}$$

satisfies

$$b(N)S_N(n) + a(N+1)S_{N+1}(n) + S_{N+2}(n) = c(N+2),$$
$$0 \leqslant N = n, n+1, n+2, \ldots, \tag{4.16}$$

with initial conditions

$$S_n(n) = c(n), \qquad S_{n+1}(n) = -a(n)c(n) + c(n+1). \tag{4.17}$$

Proof See Theorem 7.3. ■

Letting $n = 0$ gives

Corollary S_N *satisfies*

$$b(N)S_N + a(N+1)S_{N+1} + S_{N+2} = c(N+2),$$
$$S_0 = c(0), \quad S_1 = c(1) - c(0)a(0). \quad ■ \tag{4.18}$$

The two previous theorems' results are particularly welcome when only a few values of the sequence $\{w(n)\}$ are required. For these values of n $y_N(n)$ can be computed for increasing values of N, $N = 0, 1, 2, \ldots$, from (13) and (14) while from (18) S_N can be computed without the necessity of computing the sum (5) at each step.

The convergence properties of this algorithm are described in the two following theorems.

Theorem 4.2 *Let (1) have a minimal solution* $w(n)$ *for which*

$$S := \sum_{k=0}^{\infty} c(k)w(k) \neq 0.$$

Then the Miller algorithm (3)–(5) *converges to* $w(n) \Leftrightarrow$

$$\lim_{N \to \infty} \tau_N S_N^* = 0, \tag{4.19}$$

where

$$\tau_N := \frac{w(N+1)}{y(N+1)}, \qquad S_N^* := \sum_{k=0}^{N} c(k)y(k), \tag{4.20}$$

$y(n)$ *any dominant solution of the equation.*

Proof Let

$$U_N := \sum_{k=N+1}^{\infty} c(k)w(k).$$

A direct computation using (4) and (5) gives

$$w_N(n) - w(n) = \frac{-S\tau_N y(n) + w(n)(U_N + \tau_N S_N^*)}{s - U_N - \tau_N S_N^*}. \tag{4.21}$$

If the limit (19) exists, then $w_N(n)$ is defined for n sufficiently large and $w_N(n) \to w(n)$. The converse is obvious. ■

The following corollary is useful in many applications.

Corollary Let (1) have a minimal solution $w(n)$ with $w(0) \neq 0$.
Then the simplified Miller algorithm defined by (3), (7) converges to $w(n)$.
Conversely, when the algorithm converges in the sense that

$$\lim_{N \to \infty} y_N(n)/y_N(0) = w(n)$$

exists, then $w(n)$ is a minimal solution of the equation.

Proof The proof of the first part of the corollary is direct.
To show the second part, choose a fundamental set $\{y^{(h)}(n)\}$ with the properties

(i) $y^{(2)}(n)$ does not vanish for $n > n_0$;
(ii) $r(n) := y^{(1)}(n)/y^{(2)}(n)$ does not approach a finite nonzero constant C.

By A-IV such a $y^{(2)}(n)$ exists. Also, if $r(n) \to C$, choose instead the set $\{y^{(1)}(n) - Cy^{(1)}(n), y^{(2)}(n)\}$ which will be linearly independent by Milne–Thomson (1960, p. 360).
I have

$$\frac{y_N(n)}{y_N(0)} = \frac{y^{(1)}(n) - r(N+1)y^{(2)}(n)}{y^{(1)}(0) - r(N+1)y^{(2)}(0)} .$$

Either $r(n) \to \pm\infty$ or $r(n) \to 0$ if convergence is to occur. In the first case convergence is to $y^{(2)}(n)/y^{(2)}(0)$ which, indeed, is a minimal solution.
In the second case we must have $y^{(1)}(0) \neq 0$, else $y^{(1)}(n) \equiv 0$ and $\{y^{(h)}(n)\}$ is not a fundamental set. So convergence is to $y^{(1)}(n)$, which is the minimal solution. ∎

The convergence of the Miller algorithm also depends intimately on the speed of convergence of the normalization series $\sum c(k)w(k)$, or, more accurately, on the decay rate of the coefficients $c(k)$. In fact if $w(n)$ is a minimal and $y(n)$ a dominant solution in most instances the relative error

$$E_N^*(n) := \frac{w_N(n) - w(n)}{w(n)},$$

is $O(w(N+1)c(N)y(N)/y(N+1))$ for the Miller algorithm and $O(w(N+1)/y(N+1))$ for the simplified algorithm, as can be inferred, under appropriate conditions, from (20), (21). The latter is usually much smaller. Of course the price to be paid in the simplified algorithm is that $w(0)$ must be known.
When the equation is of Birkhoff type (i.e., $a(n)$, $b(n)$ have appropriate asymptotic series, see Section B.2) a much stronger statement about

convergence can be made. The theory of linear difference equations discussed in Section B.2 shows that in such a case there is a fundamental set of solutions having the behavior

$$y^{(h)}(n) \sim e^{Q_h(\rho;n)} S_h(\rho, n), \qquad n \to \infty.$$

(See Appendix B for the notation in this equation.)

Wimp (1969) demonstrates the following.

Theorem 4.3 *Let equation (1) be of the Birkhoff type, let $w(n)$ be a minimal solution of the equation, let*

$$c(k) = O(e^{Q(k)} k^{\alpha}),$$

$Q(k)$ of the form (B.11) and let

$$e^{Q(k)} k^{\alpha+1} \ln k \, w(k) = o(1).$$

Then the computation of $w(n)$ by backward recursion based on (3)–(5) converges.

Proof See Wimp (1969). ■

Sources for much of the material to follow, as well as good general surveys of many computational aspects of three-term recurrences, are in Gautschi (1967, 1972a).

When the simplified algorithm converges an interesting property of a related continued fraction results.

Theorem 4.4 *Let the limit*

$$\lim_{N \to \infty} \frac{y_N(n)}{y_N(0)} := w(n) \neq 0, \qquad n \geqslant 0,$$

exist.

Then

$$\frac{w(n+1)}{w(n)} = \frac{-1}{a(n)-} \frac{b(n)}{a(n+1)-} \frac{b(n+1)}{a(n+2)-} \cdots, \qquad n \geqslant 0. \tag{4.22}$$

Proof Note that for each n, $y_N(n) \neq 0$, N sufficiently large. In fact if $y_N(0) = 0$ for an infinite number of values of N, then $w(n)$ is not defined; clearly $y_N(n) \neq 0$, $n > 0$, for N sufficiently large.

Now let $\tau_N(n) := y_N(n+1)/y_N(n)$. From (3) I find for N sufficiently large, n fixed,

$$\tau_N(n) = \frac{-1}{a(n)+b(n)\tau_N(n+1)}, \tag{4.23}$$

or

$$\tau_N(n) = \frac{-1}{a(n)-} \frac{b(n)}{a(n+1)-} \frac{b(n+1)}{a(n+2)-} \cdots \frac{b(N-2)}{a(N-1)}, \tag{4.24}$$

by repeated application of (23), since $\tau_N(N) = 0$. (Note $a(N-1) \neq 0$ or else $y_N(n) \equiv 0$, $0 \leqslant n \leqslant N-1$.) From the above expression, the theorem follows on letting $N \to \infty$. ■

A criterion for the existence of a minimal solution can be based on the continued fraction (22).

Theorem 4.5 (Pincherle (1894)) *The continued fraction (22) with $n = 0$ converges $\Leftrightarrow$ (1) has a minimal solution $\mu(n)$ with $\mu(0) \neq 0$.*

When the fraction does converge the relationship (22) holds (with $w = \mu$) for each n for which $\mu(n) \neq 0$.

Proof $\Rightarrow$: The numerator approximants $P(n)$ and the denominator approximants $Q(n)$ of the continued fraction both satisfy

$$y(n) = a(n-1)y(n-1) - b(n-2)y(n-2), \qquad n \geqslant 2, \tag{4.25}$$

with

$$P(0) = 0, \quad P(1) = -1; \qquad Q(0) = 1, \quad Q(1) = a(0). \tag{4.26}$$

Let $\{y^{(h)}(n)\}$ be a fundamental set for (1). The equation (25) is essentially the adjoint equation and has solutions

$$(-1)^{n+h+1} y^{(h)}(n+1)/D(n);$$

see A-VI. Thus the fact that the ratio of two solutions of (25) approaches a limit implies the ratio of two solutions of (1) approaches a limit; again, call these solutions $\{y^{(h)}(n)\}$. Thus

$$\lim_{n \to \infty} \frac{y^{(1)}(n)}{y^{(2)}(n)} = c.$$

Define

$$\mu(n) := y^{(1)}(n) - c y^{(2)}(n).$$

The initial conditions (26) guarantee that $\{y^{(h)}(n)\}$ are linearly independent. Let $y(n) = c_1 y^{(1)}(n) + c_2 y^{(2)}(n)$ be any other solution of (1). Then

$$\lim_{n \to \infty} \frac{\mu(n)}{y(n)} = \lim_{n \to \infty} \frac{[y^{(1)}(n)/y^{(2)}(n)] - c}{c_1\{[y^{(1)}(n)/y^{(2)}(n)] - c\} + c_2 + c_1 c} = 0,$$

unless $c_2 + c_1 c = 0$. But if this is true, $y(n)$ is a multiple of $\mu(n)$.

$\Leftarrow$: Obvious. ■

When the coefficients $a(n)$, $b(n)$ of the equation (1) are negative the continued fraction (22) provides a very clean criterion for convergence of $y_N(n)/y_N(0)$.

Theorem 4.6 *Let $a(n), b(n) < 0$, $n \geq 0$. Then the limit*

$$\lim_{N \to \infty} \frac{y_N(n)}{y_N(0)},$$

exists for $n = 1 \Leftrightarrow$ it exists for all $n \geq 1$ and this happens $\Leftrightarrow$ the series

$$\sum \left| \frac{a(k)a(k+1)}{b(k)} \right|^{1/2}, \tag{4.27}$$

diverges.

Proof By Pringsheim's theorem (Khovanskii 1963, p. 45) the continued fraction (22) converges $\Leftrightarrow$ the series

$$\sum_{k=2}^{\infty} \left| \frac{a(n+k-2)a(n+k-1)}{b(n+k-2)} \right|^{1/2},$$

is divergent.

But this series diverges with (27). Thus $\lim_{N \to \infty} y_N(n+1)/y_N(n)$ exists for one $n \in Z^0 \Leftrightarrow$ it exists for all $n \in Z^0$ and this happens $\Leftrightarrow$ (27) diverges. (Note that $y_N(n) \neq 0$, $0 \leq n \leq N$.) I have

$$\frac{y_N(n)}{y_N(0)} = \prod_{j=0}^{n-1} \frac{y_N(j+1)}{y_N(j)}.$$

If any $y_N(j+1)/y_N(j)$ converges they all do and the quantity on the left converges. ∎

This theorem does not reveal whether the simplified Miller algorithm converges to the *desired* solution. The following is more or less a negative criterion.

Theorem 4.7 *Let*

$$a(n), b(n) < 0, \qquad n \geq 0.$$

If the simplified Miller algorithm converges it converges to a solution $w(n)$ of the recurrence having the property

$$\frac{1}{|a(n)| + \dfrac{b(n)}{a(n+1)}} \leq \frac{w(n+1)}{w(n)} \leq \frac{1}{|a(n)|}. \tag{4.28}$$

Proof This follows from the representation (22) and the inequality (Khovanskii (1963, p. 7 (1.10))). ∎

Example 4.2 For the recursion (8) satisfied by $I_n(x)$, $a(n) = -2(n+1)/x$, $b(n) = -1$, and the series (27) diverges. From the asymptotic estimates (11)–(12) we conclude immediately that the algorithm converges to $I_n(x)$. ∎

Example 4.3 Consider the recurrence (2.30) satisfied by $w(n) =$ i^n erfc x; here $a(n) = -2x$, $b(n) = -2(n+2)$. The series (27) diverges so the algorithm converges. However, unless we have some *a priori* knowledge of the asymptotic behavior of $w(n)$ there is no easy way to decide to which solution the Miller algorithm converges. ∎

As the previous example shows, it is difficult, lacking any additional information, to characterize the limit function of the Miller algorithm by a cursory examination of the difference equation. Even when a great deal of information is available about the coefficients $a(n)$ and $b(n)$—for example, the fact that they are rational functions of n—it may be very difficult to identify $w(n)$.

The Olver growth theorems given in Section B.3 sometimes allow one to decide whether a solution having a prescribed initial value is a minimal solution and thus to conclude that the Miller algorithm converges to that solution. In effect the theorems allow us to match the given solution with its asymptotic expansion. Example B.4 demonstrates how the theorems can be used.

4.2 Reduction of the error by the use of asymptotic information

Gautschi (1967) and Scraton (1972) have observed that when asymptotic information is available about the desired solution of the difference equation this information may often be put to advantage.

Suppose

$$w(n) = \phi(n)[1 + o(1)], \qquad n \to \infty, \quad \phi(n) \neq 0,$$

where $\phi(n)$ is known. Instead of taking $y_N(N+1) := 0$, $y_N(N) := 1$, take

$$y_N(N+1) := 1, \qquad y_N(N) := \frac{\phi(N)}{\phi(N+1)},$$

and compute as usual.

The following example is taken from Scraton's paper.

Example 4.4

$$w(n) := \frac{1}{\sqrt{\pi}} \int_0^\infty \frac{e^{-t} t^{n-1/2}}{(t+\frac{1}{2})^n} \, dt, \tag{4.29}$$

Table 4.2   Error in the tabulation of $w(n)$ by Miller's algorithm (upper part) and the revised algorithm (lower part).

		$n = 1$	$n = 2$	Error $\times 10^8$ $n = 3$	$n = 4$	$n = 5$
	10	$-38\,060$	$-76\,120$	$-123\,695$	$-183\,957$	$-259\,680$
	15	$-5\,427$	$-10\,854$	$-17\,638$	$-26\,230$	$-37\,028$
Miller N	20	$-1\,035$	$-2\,070$	$-3\,364$	$-5\,002$	$-7\,062$
	25	-239	-477	-776	$-1\,153$	$-1\,628$
	30	-63	-126	-205	-305	-430
	10	-206	-412	-669	-995	$-1\,404$
	15	-18	-36	-58	-87	-123
Revised N	20	-2	-5	-8	-12	-17
	25	0	-1	-1	-2	-3
	30	0	0	0	0	-1

satisfies

$$(n + \tfrac{1}{2})y(n) - 2(n + 1)y(n + 1) + (n + 1)y(n + 2) = 0, \tag{4.30}$$

and has the behavior

$$w(n) = Ke^{-\sqrt{(2n)}}[1 + O(n^{-1/2})],$$

see Section 5.2; $w(n)$ is minimal. Since

$$w(n)/w(n + 1) = 1 + (2n)^{-1/2} + (4n)^{-1} + O(n^{-3/2})$$

(as is easily determined by direct substitution of a series of the form B.13 with $\rho = 2$ into the equation) we take

$$y_N(N + 1) := 1, \qquad y_N(N) := 1 + (2N)^{-1/2} + (4N)^{-1}.$$

Table 4.2 is self-explanatory. ■

4.3   Error analysis, the simplified algorithm: case of negative coefficients

The *truncation error* of the algorithm is the error in exact arithmetic, i.e., $|w(n) - w_N(n)|$. When the difference equation has *real negative coefficients* the error study of the simplified Miller algorithm is very elegant. The analysis and numerical data to follow are due to Tait (1967). Note that if $a(n) > 0$ while $b(n) < 0$ then the substitution $y(n) = (-1)^n z(n)$ brings the equation into the desired form. Most of the second-order homogeneous difference equations that arise in practical applications are amenable to this analysis.

I shall actually consider a slight generalization of the algorithm defined by (7), i.e., let $y_N(n)$ be defined by

$$y_N(n) := \begin{cases} \gamma, & n = N+1, \\ 1, & n = N, \end{cases}$$

and $\mathcal{L}[y_N(n)] = 0$, $n = N-1, N-2, \ldots, 0$. Allowing $y_N(N+1)$ to be other than 0 can give rise to certain computational advantages.

Again, let $w(n)$ be the solution to be computed. Define, as usual,

$$E_N(n) := w_N(n) - w(n), \qquad E_N^*(n) := E_N(n)/w(n), \qquad \textbf{(4.31)}$$

$w_N(n)$ as in (7).

Theorem 4.8   *Let $\{y_N^{(h)}(n)\}$, $h = 1, 2$, be a fundamental set for $\mathcal{L}[y(n)] = 0$ with the property*

$$y_N^{(1)}(n) := \begin{cases} 0, & n = N+1; \\ 1, & n = N; \end{cases} \qquad y_N^{(2)}(n) := \begin{cases} 1, & n = N+1; \\ 0, & n = N. \end{cases}$$

Assume $y_N^{(1)}(n) \neq 0$, $0 \leq n \leq N$, and $w_N(0) \neq 0$. Then

$$E_N(n) = \frac{(\gamma w(N) - w(N+1))}{y_N(0)} \, y_N^{(1)}(n) y_N^{(1)}(0) \sum_{k=0}^{n-1} T_k, \qquad 1 \leq n \leq N,$$

$$\textbf{(4.32)}$$

where

$$T_k := \frac{b(n-k-1)b(n-k) \cdots b(N-1)}{y_N^{(1)}(n-k)y_N^{(1)}(n-k-1)}. \qquad \textbf{(4.33)}$$

Proof   We have

$$w(n) = w(N)y_N^{(1)}(n) + w(N+1)y_N^{(2)}(n),$$

$$y_N(n) = y_N^{(1)}(n) + \gamma y_N^{(2)}(n),$$

so by direct substitution

$$E_N(n) = \frac{(\gamma w(N) - w(N+1))}{y_N(0)} \, y_N^{(1)}(0) y_N^{(1)}(n) \left(\frac{y_N^{(2)}(n)}{y_N^{(1)}(n)} - \frac{y_N^{(2)}(0)}{y_N^{(1)}(0)} \right).$$

However, by (A.12),

$$\frac{y_N^{(2)}(n)}{y_N^{(1)}(n)} = \frac{y_N^{(2)}(n-1)}{y_N^{(1)}(n-1)} + \frac{b(n-1)b(n) \cdots b(N-1)}{y_N^{(1)}(n)y_N^{(1)}(n-1)},$$

so the result follows by induction. ∎

Corollary

(i) $\left|\dfrac{T_{k+1}}{T_k}\right|<1 \Leftrightarrow \left|\dfrac{y_N^{(1)}(n-k-2)}{y_N^{(1)}(n-k)}\right|>|b(n-k-2)|,$

(ii) $\left|\dfrac{T_{k+1}}{T_k}\right|\leqslant\alpha<1 \Rightarrow |E_N(n)|\leqslant(1-\alpha)^{-1}$

$$\times\left|\frac{(\gamma w(N)-w(N+1))y_N^{(1)}(0)}{y_N(0)y_N^{(1)}(N-1)}\,b(n-1)b(n)\cdots b(N-1)\right|,$$

$$0\leqslant k\leqslant n-2. \quad\blacksquare$$

I now make the following assumptions:

(A) $w(n)$ is positive nonincreasing;
(B) $a(n), b(n)<0,\ 0\leqslant n\leqslant N-1.$

Note that (B) guarantees that $y_N^{(1)}(n),\ y_N^{(2)}(n)>0$. The use of equation (1) with n replaced by $(n-k-2)$ shows that Corollary (ii) holds. Obviously T_k alternates in sign and is decreasing in magnitude.

For our error studies it is sufficient to take $\gamma=0$ for the formula (32) displays explicitly how the error depends on γ. Thus $y_N^{(1)}(n)\equiv y_N(n)$.

Now $E_N(0)=0$,

$$E_N(1)=\frac{(-1)^{N-1}w(N+1)}{y_N(0)}\prod_{j=0}^{N-1}[-b(j)],$$

and from the difference equation

$$E_N(2)=\frac{(-1)^{N-2}w(N+1)[-a(0)]}{y_N(0)}\prod_{j=1}^{N-1}[-b(j)].$$

By induction the following may be established.

Theorem 4.9   *Under the stated conditions*

(i) $0\leqslant\dfrac{-a(n-2)w(N+1)}{y_N(n-2)}\displaystyle\prod_{j=n-1}^{N-1}[-b(j)]\leqslant(-1)^{N-n}E_N(n)$

$\qquad\qquad \leqslant\dfrac{w(N+1)}{y_N(n-1)}\displaystyle\prod_{j=n-1}^{N-1}[-b(j)],\qquad 2\leqslant n\leqslant N;$ (4.34)

(ii) $0\leqslant(-1)^{N-n}E_N^*(n)\leqslant\dfrac{w(N+1)\prod_{j=n-1}^{N-1}[-b(j)]}{w(N)y_N(n)y_N(n-1)},\qquad n\geqslant2.$ (4.35)

Remark   Note that (i) implies that $E_N(n)$ *alternates in sign* (with n).

Proof   Both $y_N^{(h)}(n)$ are nondecreasing as n decreases from N to $m-1$.

Assume

$$y_N(n)y_N(n-1) \geqslant -a(N-1) \prod_{j=n-1}^{N-2} [a(j)a(j+1)-b(j)]. \tag{4.36}$$

Then

$$\begin{aligned} y_N(n-1)y_N(n-2) &= -a(n-2)y_N(n-1)^2 - b(n-2)y_N(n)y_N(n-1) \\ &\geqslant -a(n-2)y_N(n-1)[-a(n-1)y_N(n)] \\ &\quad - b(n-2)y_N(n)y_N(n-1) \\ &= [a(n-2)a(n-1)-b(n-2)]y_N(n)y_N(n-1), \end{aligned}$$

and this provides the statement (36) with $n \to n-1$. Also the statement is true for $n = N-1$. Thus the induction is completed to demonstrate (36) for $m-1 \leqslant n \leqslant N-1$. Now $w(N) \geqslant -a(N)w(N+1)$. Using this and (36) in Theorem 9 (ii) gives the theorem. ■

Corollary Let the conditions of Theorem 9 hold and $a(n) \leqslant -1$ for $n \geqslant m-1$ for some $m \geqslant 1$ and let $b(N) < 0$. Then

$$0 \leqslant (-1)^{N-n} E_N^*(n) \leqslant \frac{\prod_{j=n-1}^{N-1}[-b(j)]}{a(N)a(N-1)\prod_{j=n-1}^{N-2}[a(j)a(j+1)-b(j)]},$$
$$m-1 \leqslant n \leqslant N-2. \quad ■ \quad (4.37)$$

The usefulness of equation (37) results from the fact that the quantity on the right depends only on the known coefficients of the equation.

Example 4.5 (Bessel functions) Consider the computation of $I_n(x)$ from equation (8). We get

$$0 \leqslant (-1)^{N-n} E_N^*(n) \leqslant \frac{x^2/4N^2}{\prod_{j=n}^{N-1}[(4j(j+1)/x^2)+1]}, \qquad n \geqslant \frac{x}{2}. \tag{4.38}$$

Denote the denominator of the right-hand side by $V(N)$.
   A simple computation shows $V(N)$ satisfies

$$V(N) \approx \left(\frac{4}{x^2}\right)^{N-n} \frac{\Gamma(N)\Gamma(N+1)}{\Gamma(n)\Gamma(n+1)} \prod_{j=n}^{\infty}\left[1+\frac{x^2}{4j(j+1)}\right], \qquad N \to \infty.$$

Thus convergence is

$$O\left(\left(\frac{x^2}{4}\right)^N \Big/ N(N!)^2\right).$$

The data in Table 4.3 illustrate a typical computation. ■

There is one more useful result concerning $E_N(n)$, $E_N^*(n)$.

Table 4.3 Computation of $I_n(x)$; $x = 2$; $N = 10$.

n	$y_N(n)$	$w_N(n)$	$w(n)$	bound $\|E_N^*(n)\|$	$\|E_N^*(n)\|$
11	0	—	—	—	—
10	1	—	—	—	—
9	10	3.043 890 744 (-6)	3.044 185 903 (-6)	1.099 (-4)	0.9695 (-4)
8	91	2.769 940 577 (-5)	2.769 936 951 (-5)	1.506 (-6)	1.309 (-6)
7	738	2.246 391 369 (-4)	2.246 391 420 (-4)	2.641 (-8)	2.207 (-8)
6	5 257	1.600 173 364 (-3)	1.600 173 364 (-3)	6.142 (-10)	—
⋮	⋮	⋮	⋮	⋮	⋮
0	7 489 051	2.279 585 302	2.279 585 302	0	0

Theorem 4.10

(i)   Let (A), (B) hold and $a(n) \leqslant b(n)$. Then

$$|E_N(n)| \leqslant |E_N(n+1)|; \qquad |E_N^*(n)| \leqslant |E_N^*(n+1)|, \qquad 1 \leqslant n \leqslant N-1;$$

(ii)   Let (A), (B) hold. Then

$$|E_N^*(n-1)| \leqslant |E_N^*(n+1)|, \qquad 1 \leqslant n \leqslant N-1.$$

Proof   Left to the reader. ■

When $a(n)$, $b(n)$ are not necessarily negative then, as might be expected, the analysis is not so simple nor the convergence so rapid. The following result is, however, useful. Again we take $\gamma = 0$ and assume $b(n) \neq 0$. Let

$$\alpha := \sup \left| \frac{T_{k+1}}{T_k} \right|, \qquad 0 \leqslant k \leqslant n-2.$$

Theorem 4.11   Let

$$-a(k) > 1 + |b(k)|, \qquad\qquad 0 \leqslant k \leqslant N-1,$$

and

$$-a(k) > \max\left[1 + |b(k)|, 2\,|a(k)|\right], \qquad 0 \leqslant k \leqslant n-2.$$

Then

$$|E_N(n)| \leqslant w(N+1)/(1-\alpha) y_N^{(1)}(n-1) \, |b(n-1) \cdots b(N-1)|. \qquad (4.39)$$

Proof   See Tait, *ibid*. ■

The round-off error can be handled similarly. It is not so easy to discuss the *relative* round-off error as was done for the first-order equation, cf.

Theorem 3.2. I shall instead adopt the tactic of examining the propagation of a single error, ε, introduced at some stage in the computations. Let m be an integer $< N$ and consider $z_N(n)$ defined by

$$
z_N(n) := \begin{cases} 0, & n = N+1, \\ 1, & n = N, \end{cases}
$$

and

$$
z_N(n) = -a(n)z_N(n+1) - b(n)z_N(n+2) + \varepsilon_N(n),
$$

$$
n = N-1, N-2, \ldots, 0,
$$

where

$$
\varepsilon_N(n) := \varepsilon \delta_{mn}.
$$

Then

$$
z_N(n) = y_N(n) + \delta(n),
$$

where $\delta(n)$ satisfies the same equation as $z_N(n)$. Using the particular solution in (A.30) gives

$$
\delta(n) = -\varepsilon \sum_{k=n}^{N} [y_N(k+1)y_N^{(2)}(n) - y_N^{(2)}(k+1)y_N(n)] \frac{\delta_{mk}}{D(k)}
$$

$$
= \frac{\varepsilon[y_N^{(2)}(m+1)y_N(n) - y_N(m+1)y_N^{(2)}(n)]}{D(m)}, \qquad 0 \leqslant n \leqslant m.
$$

Let, as before,

$$
E_N(n) := w_N(n) - w(n),
$$

and

$$
\hat{E}_N(n) := \begin{cases} \dfrac{w(0)}{z_N(0)} y_N(n) - w(n), & n > m, \\[2mm] \dfrac{w(0)}{z_N(0)} z_N(n) - w(n), & 0 \leqslant n \leqslant m. \end{cases}
$$

Tait shows that

$$
\hat{E}_N^*(n) := \frac{\hat{E}_N(n)}{w(n)} = E_N^*(n) - \varepsilon(1 + K_N(n))y_N(m+1)
$$

$$
\times \sum_{k=0}^{n-1} \frac{b(n-k-1)b(n-k) \cdots b(m-1)}{y_N(n-k)y_N(n-k-1)},
$$

where

$$
K_N(n) := \frac{y_N(0)}{z_N(0)} E_N^*(n) + \left(\frac{y_N(0)}{z_N(0)} - 1 \right).
$$

Since

$$\frac{y_N(0)}{z_N(0)} = \left(1 + \frac{\delta(0)}{y_N(0)}\right)^{-1},$$

$K_N(n)$ is known.

Example 4.6 Consider the previous example with $x = 2$, $N = 10$. Write

$$\hat{E}_N^*(n) := E_N^*(n) + W(n).$$

Using the previous results it is readily established that

$$|W(n)| \leqslant \frac{|\varepsilon|\, y_N(m+1)\, |b(n-1) \cdots b(m-1)|\, (1 + |E_N^*(n)|)}{y_N(n) y_N(n-1)(1 - |\varepsilon|/y_N(m))}.$$

Now let the value 10 for $y_N(9)$ be replaced by 9, i.e., take $\varepsilon = -1$. Table 4.4, also taken from Tait, displays the results obtained by using the bounds

$$|W(n)| \leqslant \frac{10(1 + |E_N^*(n)|)}{9\,|a(N-1)|\,\prod_{j=n}^{N-1}(a(j-1)a(j) - b(j-1))},$$

and

$$|\hat{E}_N^*(n)| \leqslant |E_N^*(n)| + |W(n)|. \quad \blacksquare$$

The above analysis can be generalized considerably. For instance by the additivity of particular solutions we find that if errors $\varepsilon_N(n)$ are introduced at each stage of the analysis, then

$$\delta(n) = -\sum_{k=n}^{N} \frac{\varepsilon_N(k)}{D(k)} [y_N(k+1) y_N^{(2)}(n) - y_N^{(2)}(k+1) y_N(n)],$$

and the total contribution of the error to $y_N(n)$ can be assessed by a

Table 4.4

n	$z_N(n)$	$z_n(n)w(0)/z_N(0)$	$|\hat{E}_N(n)|$	$|\hat{E}_N(n)|$ bound
11	0	—	—	—
10	1	—	—	—
9	9	3.040 228 070 (−6)	1.300 1 (−3)	1.332 (−3)
8	82	2.769 985 575 (−5)	1.755 4 (−5)	1.824 (−5)
7	665	2.246 390 741 (−4)	3.022 6 (−7)	3.200 (−7)
6	4 737	1.600 173 374 (−3)	6.249 3 (−9)	7.440 (−9)
.	.	.	.	.
.	.	.	.	.
.	.	.	.	.
0	6 748 266	2.279 585 302	0	0

straightforward summation. The contribution to the error in $w_N(n)$ is not so clear because the normalization of $y_N(n)$ makes the process nonlinear.

Note that the alternative algorithm with normalization $S = \sum c(k)w(k)$ can easily be incorporated into the previous analysis provided only

$$\sum_{k=0}^{N} c(k)E_N(k) = o(1), \qquad N \to \infty,$$

see Tait (1967) for details.

4.4 Error analysis, the general algorithm

The general Miller algorithm with unrestricted coefficients is more of a challenge. Olver (1964), however, has shown that by making a few simple estimates of the solutions of the equation useful error bounds may still be obtained.

Let the equation have a minimal solution $u(n) = y^{(1)}(n)$, the solution to be computed, and let $y^{(2)}(n)$ be any solution dominating $w(n)$. Define

$$\rho := |w(N+1)/w(N)|,$$

$$c := \max_{0 \leqslant k \leqslant N} |c(k)|,$$

$$\tau := \left| \frac{1}{cw(N)} \sum_{k=N+1}^{\infty} c(k)w(k) \right|,$$

$$a := c \max_{0 \leqslant k \leqslant N} |w(k)|.$$

Let σ be the least quantity such that the inequalities

$$\left| \frac{y^{(2)}(n)}{y^{(2)}(N+1)} \right| \leqslant \sigma^{N+1-n}, \qquad 0 \leqslant n \leqslant N,$$

hold and let

$$S_N := \sum_{k=0}^{N} c(k)y_N(k).$$

Then if σ, ρ are small (such that, e.g., $\rho\sigma < 1$, $\sigma < 1$),

$$|E_N(n)| \leqslant \frac{(1-\rho\sigma)^{-1}[(\sigma^{N-n}+(1-\sigma)^{-1}a)\rho\sigma + a\tau]}{|S_N|}, \qquad 0 \leqslant n \leqslant N.$$

$$\textbf{(4.40)}$$

Note that $y^{(2)}(n)$ may be generated by using the difference equation in the forward direction. An application of this (to $J_{n+\nu}$) will be given in Section 5.7.

Olver (1964) provides simple estimates for round-off error and gives numerical examples.

The previous estimates may, depending on the problem, be very imprecise but to do better one must work much harder. Oliver (1967), Gautschi (1967, 1972a), and Mattheij and van der Sluis (1976) have all studied the error of this algorithm. The last study is the most ambitious to date and the authors attack the problems of rounding and truncation error for both the scalar equation and the 2×2 system with indications of how their studies may be generalized to higher-order systems. Since the estimates are very technical the reader is referred to the paper itself for details.

4.5 Algorithms based on continued fractions

When N becomes large in the Miller algorithm so, in general, do the quantities $y_N(n)$ and machine overflow becomes a very real possibility. However, the quantities $y_N(n+1)/y_N(n)$ are usually moderate in size, so it is preferable to compute these ratios.

Gautschi (1967) has analyzed an algorithm for doing this which is mathematically equivalent to the Miller algorithm but for which overflow can be avoided.

Let

$$r(n) := \frac{w(n+1)}{w(n)}, \qquad s(n) := \frac{1}{w(n)} \sum_{k=n+1}^{\infty} c(k)w(k),$$

where $w(n)$ is the desired solution of the equation.

From the difference equation we get

$$r(n) = \frac{-1}{a(n)+b(n)r(n+1)}, \tag{4.41}$$

under appropriate conditions. Also

$$s(n) = \frac{1}{w(n)}\left[c(n+1)w(n+1) + \sum_{k=n+2}^{\infty} c(k)w(k)\right],$$

so

$$s(n) = r(n)(c(n+1)+s(n+1)). \tag{4.42}$$

Also

$$w(0) = S/(c(0)+s(0)). \tag{4.43}$$

The algorithm results from truncating the infinite continued fraction and infinite series representing $r(n)$ and $s(n+1)$. In fact let

$$r_N(N):=0; \qquad r_N(n):=\frac{-1}{a(n)-}\frac{b(n)}{a(n+1)-}\cdots\frac{b(N-2)}{a(N-1)},$$

$$s_N(N):=0; \qquad s_N(n):=\frac{1}{y_N(n)}\sum_{k=n+1}^{N}c(k)y_N(k)$$

$$=\sum_{k=n+1}^{N}c(k)r_N(n)r_N(n+1)\cdots r_N(k-1).$$

The algorithm is then defined by the following computational format:

$$\left.\begin{array}{l}
r_N(N)=0; \quad r_N(n)=\dfrac{-1}{a(n)+b(n)r_N(n+1)}, \quad n=N-1,N-2,\ldots,0; \\[2mm]
s_N(N)=0; \quad s_N(n)=r_N(n)(c(n+1)+s_N(n+1)), \\[2mm]
\hspace{6cm} n=N-1,N-2,\ldots,0; \\[2mm]
w_N(0)=\dfrac{S}{c(0)+s_N(0)}; \quad w_N(n)=r_N(n-1)w_N(n-1), \quad n=1,2,\ldots,N.
\end{array}\right\}$$

$$(4.44)$$

As with the Miller algorithm an initial value of N is chosen for the computations. The algorithm (44) is particularly useful when large values of N are required to define $w(n)$ accurately.

In the simplified Miller algorithm ($c(0)=1$, $c(k)=0$, $k>0$) all the $s_N(n)$ vanish so the computation of $s_N(n)$ in (44) may be omitted. In this case, the value of $w(0)$ must be known. It is clear that this algorithm is precisely equivalent to the usual Miller algorithm and thus converges when the conditions of Theorem 2 are satisfied.

If no *a priori* knowledge is available about a starting value of N, the previous procedure has to be repeated for increasing values of N until the required accuracy is achieved. One might think there would be some way of employing the duality Theorem 1 to avoid repeating similar computations. Indeed there is and the corresponding algorithm is due jointly to Gautschi (1967) and Shintani (1965).

Note that one can compute $r(n)$, $s(n)$ by recursion for $0\leqslant n\leqslant K$, and thus $w(n)$, for $0\leqslant n\leqslant K$, once $r(K)$, $s(K)$ are known for some value of K. This is accomplished by means of the relationships (41), (42), (43), respectively.

Let's first discuss the computation of $r(K)$ for some fixed value of K. To do this we exploit the connection between continued fractions and series. The necessary formulas can be found in Wall (1948, p. 17ff.). Suppose we are given the continued fraction

$$K:=\frac{\alpha(1)}{\beta(1)+}\frac{\alpha(2)}{\beta(2)+}\frac{\alpha(3)}{\beta(3)+}\cdots.$$

Let

$$\frac{P(n)}{Q(n)} := \frac{\alpha(1)}{\beta(1)+} \frac{\alpha(2)}{\beta(2)+} \cdots \frac{\alpha(n)}{\beta(n)}.$$

Then

$$\frac{P(n)}{Q(n)} = \sum_{k=1}^{n} \rho(1)\rho(2) \cdots \rho(k),$$

where $\rho(k)$ satisfies

$$1+\rho(k+1) = \frac{1}{1+\dfrac{\alpha(k+1)}{\beta(k)\beta(k+1)}(1+\rho(k))}, \qquad 2 \leqslant k \leqslant n-1,$$

with

$$\rho(1) = \alpha(1)/\beta(1); \qquad 1+\rho(2) = \left[1+\frac{\alpha(2)}{\beta(1)\beta(2)}\right]^{-1}.$$

The above formulas can be used to generate approximants to the continued fraction, $P(n)/Q(n)$.

Let

$$u(1) := 1, \quad u(k) := 1+\rho(k), \qquad k \geqslant 2,$$
$$v(k) := \rho(1)\rho(2) \cdots \rho(k), \qquad k \geqslant 1,$$
$$V(k) := \sum_{i=1}^{k} v(i), \qquad k \geqslant 1.$$

Then $V(k) = P(k)/Q(k)$ and

$$u(k+1) = \left[1+\frac{\alpha(k+1)u(k)}{\beta(k)\beta(k+1)}\right]^{-1}, \qquad k \geqslant 1,$$
$$v(k+1) = v(k)[u(k+1)-1], \qquad k \geqslant 1,$$
$$V(k+1) = V(k)+v(k+1), \qquad k \geqslant 1,$$

with

$$u(1) = 1; \qquad v(1) = V(1) = \alpha(1)/\beta(1).$$

We now apply this to the continued fraction (22).
Take K fixed $\geqslant 0$, and let

$$\left. \begin{aligned} &u(1) := 1, \qquad v(1) := V(1) = -1/a(K); \\ &u(k+1) = \left[1-\frac{b(K+k-1)u(k)}{a(K+k-1)a(K+k)}\right]^{-1}, \qquad k \geqslant 1, \\ &v(k+1) = v(k)[u(k+1)-1], \qquad k \geqslant 1, \\ &V(k+1) = V(k)+v(k+1), \qquad k \geqslant 1. \end{aligned} \right\} \qquad \textbf{(4.45)}$$

Then $r(K)$ may be computed to the desired accuracy, assuming the conditions of Theorem 1 are satisfied, by taking k sufficiently large, i.e.,

$$\lim_{k \to \infty} V(k) = r(K).$$

Now consider the computation of $s(K)$. The quantities $s_N(K)$, $N \geqslant K$, may be computed by recursion also. To do this I use the formulas (22) and (24). Define

$$\rho_N(K) := y_N(K)/y_{N+1}(K).$$

A straightforward application of these equations gives the format:

$$\left.\begin{aligned}
&\rho_{K-1}(K) = 0; \qquad y_K(K) = 1, \qquad y_{K-1}(K) = 0; \\
&s_K(K) = s_{K-1}(K) = 0; \\
&\rho_{N-1}(K) = -[a(N-1) + b(N-2)\rho_{N-2}(K)]^{-1}, \\
&\qquad\qquad\qquad\qquad N = K+1, K+2, \ldots; \\
&y_N(K) = -a(N-1)y_{N-1}(K) - b(N-2)y_{N-2}(K), \\
&\qquad\qquad\qquad\qquad N = K+1, K+2, \ldots; \\
&s_N(K) = -\rho_{N-1}(K)[a(N-1)s_{N-1}(K) \\
&\qquad + \rho_{N-2}(K)b(N-2)s_{N-2}(K)] + \frac{c(N)}{y_N(K)}, \\
&\qquad\qquad\qquad\qquad N = K+1, K+2, \ldots.
\end{aligned}\right\} \quad \textbf{(4.46)}$$

For N sufficiently large $s_N(K)$ will be as close as one desires to $s(K)$, i.e.,

$$\lim_{N \to \infty} s_N(K) = s(K).$$

To review the procedure I tabulate the steps involved in the algorithm:

(1) An integer $K > 0$ is chosen; this means the desired solution $w(n)$ will be computed for the $K+2$ successive values $0 \leqslant n \leqslant K+1$;
(2) In equation (45) take $k = 1, 2, \ldots$, until $V(k)$ defines $r(K)$ to the desired number of places;
(3) Take $N = K+1, K+2, \ldots$, in equation (46) until $s_N(K)$ defines $s(K)$ to the desired number of places;
(4) Compute $r(K-1), r(K-2), \ldots, r(0)$ from (41);
(5) Compute $s(K-1), s(K-2), \ldots, s(0)$ from (42);
(6) Compute $w(0)$ from (43);
(7) Compute $w(1), w(2), \ldots, w(K+1)$ from $w(n+1) = r(n)w(n)$.

Van der Cruyssen (1981) has analyzed error propagation in this algorithm.

Gautschi and Slavik (1978) and Amos (1974) have discussed the computation of the modified Bessel function ratios $r_\nu(x):=I_\nu(x)/I_{\nu-1}(x)$ by means of continued fraction algorithms, the former reference comparing the convergence properties of two different continued fraction representations for r_ν,

$$r_\nu(x):=\cfrac{1}{\dfrac{2\nu}{x}+}\cfrac{1}{\dfrac{2(\nu+1)}{x}+}\cfrac{1}{\dfrac{2(\nu+2)}{x}+}\cdots, \qquad x, \nu>0,$$

and

$$r_\nu(x):=\cfrac{x}{2\nu+x-}\cfrac{2x(\nu+\frac{1}{2})}{2\nu+1+2x-}\cfrac{2x(\nu+\frac{3}{2})}{2\nu+2+2x-}\cdots, \qquad x, \nu>0,$$

known as the Gauss and Perron continued fractions, respectively. Asymptotically the Gauss continued fraction converges twice as fast as Perron's. Unless very high accuracies are desired, it is the initial convergence behavior that matters and the authors conclude that if $x \gg \nu$ the Perron fraction has distinct computational advantages.

In the second reference it is shown, for example, that

$$r_{\nu+1}(x)\leqslant\frac{x}{\nu+(x^2+\nu^2)^{1/2}}\leqslant r_\nu(x), \qquad \nu, x\geqslant 0,$$

and this estimate can be used to determine an efficient value of r_ν with which to begin backward recursion in the algorithm

$$r_\nu(x)=\cfrac{1}{\dfrac{2\nu}{x}+r_{\nu+1}(x)},$$

which is the formalization of Gauss' continued fraction.

Gautschi (1970) discusses the application of continued fractions to the complex error function.

The literature on computation with continued fractions is vast. The survey by Blanch (1964) is probably the best reference to consult for an overview of the subject. Significant advances have also been made in analyzing the truncation error for continued fractions, for instance, Gautschi (1982), Field (1978), Field and Jones (1972), Jones and Thron (1968, 1971), Jones and Snell (1969), Sweezy and Thron (1967), Henrici and Pfluger (1966), and Merkes (1966).

Van der Cruyssen (1979a) has generalized this algorithm by using the concept of a *generalized continued fraction*. The resulting algorithm is useful for solving recurrences of order σ, $\sigma\geqslant 2$, for which there exist $\sigma-1$ linearly independent solutions that are dominated by any solution not lying in the subspace spanned by these solutions. (This means, according to Definition 2.8, that the equation possesses a dominant solution.) His definitions and main results are easy to state.

Definition 4.1 *A generalized continued fraction (GCF) of dimension σ is a $(\sigma+1)$-tuple*

$$W = (\mathbf{a}^{(1)}, \mathbf{a}^{(2)}, \ldots, \mathbf{a}^{(\sigma)}; \mathbf{b}),$$

with elements $\in \mathscr{R}_S$ (i.e., real sequences) and a convergence structure defined as follows.
Let $\mathbf{A}^{(1)}, \mathbf{A}^{(2)}, \ldots, \mathbf{A}^{(\sigma)}, \mathbf{B} \in \mathscr{R}_S$ satisfy the recurrence

$$y(\sigma+n+1) = b(n)y(\sigma+n) + a^{(\sigma)}(n)y(\sigma+n-1) + \cdots + a^{(1)}(n)y(n),$$
$$\text{(4.47)}$$

with initial conditions

$$A^{(i)}(k) = \delta_{ki}, \qquad 0 \leq k \leq \sigma, \quad 1 \leq i \leq \sigma;$$
$$B(k) = 0, \qquad 0 \leq k \leq \sigma-1;$$
$$B(\sigma) = 1.$$

The GCF $(\mathbf{a}^{(1)}, \mathbf{a}^{(2)}, \ldots, \mathbf{a}^{(\sigma)}; \mathbf{b})$ is said to converge *if all the limits*

$$d(i) = \lim_{n \to \infty} A^{(i)}(n)/B(n), \qquad 1 \leq i \leq \sigma,$$

exist. The value $K(\mathbf{a}^{(1)}, \mathbf{a}^{(2)}, \ldots, \mathbf{a}^{(\sigma)}; \mathbf{b})$ is then defined to be the σ-tuple $(d(1), d(2), \ldots, d(\sigma)) \in \mathscr{R}^{\sigma}$. ∎

(This concept is due to de Bruijn. For a further discussion of GCF, see *ibid*, the previous references.)
Van der Cruyssen establishes

Theorem 4.12 *A GCF converges iff there exist σ linearly independent solutions $y^{(h)}(n)$, $1 \leq h \leq \sigma$, of (47) such that:*

(i) $y^{(h)}(n)$, $1 \leq h \leq \sigma$, *is dominated by any solution*

$$y(n) \notin \text{lin}\,[y^{(1)}(n), \ldots, y^{(\sigma)}(n)].$$

(ii) $\begin{vmatrix} y^{(1)}(0) & \cdots & y^{(\sigma)}(\sigma-1) \\ \vdots & & \vdots \\ y^{(\sigma)}(0) & \cdots & y^{(\sigma)}(\sigma-1) \end{vmatrix} = 0.$ ∎

Now suppose we wish to compute (approximate) values of a function

$$w(n) \in \text{lin}\,[y^{(1)}(n), \ldots, y^{(\sigma)}(n)]$$

given the initial values $w(0), w(1), \ldots, w(\sigma-1)$. Pick $N \gg 0$ and define $r_N^{(h)}(n)$, $1 \leqslant h \leqslant \sigma$, by

$$r_N^{(1)}(N) = r_N^{(2)}(N) = \cdots = r_N^{(\sigma)}(N) = 0;$$

$$r_N^{(1)}(n) = a^{(1)}(n)/(b(n) + r_N^{(\sigma)}(n+1)), \qquad n = N-1, N-2, \ldots, 0;$$

$$r_N^{(h)}(n) = (a^{(i)}(n) + r_N^{(h-1)}(n+1))/(b(n) + r_N^{(\sigma)}(n+1)), \qquad 1 \leqslant h \leqslant \sigma,$$
$$n = N-1, N-2, \ldots, 0.$$

Now put

$$w_N(0) = w(0), \qquad w_N(1) = w(1), \ldots, w_N(\sigma-1) = w(\sigma-1),$$

and

$$w_N(n+\sigma) = -\sum_{j=1}^{\sigma} r_N^{(j)}(n) w_N(n+j-1), \qquad 0 \leqslant n \leqslant N-\sigma-1.$$

Van der Cruyssen establishes that if the conditions of Theorem 12 are satisfied and, in addition,

$$\begin{vmatrix} y^{(1)}(k) & \cdots & y^{(1)}(k+\sigma-1) \\ \vdots & & \vdots \\ y^{(\sigma)}(k) & \cdots & y^{(\sigma)}(k+\sigma-1) \end{vmatrix} \neq 0, \qquad k \geqslant 1,$$

then

$$\lim_{N \to \infty} w_N(n) = w(n), \qquad n \geqslant 0.$$

Example 4.7 Consider the recurrence

$$y(n+3) = \frac{(4n+22)(n+3)}{(n+4)(n+5)} y(n+2) - \frac{(5n+29)(n+2)}{(n+4)(n+5)} y(n+1)$$
$$+ \frac{2(n+1)(n+6)}{(n+4)(n+5)} y(n),$$

which has a fundamental set

$$y^{(1)}(n) = 1, \qquad y^{(2)}(n) = \frac{n}{n+1}, \qquad y^{(3)}(n) = 2^n.$$

Here $\sigma = 2$. The conditions are satisfied, and the previous algorithm will compute successfully any solution in $\lin[1, n/(n+1)]$. ∎

An algorithm analogous to the one based on a three-term recurrence that utilizes normalization relationships rather than initial values can be formulated in a straightforward manner.

4.6 Minimal solutions and orthogonal polynomials

Gautschi (1981) has observed that there is an intimate connection between the existence of minimal solutions for the three-term recurrence (1) and the determinacy of a related moment problem. Since his work has important implications for Gaussian quadrature, I shall present his results in detail.

To maintain consistency with the standard treatises on orthogonal polynomials and continued fractions I shall write the recurrence

$$y(n+1) = (z - a(n))y(n) - b(n)y(n-1), \qquad n \geq 1. \tag{4.48}$$

Obviously (1) may be written in the form (48) in many ways, the role of the variable z being discretionary.

Let $\mu(t)$ be a nondecreasing function with infinitely many points of increase (see Henrici (1977, vol. 2, p. 579)) defined on an interval $[a, b]$, finite or infinite, and let the moments

$$m(n) := \int_a^b t^n \, d\mu(t), \qquad n \geq 0,$$

be defined.

Let $p_n(z)$ be the monic polynomials orthogonal on $[a, b]$ with respect to $d\mu(t)$,

$$\int_a^b p_n(t)p_m(t) \, d\mu(t) = 0, \qquad n \neq m.$$

As is well known p_n satisfies the recurrence

$$\left. \begin{aligned} p_{n+1}(z) &= (z - a(n))p_n(z) - b(n)p_{n-1}(z), \qquad n \geq 0, \\ p_{-1}(z) &:= 0, \quad p_0(z) := 1, \qquad a(n) \text{ real}, \quad b(n) > 0. \end{aligned} \right\} \tag{4.49}$$

The polynomials

$$q_n(z) := \int_a^b \frac{p_n(z) - p_n(t)}{z - t} \, d\mu(t), \qquad n \geq 0, \tag{4.50}$$

are called the polynomials *associated* with the $p_n(z)$. They also satisfy (49) for $n \geq 1$, but with

$$q_0(z) := 0, \qquad q_1(z) := m(0).$$

Clearly $p_n(z)$, $q_n(z)$ are linearly independent solutions of (48).

Now let

$$F(z) := \int_a^b \frac{d\mu(t)}{z - t}, \qquad z \notin [a, b]. \tag{4.51}$$

$F(z)$ then possesses the 'formal' series in powers of $1/z$,

$$F(z) \sim \sum_{k=0}^{\infty} \frac{m(k)}{z^{k+1}}, \quad . \tag{4.52}$$

and corresponding to this series is the continued fraction

$$F(z) \sim \frac{b(0)}{z - a(0) -} \frac{b(1)}{z - a(1) -} \frac{b(2)}{z - a(2) -} \cdots, \qquad b(0) = m(0), \tag{4.53}$$

where $a(n)$, $b(n)$ are the coefficients in (49); see Perron (1957, Theorem 4.1). By corresponding I mean that when the nth convergent to the continued fraction (53) is developed in a power series in $1/z$ it agrees with the series (52) to n terms; see Henrici (1977, vol. 2, p. 518). Furthermore, the nth convergent to (53) is

$$\frac{b(0)}{z - a(0) -} \frac{b(1)}{z - a(1) -} \cdots \frac{b(n-1)}{z - a(n-1)} = \frac{q_n(z)}{p_n(z)}, \qquad n \geq 1.$$

I wish to investigate the case where the continued fraction (53) actually converges to $F(z)$,

$$F(z) := \lim_{n \to \infty} \frac{q_n(z)}{p_n(z)}, \qquad z \notin [a, b]. \tag{4.54}$$

If (54) holds, then

$$f_n(z) := F(z) p_n(z) - q_n(z), \tag{4.55}$$

is a minimal solution of (48) since (54) implies

$$\frac{f_n(z)}{p_n(z)} = F(z) - \frac{q_n(z)}{p_n(z)} \to 0, \qquad n \to \infty.$$

Because of (50) and (51) f_n can be written

$$f_n(z) = \int_a^b \frac{p_n(t)}{z - t} \, d\mu(t), \qquad z \notin [a, b], \quad n \geq 0. \tag{4.56}$$

If we define $b(0) = m(0)$, $q_{-1}(z) = -1$, then $f_n(z)$ satisfies (48) for $n \geq 0$ and the starting value is

$$f_{-1}(z) = 1.$$

Any condition which guarantees (54) will be sufficient for the existence of a minimal solution to (48), $f_n(z)$, in fact. There are a number of such conditions. For instance if $[a, b]$ is finite (54) will hold by Markov's theorem. If $[a, b] = [0, \infty]$ (54) will hold if the corresponding Stieltjes moment problem is determined. (These statements can be found in Perron (1957).) A sufficient condition for the latter is due to Carleman

(Henrici (1977, vol. 2, p. 608)):

$$\sum_{k=1}^{\infty} m(k)^{-1/2k} = \infty.$$

For the interval $[-\infty, \infty]$ (54) is assured if the corresponding Hamburger moment problem is determined. Sufficient conditions for this, also due to Carleman, are

$$\sum_{k=1}^{\infty} m(2k)^{-1/2k} = \infty \qquad \text{or} \qquad \sum_{k=1}^{\infty} b^{-1/2}(k) = \infty.$$

Since the moment problem for a finite interval is always determined (Shohat and Tamarkin (1943), p. 9) all cases are exhausted and I summarize the above observations in

Theorem 4.13 *The recurrence (48) possesses a minimal solution, given by (56), whenever the moment problem for $m(n)$ is determined.* ∎

Example 4.8 (Special Gaussian quadrature rules) Let $d\mu(t)$ be a distribution with finite or infinite support $[a, b]$, let x be real, $x \notin [a, b]$, and consider the new distribution

$$d\sigma(t) := \frac{d\mu(t)}{|x-t|}, \qquad t \in [a, b], \quad x \notin [a, b].$$

Suppose we wish to construct the monic orthogonal polynomials $\{\pi_n\}$ with respect to $d\sigma(t)$ and the corresponding Gaussian quadrature rules, all of which exist and are unique.

To do this one first determines the coefficients $\alpha(n)$, $\beta(n)$ in the recurrence for the desired polynomials

$$\pi_{n+1}(z) = (z - \alpha(n))\pi_n(z) - \beta(n)\pi_{n-1}(z), \qquad n \geqslant 0,$$

$$\pi_{-1}(z) := 0, \qquad \pi_0(z) := 1,$$

in terms of the $a(n)$, $b(n)$ of the given orthogonal polynomials and in terms of the modified moments of $d\sigma(t)$,

$$k(n) := \int_a^b \frac{p_n(t)\, d\mu(t)}{|x-t|}, \qquad n \geqslant 0.$$

$\pi_r(z)$ may then be interpreted as the characteristic polynomial of the rth leading principal minor matrix of the Jacobi matrix of order n

$$J(n) := \begin{bmatrix} \alpha(0) & \sqrt{\beta(1)} & & & (0) \\ \sqrt{\beta(1)} & \alpha(1) & \sqrt{\beta(2)} & & \\ & \cdot & \cdot & \cdot & \\ & & \cdot & \cdot & \cdot \\ & & & \cdot & \sqrt{\beta(n-1)} \\ (0) & & & \sqrt{\beta(n-1)} & \alpha(n-1) \end{bmatrix},$$

and, as is well-known, the required Gaussian nodes $\xi_{n\nu}$ are the zeros of $\pi_n(z)$, hence the eigenvalues of $J(n)$. The Gaussian weights (Christoffel numbers) $A_{n\nu}$ can be written either in terms of the $\{\pi_n(z)\}$

$$A_{n\nu} = h(n-1)[\pi_{n-1}(\xi_{n\nu})\pi'_n(\xi_{n\nu})]^{-1}, \qquad 1 \leq \nu \leq n,$$

where

$$h(n-1) := \int_a^b \pi_{n-1}^2(x)\, d\sigma(x),$$

or in terms of the spectral quantities (see Golub and Welsh (1969))

$$A_{n\nu} = k(0)K_{n\nu}^2, \qquad 1 \leq \nu \leq n,$$

where $K_{n\nu}$ denotes the first component of the normalized eigenvector $\mathbf{x}_{n\nu}$, $\mathbf{x}_{n\nu}^{\mathrm{T}}\mathbf{x}_{n\nu} = 1$, of $J(n)$ belonging to the eigenvalue $\xi_{n\nu}$.

The computation of eigenvalues and the first components of the corresponding eigenvectors can be conducted using the implicit QL algorithm; see Gautschi (1978) and the previous reference for details. At any rate the accurate determination of the modified moments can be accomplished by means of the continued fraction algorithm of Section 5 since, as is easily seen, these are identical with $f_n(x)$ (or $-f_n(x)$, if $x < a$) which is the minimal solution of the recurrence (48).

Gautschi (1981) points out that this procedure may not be the best possible in terms of efficiency but it has the advantage of being numerically stable.

For further observations on the generation of Gaussian quadrature rules see Gautschi (1979a) and the references given there. ∎

Example 4.9 Consider the recurrence

$$p_{n+1}(z) = zp_n(z) - b(n)p_{n-1}(z), \qquad n \geq 0,$$

with

$$b_n := \frac{(n+2\lambda-1)n}{4(n+\lambda)(n+\lambda-1)}, \qquad \lambda > 1.$$

Then

$$p_n(z) = \frac{n!}{2^n(\lambda)_n} C_n^\lambda(z).$$

For z off the cut $[-1, 1]$ the minimal solution of the recurrence is

$$f_n(z) := \frac{n!}{2^n(\lambda)_n} \int_{-1}^1 \frac{C_n^\lambda(t)(1-t^2)^{\lambda-1/2}}{z-t}\, dt$$

$$= \frac{(n+\lambda)_n}{2^{n+\lambda-3/2}} n!\sqrt{\pi}\, e^{i\pi(\lambda-1/2)}(z^2-1)^{(2\lambda-1)/4} Q_{n+\lambda-1/2}^{(1/2)-\lambda}(z),$$

see Erdélyi *et al.* (1954, vol. 2, p. 281 (5)). (The formula referred to is incorrect; it requires an $n!$ on the right-hand side.)

These are Legendre functions. Their computation is discussed in detail in Section 5.5. ■

Lewanowicz (1979b) has shown that modified moments of the form

$$v(n;f):=\int_a^b f(x)p_n(x)\,dx,$$

may be computed effectively by recurrence when $p_n(x)$ is the Gegenbauer polynomial $C_n^\lambda(x)$ and f satisfies a linear differential equation of order σ with polynomial coefficients

$$\sum_{k=0}^\sigma q_k(x)f^{(k)}(x)=q(x), \qquad -1\leqslant x\leqslant 1,$$

provided $v(n;q)$ is known. He shows that the order, ρ, of the recurrence he obtains is minimal and derives the interesting inequality

$$2\left[\max_{\substack{0\leqslant i\leqslant\sigma\\ q_i\neq 0}}(\partial q_i-i)\right]\leqslant\rho\leqslant 2\sigma+2\left[\max_{\substack{0\leqslant i\leqslant\sigma\\ q_i\neq 0}}(\partial q_i-i)\right].$$

(Recall $\partial r=$ degree of r, r a polynomial.)

As an example let

$$v(n):=\int_{-1}^1 e^{-\alpha/(x+1)}T_n(x)\,dx, \qquad n\geqslant 0.$$

Then f satisfies

$$(x+1)^2 f'-\alpha f=0,$$

$\sigma=1$ and the above inequality yields $2\leqslant\rho\leqslant 4$. Actually $\rho=3$ and (*ibid*) $v(n)$ satisfies

$$(n-1)(n+2)y(n)+[(3n+7)n-2\alpha(n+2)]y(n+1)$$
$$+[(3n+2)(n+3)+2\alpha(n+1)]y(n+2)$$
$$+(n+1)(n+3)y(n+3)=-8e^{-\alpha/2}, \qquad n\geqslant 0.$$

4.7 The Clenshaw averaging process

We have seen in previous sections that to compute a solution of the recurrence

$$y(n)+a(n)y(n+1)+b(n)y(n+2)=0, \tag{4.57}$$

successfully by using the recurrence in the backward direction it is required, in general, that the desired solution be minimal.

When the situation is such that every fundamental set for the equation consists of solutions having roughly the same rate of growth, so that, if there is a minimal solution it is only weakly minimal, backward recurrence, which is so desirable because it requires no initial values, will fail. In this chapter we outline an approach first, apparently, used by Clenshaw (1962) in constructing the Chebyshev coefficients for the exponential integral. In such situations the method can be used to generate *any* solution of the equation. (Other techniques can be used when the solutions of the equation are not strongly differentiated; I shall deal with these in later chapters.)

Before I formulate the algorithm I will provide an example to show where such a technique would be desirable.

Consider the recurrence

$$(n+a)(n+a+1-c)y(n)-(n+1)[2(n+a+1)-x-c]y(n+1)$$
$$+(n+1)(n+2)y(n+2)=0, \qquad\qquad (4.58)$$

which has as solutions

$$w(n):=\frac{(a)_n(a+1-c)_n}{n!}\Psi(n+a,c;-x), \qquad \frac{(a)_n}{n!}\Phi(n+a,c;-x).$$

The recurrence has a basis of solutions $\{y^{(h)}(n)\}$ having the behavior (see Section 5.2)

$$y^{(1)}(n)\approx n^{(4a-2c-3)/4}\cos(2\sqrt{(nx)}), \qquad y^{(2)}(n)\approx n^{(4a-2c-3)/4}\sin(2\sqrt{(nx)}).$$

Thus the equation has no minimal solution and the attempt to generate a solution by means of the Miller algorithm will fail. Note that (58) is a rather significant equation. Important special cases of $w(0)$ are error, exponential and logarithmic integrals.

To describe the Clenshaw procedure, I let $\{y^{(h)}(n)\}$ be a fundamental set for (57) and assume, without loss of generality, that I wish to compute $y^{(1)}(n)$.

Let constants $c^{(h)}(n)$ be given, $h = 1, 2$, $c^{(1)}(n):\equiv c(n)$ in equation (2) and define

$$S_N^{(h,j)}:=\sum_{k=0}^{N} c^{(j)}(k)y^{(h)}(k), \qquad 1\leqslant h, j\leqslant 2.$$
$$R^{(h)}(n):=T^{(h)}(n)/T^{(1)}(n) \quad \text{(see A-VI)}, \qquad h=1, 2.$$

The method supposes that *two* normalization relations are known for $y^{(1)}(n)$ and that the solutions $\{y^{(h)}(n)\}$ behave similarly.

The algorithm is described by the following theorem.

Theorem 4.14 *Let*

$$\lim_{N\to\infty} S_N^{(h,j)} = S^{(h,j)} < \infty,$$

with $|S^{(h,j)}|_1^2 \neq 0$, $S^{(1,1)} \neq 0$.
Let $y_N(n)$ *be defined as in* (3), *nonzero for* N *sufficiently large,* $0 \leqslant n \leqslant N$,

$$w_N(n) := S^{(1,1)} y_N(n) \bigg/ \sum_{k=0}^{N} c^{(1)}(k) y_N(k),$$

and let $y^{(2)}(n) \neq 0$ *for* n *sufficiently large.*
Then for N_1 *sufficiently large we can determine* $N_2 > N_1$ *such that the system of equations*

$$\sum_{\nu=1}^{2} \pi_\nu \sum_{k=0}^{N_\nu} c^{(2)}(k) w_{N_\nu}(k) = S^{(1,2)},$$

$$\pi_1 + \pi_2 = S^{(1,1)},$$

has a unique solution π_1, π_2 *(depending, of course, on* N_1, N_2).
Let $|R^{(h)}(N_j)|_1^2$ *be bounded away from zero as* $N_1 \to \infty$. *Then*

$$\lim_{N_1\to\infty} [\pi_1 w_{N_1}(n) + \pi_2 w_{N_2}(n)] = y^{(1)}(n), \qquad n \geqslant 0.$$

Proof The proof is deferred to Chapter 7 where the algorithm is given for the general σth-order equation. ∎

A minor modification of the Clenshaw method is achieved by taking $N_1 = N_2$ and computing two sequences $y_N^{(1)}(n)$, $y_N^{(2)}(n)$ corresponding to different *initial* values, say

$$y_N^{(1)}(n) = \begin{cases} 0, & n = N+1; \\ 1, & n = N; \end{cases} \qquad y_N^{(2)}(n) = \begin{cases} 1, & n = N+1; \\ 0, & n = N; \end{cases}$$

and then computing an appropriate average of the sequences to satisfy the two given normalizations. Although this modification makes the computations more efficient it may cause numerical instabilities because the equation for π_1 generally becomes ill-conditioned as N gets large.

Example 4.10 Compute the solution $w(n)$ of the recurrence

$$y(n) - 2y(n+1) + \left(\frac{n+2}{n+1}\right) y(n+2) = 0$$

having the behavior

$$w(0) = 1; \qquad \sum_{k=0}^{\infty} \frac{w(k)}{k!} = 1.$$

Table 4.5 $\pi_1 w_{N_1}(n) + \pi_2 w_{N_2}(n)$.

n	$N_1 = 4$	$N_1 = 6$	$N_1 = 8$	$N_1 = 10$
0	1	1	1	1
1	0.241 721 434 5	0.241 657 804 2	0.241 654 527 6	0.241 654 506 4
2	−0.258 278 565 5	−0.258 342 195 8	−0.258 347 236 7	−0.258 345 493 6
3	−0.505 519 043 6	−0.505 561 463 9	−0.505 563 648 2	−0.505 563 662 4
4	−0.564 569 641 4	−0.564 585 548 9	−0.564 586 368 1	−0.564 586 373 4

To do this by the Clenshaw method we take

$$c^{(1)}(k) = \delta_{k0}, \qquad c^{(2)}(k) = \frac{1}{k!}.$$

Note this is the recurrence (58) with the specializations $a = c = x = 1$.

In the computations (Table 4.5) I have taken $N_2 = 2N_1$. The entries in the last column are accurate to all figures given. ∎

For the second-order recursion the Clenshaw method does not have anything special to recommend it since a similar algorithm, based on forward recursion, is just as stable. One simply generates two linearly independent solutions by using the equation in the forward direction and then determines an appropriate linear combination of these solutions to satisfy the constraints on the desired solution.

The Clenshaw method is most useful in its application to higher-order equations; see Section 7.4.

Wimp (1967) has shown that when the equation is such that $a(n)$, $b(n)$ possess asymptotic series in powers of $n^{-1/\omega}$, ω an integer, and when members of a certain fundamental set for the equation have similar behavior, then the Clenshaw averaging process always converges.

5 Applications of the Miller algorithm to the computation of special functions

Since its inception the Miller algorithm has found a use in a wide variety of practical problems. Among these are the computation of many of the special functions of mathematical physics, particularly functions of hypergeometric type, the computation of a large number of definite integrals, especially integrals of generalized moment type, and the computation of zeros of functions and eigenvalues of certain differential operators.

This chapter is devoted to such applications. The treatment is intended to be illustrative rather than exhaustive in either scope or detail. For instance I do not generally present a complete account of the error involved. My intention is to give an overview of the multifarious applications of the algorithm and to emphasize the large number of areas in science and engineering in which the algorithm has been found to be an important computational resource. In fact the bibliographic sources for most of the work in this chapter are not generally contained in mathematics journals but rather in applied journals whose fields range the spectrum from celestial mechanics to hydrology.

5.1 The confluent hypergeometric function $\Phi(a, c; x)$

In the recurrence formula (C.16) let $z \to zb_2$ and $b_2 \to \infty$. After an obvious relabeling of parameters and variables we obtain

$$(2n+c+2)(n+a)y(n)-(2n+c+1)$$

$$\times \left\{(2a-c)+\frac{(2n+c)(2n+c+2)}{x}\right\}y(n+1)$$

$$-(2n+c)(n+c+1-a)y(n+2)=0. \quad \textbf{(5.1)}$$

One solution is

$$w(n):=y^{(1)}(n):=\frac{x^n(a)_n}{(c)_{2n}}\Phi(n+a, 2n+c; x). \quad \textbf{(5.2)}$$

Another is the function

$$y^{(2)}(n):=\frac{x^{-n}\Gamma(2n+c-1)}{\Gamma(n+c-a)}\Phi(a+1-c-n, 2-c-2n; x). \quad \textbf{(5.3)}$$

An elementary argument shows

$$y^{(1)}(n) \sim \frac{\Gamma(c)}{\Gamma(a)} 2^{(1/2)-c} e^{x/2} \left(\frac{ex}{4}\right)^n n^{a-c-n} \left[1 + \frac{c_1}{n} + \cdots\right],$$ (5.4)

while

$$y^{(2)}(n) \sim 2^{c-3/2} e^{x/2} \left(\frac{4}{ex}\right)^n n^{a-1+n} \left[1 + \frac{d_1}{n} + \cdots\right].$$ (5.5)

$y^{(1)}(n)$ is a minimal solution and the recursion is stable for $y^{(2)}(n)$.

Let us consider in detail the computation of $y^{(1)}(n)$ by Miller's algorithm.

A number of normalization relationships are possible; see Luke (1969, vol. 2, Ch. 9). A convenient one is

$$S = c - 1 = \sum_{k=0}^{\infty} \frac{(-1)^k}{k!} (c-1)_k (c+2k-1) w(k).$$

(An obvious modification must be made of this formula when $c = 1$.) The algorithm is not defined when any of the quantities c, $c+1-a$, a are negative integers or zero. However in such cases a meaningful computational scheme may be obtained by a change of dependent variable in the equation. I leave such details to the reader.

Theorem 4.2 is applicable and we find the Miller algorithm converges with great rapidity. In fact formula (4.21) furnishes an error estimate for exact arithmetic. Let $y(n) = y^{(2)}(n)$. The series $\sum_{k=0}^{n} c(k) y(k)$ diverges so rapidly that the last term serves as the leading term of its asymptotic series. Using (4), (5) we find

$$\tau_N = \frac{w(N+1)}{y(N+1)} = O\left(\left(\frac{ex}{4}\right)^{2N} N^{-2N-c-1}\right),$$

$$U_N + \tau_N S_N^* \approx w(N+1) c(N+1) \left[1 + \frac{c(N) y(N)}{c(N+1) y(N+1)}\right]$$

$$= w(N+1) c(N+1) [1 + O(N^{-1})].$$

Thus

$$E_N^*(n) \approx \frac{2^{-(1/2)-c}}{\Gamma(a)} xe^{x/2} \left(\frac{ex}{4}\right)^N (-1)^{N+1} N^{a-N-2}, \qquad N \to \infty.$$

This result can be compared with the error bound produced by the Tait theory, Corollary, Theorem 4.9. This result is applicable provided $a(n) \leqslant -1$ and $b(n) < 0$. These conditions are certainly met when n is sufficiently large, a, c real and $x > 0$. The Tait error bound is rather complicated.

If not all members of the sequence $\{y^{(1)}(n)\}$ are desired the Shintani form of the algorithm should be used; see Theorem 4.1.

When $c = 2a$, the recursion reduces to the one satisfied by the modified Bessel functions,

$$w(n) = \left(\frac{4}{x}\right)^{a-1/2} e^{x/2}\Gamma(a+\tfrac{1}{2})I_{n+a-1/2}\left(\frac{x}{2}\right).$$

A great many other higher transcendental functions may be expressed in terms of Φ. For a partial list consult Erdélyi et al. (1953, Section 6.9.2). Currently of great interest are the *Coulomb wave functions*

$$C_n(\eta) := 2^n e^{-\pi\eta/2}\,|\Gamma(1+n+i\eta)|/(2n+1)!,$$

$$F_n(\eta,\rho) := C_n(\eta)\rho^{n+1}e^{-i\rho}\Phi(n+1-i\eta, 2n+2; 2i\rho),$$

(see Abramowitz and Stegun (1964, Ch. 14)), where η, ρ are real, $\rho > 0$. The function F_n is easily shown from our previous work to be the minimal solution of the recurrence

$$(n+2)[(n+1)^2+\eta^2]^{1/2}y(n) - (2n+3)\left[\eta + \frac{(n+1)(n+2)}{\rho}\right]y(n+1)$$

$$+ (n+1)[(n+2)^2+\eta^2]^{1/2}y(n+2) = 0.$$

5.2   The confluent hypergeometric function $\Psi(a, c; x)$

One way of computing this function is to use the representation

$$\Psi(a,c;x) = \frac{\Gamma(1-c)}{\Gamma(a+1-c)}\Phi(a,c;x) + \frac{\Gamma(c-1)}{\Gamma(a)}x^{1-c}\Phi(a+1-c, 2-c; x).$$

A single-valued branch of this function is defined by choosing an appropriate principal branch of x^{1-c}. The parameter c, however, must be nonintegral if this formula is to be used.

Another approach, not suffering from this disadvantage, is to start with the recurrence (see Wimp (1974))

$$(n+a)(n+a+1-c)y(n) - (n+1)[2(n+a+1)+x-c]y(n+1)$$

$$+ (n+1)(n+2)y(n+2) = 0. \quad \textbf{(5.6)}$$

Solutions of this equation are

$$w(n) = y^{(1)}(n) = \frac{x^a(a)_n(a+1-c)_n}{n!}\Psi(n+a, c; x),$$

and

$$y^{(2)}(n) = \frac{(a)_n}{n!}\Phi(n+a, c; x).$$

We require $a, a+1-c \neq 0, -1, -2, \dots$. However, these restrictions may be removed by renormalization when necessary.

Formulas in Erdélyi *et al.* ((1953), vol. 1, 6.4, 6.6) show

$$y^{(1)}(n) = \frac{x}{n!} D^n[x^{n+a-1}\Psi(a, c; x)],$$

$$y^{(2)}(n) = \frac{x^{1-a}}{n!} D^n[x^{n+a-1}\Phi(a, c; x)].$$

These, used with the results (*ibid*, 6.7) and A-V show the Casorati determinant for the system is

$$D(n) := y^{(1)}(n)y^{(2)}(n+1) - y^{(2)}(n)y^{(1)}(n+1)$$
$$= \frac{D(0)(a)_n(a+1-c)_n}{n!\,(n+1)!}, \tag{5.7}$$

and

$$D(0) = \frac{\Gamma(c)}{\Gamma(a)} x^{a+1-c} e^x.$$

The asymptotic analysis of the recurrence (6) is given in Section B.2. We have

$$y^{(1)}(n) \sim K(x)n^{(4a-2c-3)/4}e^{-2\sqrt{(nx)}}\{1 + c_1 n^{-1/2} + c_2 n^{-1} + \cdots\},$$

$$K(x) = \frac{\sqrt{\pi}\,e^{x/2}x^{(4a-2c+1)/4}}{\Gamma(a)\Gamma(a+1-c)}, \qquad n \to \infty, \qquad |\arg x| < \pi. \tag{5.8}$$

A result in Slater ((1960), p. 80, 4.6.42) shows

$$y^{(2)}(n) \sim L(x)n^{(4a-2c-3)/4}e^{2\sqrt{(nx)}}\{1 - c_1 n^{-1/2} + c_2 n^{-1} + \cdots\},$$

$$L(x) = \frac{\Gamma(c)e^{x/2}x^{(1-2c)/4}}{2\sqrt{\pi}\,\Gamma(a)}, \qquad n \to \infty, \quad |\arg x| < \pi. \tag{5.9}$$

For the preceding estimates to hold it is, strictly speaking, necessary that none of the quantities a, c, $a+1-c$ be negative integers or zero. If this is not the case the analysis must be modified. However, doing so is quite straightforward, and details are left to the reader. For $|\arg x| < \pi$, $y^{(1)}(n)$ is a minimal solution and may be computed by the Miller algorithm. (Note $y^{(2)}(n)$ is then a solution for which the equation is stable at any point in the forward direction by Theorem 2.2.)

A normalization relation is provided by

$$1 = \sum_{k=0}^{\infty} y^{(1)}(k). \tag{5.10}$$

As in the previous section we can use formula (4.21) to obtain an estimate for the relative error of the Miller algorithm based on the normalization (10). The necessary asymptotic estimates are easily made and we have

$$E_N^*(n) \approx \frac{2\sqrt{\pi}\, e^{x/2} (Nx)^{(4a-2c-1)/4}}{\Gamma(a)\Gamma(a+1-c)} e^{-2\sqrt{(Nx)}}. \tag{5.11}$$

The right-hand side is not a very good approximation to $E_N^*(n)$ unless $\sqrt{(x/N)}$ is small; this is a consequence of the fact that otherwise the leading term on the right-hand side of (8) approximates $y^{(1)}(n)$ poorly. A better approximation to $y^{(1)}(n)$ can be obtained using the method of steepest descents on the integral representation of $y^{(1)}(n)$. The result is

$$y^{(1)}(n) \approx \frac{\sqrt{(2\pi)} x^a n^{a-c-1/2} e^{-xt_0} t_0^{n+a}}{\Gamma(a)\Gamma(a+1-c)\sqrt{(1+2t_0)}} (1+t_0)^{c-a-n}, \qquad n \to \infty,$$

$$t_0 = \tfrac{1}{2}[-1+\sqrt{(1+4n/x)}].$$

When the minus sign is affixed to the radical in the equation for t_0 the result provides an asymptotic formula for a constant multiple of $y^{(2)}(n)$. These estimates, in turn, can be used to obtain a more accurate asymptotic formula for $E_N^*(n)$.

Convergence for the simplified algorithm is considerably stronger, being $O(e^{-4\sqrt{(nx)}})$.

Example 5.1 Mori (1980) has shown the functions

$$w(n) \equiv w(n, a, c, x) := \frac{x^a (a)_n (a+1-c)_n}{n!} \Psi(n+a, c; x),$$

are useful in evaluating infinite integrals. I shall give an example.
Let

$$F(z) := z e^z E_1(z) = \int_0^\infty e^{-t} \left(1+\frac{t}{z}\right)^{-1} dt, \tag{5.12}$$

where z is complex, $|\arg z| < \pi$.
Let

$$t := \phi(u) := \frac{2\lambda u}{1-u}; \qquad u = \phi^{-1}(t) = \frac{t}{t+2\lambda}, \tag{5.13}$$

where λ is a fixed real number, $\mathrm{Re}\, z > \lambda$. The function (13) maps the right half of the t-plane onto $|u| < 1$. Substituting (13) in (12) gives

$$F(z) = \int_0^1 f(z, \phi(u)) e^{-\phi(u)} \phi'(u)\, du, \tag{5.14}$$

$$f(z, \phi(u)) = \sum_{k=0}^\infty c(k, z) u^k, \tag{5.15}$$

where

$$c(k, z) := \begin{cases} 1, & k = 0; \\ -\dfrac{2\lambda}{z}\left(1 - \dfrac{2\lambda}{z}\right)^{k-1}, & k \geq 1. \end{cases} \tag{5.16}$$

Since the pole $t = -z$ of $(1 + t/z)^{-1}$ is mapped into $u = 1/(1 - 2\lambda/z)$, the circle of convergence of the power series (15) is $U_r(u)$, $r = |1 - 2\lambda/z|^{-1}$. The range of integration in (14) is strictly within this circle, so the substitution of the series (15) into the integral and termwise integration is justified. We get

$$
\begin{aligned}
F(z) &= \sum_{k=0}^{\infty} c(k, z) \int_0^1 u^k e^{-\phi(u)} \phi'(u)\, du \\
&= \sum_{k=0}^{\infty} c(k, z) \int_0^{\infty} [\phi^{-1}(t)]^k e^{-t}\, dt \\
&= \sum_{k=0}^{\infty} c(k, z) \psi(k),
\end{aligned}
\tag{5.17}
$$

$$\psi(k) := \int_0^{\infty} \left(\frac{t}{t + 2\lambda}\right)^k e^{-t}\, dt = 2\lambda k!\, \Psi(k + 1, 2; 2\lambda).$$

Because $a + 1 - c = 0$ the recursion (6) for Ψ must be modified a bit. This however is straightforward and we find that $\psi(n)$ satisfies

$$y(n) - 2\left[1 + \frac{\lambda}{(n+1)}\right] y(n+1) + y(n+2) = 0. \tag{5.18}$$

$\psi(n)$ behaves like

$$\psi(n) \sim e^{\lambda} \sqrt{\pi}\, (2\lambda n)^{1/4} e^{-2\sqrt{(2\lambda n)}}[1 + c_1 n^{-1/2} + c_2 n^{-1} + \cdots], \qquad n \to \infty,$$

and a linearly independent solution has a representation obtainable from this by letting $n \to n e^{2\pi i}$. Thus the Miller algorithm based on the recursion (18) and

$$\psi(0) = 1,$$

converges to $\psi(k)$.

Furthermore we can get an idea of how the series (16) converges:

$$|c(k, z)\psi(k)| < \left|\frac{2\lambda}{z}\right| \left|1 - \frac{2\lambda}{z}\right|^{k-1} e^{\lambda} \sqrt{\pi}\, (2\lambda k)^{1/4} e^{-2\sqrt{(2\lambda k)}} \left[1 + \frac{C}{k^{1/2}}\right].$$

Ordinarily λ will be fixed and the values $\psi(k)$ will be stored.

An even more satisfactory approach is to evaluate the series (17) by truncating and then using the Clenshaw nesting procedure. This does not require the storage of the individual $\psi(k)$ (see Section 10.2).

Obviously the same procedure can be used to evaluate any integral of the form

$$\int_0^\infty f(z;t)e^{-t}\,dt,$$

when

$$f(z;\phi(u))=\sum_{k=0}^\infty c(k,z)u^k,$$

and f is analytic in z and t and the previous series has a radius of convergence >1. ∎

Mori (1980) gives these and other generalizations as well as applications to the functions $\mathrm{Si}\,(z)$, $\mathrm{Ci}\,(z)$, $\psi(z)$, $\ln\Gamma(z)$, $\mathrm{erf}\,z$, $\Gamma(a,z)$ and an error analysis.

The exponential integral

$$F(z)=ze^z E_1(z),$$

may also be computed for $|\arg z|<\pi$ directly by applying Miller's algorithm to the equation

$$y(n)-\left[2+\frac{z}{n+1}\right]y(n+1)+y(n+2)=0,$$

with normalization $y(0)=1$. Then

$$F(z)\sim 1-w_N(1).$$

Example 5.2 (The computation of $K_\nu(x)$) The modified Bessel functions $K_{n+\nu}(x)$ may be computed from the recursion formula (6). Let $a\to\nu+1/2$, $c\to 1+2\nu$, $x\to 2x$. Then

$$w(n)=\sqrt{\left(\frac{2}{\pi}\right)}\frac{x}{n!}D^n\{x^{n-1/2}e^x K_\nu(x)\},$$

and the recurrence becomes

$$(n+\nu+\tfrac{1}{2})(n-\nu+\tfrac{1}{2})y(n)-2(n+1)(n+x+1)y(n+1)$$
$$+(n+1)(n+2)y(n+2)=0. \quad \textbf{(5.19)}$$

By making use of a generating function we may express $w(n)$ in terms of $K_\nu(x), K_{\nu\pm1}(x),\ldots,K_{\nu\pm n}(x)$.

Since

$$K_\nu(x)=\tfrac{1}{2}\int_0^\infty e^{-x\phi/2}t^{-\nu-1}\,dt,\qquad \phi:=t+t^{-1},\quad \mathrm{Re}\,x>0,$$

we have

$$w(n)=\frac{x}{\sqrt{(2\pi)}n!}\int_0^\infty\frac{d^n}{dx^n}[x^{n-1/2}e^{x(1-\phi/2)}]t^{-\nu-1}\,dt.$$

Now use the Rodrigues representation for the Laguerre polynomial $L_n^{(-1/2)}(x)$ in the form

$$D^n[x^{n-1/2}e^{\sigma x}] = n!\, e^{\sigma x} x^{-1/2} L_n^{(-1/2)}(-\sigma x),$$

so

$$w(n) = \sqrt{\left(\frac{x}{2\pi}\right)}e^x \int_0^\infty e^{-x\phi/2} L_n^{(-1/2)}\left(\left(\frac{\phi}{2}-1\right)x\right)t^{-\nu-1}\,\mathrm{d}t.$$

Define $c_{nk}(x)$ by

$$L_n^{(-1/2)}\left(-x+\frac{x}{2}(t+t^{-1})\right) = c_{n0}(x) + \sum_{k=1}^n c_{nk}(x)(t^k + t^{-k}).$$

Then

$$w(n) = \sqrt{\left(\frac{2x}{\pi}\right)}e^x \sum_{k=0}^n \frac{\varepsilon_k}{2} c_{nk}(x)[K_{\nu-k}(x) + K_{\nu+k}(x)].$$

Using the formula

$$L_n^{(-1/2)}(y) = \frac{(1/2)_n}{n!}\,\Phi(-n, \tfrac{1}{2};\, y),$$

we find, for $n = 0$,

$$w(0) = \sqrt{\left(\frac{2x}{\pi}\right)}e^x K_\nu(x),$$

and for $n = 1$,

$$w(1) = \sqrt{\left(\frac{2x}{\pi}\right)}e^x[(x+\tfrac{1}{2}-\nu)K_\nu(x) - xK_{\nu+1}(x)].$$

For the error analysis I assume ν, x real and that the computation is based on the normalization (10).

Equation (11) gives

$$E_N^*(n) \approx \frac{2\cos(\nu\pi)}{\sqrt{\pi}}\,e^x (2Nx)^{-1/4}e^{-2\sqrt{(2Nx)}},$$

$$N \to \infty, \qquad |\arg x| < \pi, \qquad \nu \neq \pm\tfrac{1}{2}, \pm\tfrac{3}{2}, \ldots.$$

For a numerical example take $x = 0.2$, $\nu = 0$. Standard tables give

$$w(0) = 0.763\,875\,353\,4, \qquad w(1) = 0.118\,413\,359\,1,$$

while

$$w_{100}(0) = 0.763\,876\,603\,5, \qquad w_{100}(1) = 0.118\,413\,552\,9.$$

Thus

$$E^*_{100}(0) \cong E^*_{100}(1) = 1.64 \ (-6).$$

The above formula gives $E^*_{100}(n) \sim 1.76 \ (-6)$.

Once $K_\nu, K_{\nu+1}$ are computed from $w(0)$ and $w(1)$, $K_{\nu+n}$ may be computed in a stable fashion by using the difference equation for $K_{\nu+n}$ in the forward direction, i.e.,

$$y(n) + \frac{2(n+\nu+1)}{x} y(n+1) - y(n+2) = 0.$$

A more rapidly convergent method for computing $K_\nu(x)$ based on a third-order difference equation is given in Example 7.3.

Temme (1975) elaborates on this method of computing K_ν, and gives numerics and an ALGOL procedure. ■

Example 5.3 (Incomplete Gamma function)   When $a = c$ in the recurrence (6), the normalization relation (10), which, as previously pointed out, adversely affects the convergence of the algorithm, can be dispensed with since the simplified Miller algorithm can be used. This is a consequence of the fact that in the case $a = c$ the functions $y^{(2)}(0)$, $y^{(2)}(1)$ are known. The recurrence becomes

$$(n+a)y(n) - (2n+a+2+x)y(n+1) + (n+2)y(n+2) = 0, \qquad n \geqslant 0.$$

Then

$$w(0) = y^{(1)}(0) = x^a e^x \Gamma(1-a, x)$$

$$= x^a e^x \int_x^\infty e^{-t} t^{-a} \, dt, \qquad x > 0.$$

The Casorati determinant (7) and the relations

$$y^{(2)}(0) = e^x, \qquad y^{(2)}(1) = (x+a)e^x,$$

give

$$(x+a)w(0) - w(1) = x.$$

Thus if $y_N(n) \approx K_N w(n)$, $N \to \infty$, we have

$$K_n \approx \frac{(x+a)y_N(0) - y_N(1)}{x},$$

for example,

$$w(0) \approx x y_N(0)/((x+a)y_N(0) - y_N(1)), \qquad N \to \infty.$$

This discussion has been necessarily brief. For a much more detailed description of the computation of the incomplete gamma function, see Gautschi (1979b, c).

For methods of computing other incomplete gamma functions, see Examples 11.1, 11.2. ■

5.3 The Gaussian hypergeometric function

The necessary recursion formula, derived in Appendix C, is

$$
\begin{aligned}
(2n+c+2)(n+a)(n+b)y(n) & \\
+\{(2n+c+1)\langle(2n+c+2)(n+c-a)(n+c-b) & \\
-(2n+c)(n+c+1-a)(n+c+1-b)\rangle & \\
+z^{-1}(2n+c)(2n+c+1)(2n+c+2)\}y(n+1) & \\
+(2n+c)(n+c+1-a)(n+c+1-b)y(n+2) = 0, & \quad \textbf{(5.20)}
\end{aligned}
$$

where none of the quantities a, b, c, $c+1-a$, $c+1-b$ are negative integers or zero. (As is the case with the confluent hypergeometric function, these values correspond to either finite or singular cases of the function.)

Solutions of this recurrence are

$$
w(n):=y^{(1)}(n):=\frac{(-z)^n(a)_n(b)_n}{(c)_{2n}} F\!\left(\begin{matrix} n+a, & n+b \\ & 2n+c \end{matrix} \,\middle|\, z\right), \quad \textbf{(5.21)}
$$

and

$$
y^{(2)}(n):=\frac{(-z)^n\Gamma(2n+c-1)}{\Gamma(n+c-a)\Gamma(n+c-b)} F\!\left(\begin{matrix} 1+a-c-n, & 1+b-c-n \\ & 2-c-2n \end{matrix} \,\middle|\, z\right).
$$
$$
\textbf{(5.22)}
$$

Birkhoff series are easily obtained for the recurrence and a famous result of Watson (Erdélyi *et al.* (1953, vol. 1, p. 77 (16))) allows us to make the identification

$$
\left.
\begin{aligned}
w(n) &\sim K(z)\lambda^n n^{a+b-c-1/2}\left[1+\frac{c_1}{n}+\frac{c_2}{n^2}+\cdots\right], \\
K(z)&:=\sqrt{\pi}\,Z^{c-1}(1-z)^{(c-a-b)/2-1/4}\Gamma(c)/\Gamma(a)\Gamma(b), \\
\lambda&:=-zZ^2; \quad Z:=\frac{1-\sqrt{(1-z)}}{z}.
\end{aligned}
\right\}
\quad \textbf{(5.23)}
$$

This holds for $z \in \Omega$, $\Omega:=\mathscr{C}-[1,\infty)$. The substitution $n \to -n$, $a \to 1+a-c$, $b \to 1+b-c$, $c \to 2-c$ leaves (20) unaltered and sends $y^{(1)}$ into a constant multiple of $y^{(2)}$. Thus, for some K',

$$
y^{(2)}(n) \sim K'(z)\lambda^{-n} n^{a+b-c-1/2}\left[1+\frac{d_1}{n}+\frac{d_2}{n^2}+\cdots\right]. \quad \textbf{(5.24)}
$$

We can write

$$\lambda = -[\tau - \sqrt{(\tau^2 - 1)}]^2, \quad \tau := z^{-1/2}, \quad \tau \notin [0, 1],$$

and so $|\lambda| < 1$. Thus $w(n)$ is a minimal solution of the equation and the Miller algorithm for $w(n)$ will converge provided a suitable normalization relation can be found. A whole class of these is contained in the following expansion (see Luke (1969, vol. 2, p. 7. (13))):

$$(1 - z\omega)^{-a} = \frac{1}{\Gamma(c)} \sum_{k=0}^{\infty} \frac{\Gamma(k+c-1)}{k!}$$
$$\times (c + 2k - 1) w(k) F\left(\begin{matrix} -k, & k+c-1 \\ & b \end{matrix} \;\middle|\; \omega\right), \quad (5.25)$$

with obvious conditions on the parameters. It is most convenient in (25) to take $\omega = 0$ (although, as Gautschi points out in connection with the computation of Legendre functions, other choices may confer more computational stability). We find

$$S = 1 = \frac{1}{\Gamma(c)} \sum_{k=0}^{\infty} \frac{\Gamma(k+c-1)}{k!} (c + 2k - 1) w(k). \quad (5.26)$$

A formula for the relative error of the Miller algorithm based on the normalization (26) can be found by applying Theorems B.5, B.6 on exponential sums to the asymptotic estimates (23), (24) and using the results in formula (4.21). We find

$$U_N \approx \frac{2K(z)\lambda^{N+1}}{\Gamma(c)(1-\lambda)} N^{a+b-3/2}, \quad N \to \infty,$$

$$\tau_N S_N^* \approx \frac{2K(z)\lambda^{N+2}}{\Gamma(c)(1-\lambda)} N^{a+b-3/2}, \quad N \to \infty.$$

Consequently

$$E_N^*(n) \approx \frac{2K(z)\lambda^{N+1}}{\Gamma(c)} N^{a+b-3/2}\left(\frac{1+\lambda}{1-\lambda}\right), \quad N \to \infty. \quad (5.27)$$

Example 5.4 Let

$$z := e^{\pi i/3} = (1 + (\sqrt{3})i)/2,$$

with $a = \frac{2}{3}$, $b = 1$, $c = \frac{4}{3}$. It is known (e.g., Erdélyi *et al.* (1953, vol. 1, p. 105 (55))) that

$$w(0) = F\left(\begin{matrix} \frac{2}{3}, & 1 \\ & \frac{4}{3} \end{matrix} \;\middle|\; e^{\pi i/3}\right) = \frac{e^{\pi i/6} 2\pi \Gamma(\frac{1}{3})}{9\Gamma(\frac{2}{3})^2}$$

$$= 0.883\,319\,376 + 0.509\,984\,679i. \quad (5.28)$$

The computation of $y(0)$ by backwards recurrence with the normaliza-

Table 5.1 Computation of $F\left(\begin{smallmatrix} \frac{2}{3}, & 1 \\ & \frac{4}{3} \end{smallmatrix}\middle| e^{\pi i/3}\right)$.

N	$w_N(0)$	$E_N(0) = w_N(0) - w(0)$
5	$0.882\,398\,541 + 0.509\,453\,036i$	$-(9.2 + 5.3i)\ (-4)$
9	$0.883\,314\,192 + 0.509\,981\,687i$	$-(5.2 + 3.0i)\ (-6)$
13	$0.883\,319\,347 + 0.509\,984\,663i$	$-(3.0 + 1.8i)\ (-8)$

tion (26) yields the results shown in Table 5.1, and $w_{15}(0)$ agrees with $w(0)$ to all the decimal places given in (28).

The series for $w(0)$ converges only conditionally, but this is irrelevant, because the Miller algorithm will work whether the Taylor series for F converges or not as long as $|\arg(1-z)| < \pi$. Formula (27) shows the convergence is exponential and in this case

$$\lambda = i(2 - \sqrt{3}), \qquad |\lambda| = 0.268. \quad \blacksquare$$

There is another approach, effective for Re $z < \frac{1}{2}$, which is interesting in its own right. It is based on the recurrence (C.20). Let $\Psi(n) = n!\,y(n)$ and make the obvious identification of parameters. The recurrence becomes

$$(z-1)y(n) + [(n+c) - z(2n+2c-a-b+1)]y(n+1)$$
$$+ z(n+c+1-a)(n+c+1-b)y(n+2) = 0, \quad (5.29)$$

which has solutions

$$w(n) := \frac{1}{\Gamma(n+c)}\, F\left(\begin{smallmatrix} a, & b \\ & n+c \end{smallmatrix}\middle| z\right),$$

$$y(n) := \frac{(-z)^{-n}\Gamma(n+c-1)}{\Gamma(n+c-a)\Gamma(n+c-b)}\, F\left(\begin{smallmatrix} a+1-c-n, & b+1-c-n \\ & 2-c-n \end{smallmatrix}\middle| z\right).$$

For $z \in \Omega$ we have

$$w(n) \sim \frac{n^{1-c}}{n!}\left[1 + \frac{c_1}{n} + \frac{c_2}{n^2} + \cdots\right].$$

It is known (Luke (1969, vol. 1, p. 241 (26))) that $y(n)$ has the behavior

$$y(n) = \frac{1}{n!}\left\{(1-z)^{c-a-b}\left(1 - \frac{1}{z}\right)^n n^{a+b-c}[1 + O(n^{-1})]\right.$$

$$\left. - \frac{\pi z^{c-1} n^{1-c}}{\Gamma(1-a)\Gamma(1-b)\sin(\pi c)}[1 + O(n^{-1})]\right\}.$$

As long as $|1 - 1/z| > 1$, $w(n)$ is a minimal solution. This requires Re $z < \frac{1}{2}$.

We can represent $w(n)$ as

$$w(n) = \frac{1}{\Gamma(b)\Gamma(n+c-b)} \int_0^1 t^{b-1}(1-t)^{n+c-b-1}(1-tz)^{-a} \, dt,$$

$$z \in \Omega, \qquad \mathrm{Re}\,(c+n) > \mathrm{Re}\,b > 0.$$

Consequently

$$\sum_{k=0}^{\infty} \lambda^k \Gamma(k+c-b) \frac{(\alpha)_k}{k!} w(k) = \frac{(1-\lambda)^{-\alpha}}{\Gamma(b)} \int_0^1 t^{b-1}(1-t)^{c-b-1}(1-tz)^{-a}$$

$$\times \left[1 + \frac{\lambda t}{1-\lambda}\right]^{-\alpha} dt.$$

Picking $\lambda = -[(1/z)-1]^{-1}$, $\alpha = -a$, produces a convergent series since $|\lambda| < 1$. Thus

$$S = \frac{(1-z)^{-a}}{\Gamma(c)} = \sum_{k=0}^{\infty} \frac{[1-(1/z)]^{-k}(c-b)_k(-a)_k}{k!} w(k). \tag{5.30}$$

The above holds for all $c \neq 0, -1, -2, \ldots$, uniformly on compact subsets of $\mathrm{Re}\,z < \frac{1}{2}$. Subject to these conditions the Miller algorithm for the computation of $w(n)$ based on (29) and (30) converges.

5.4 Associated Legendre functions

The Legendre function

$$P_\alpha^n(z) := \frac{2^{-n}\Gamma(\alpha+n+1)(z^2-1)^{n/2}}{\Gamma(\alpha-n+1)n!} F\left(\begin{matrix} 1+n+\alpha, & n-\alpha \\ & 1+n \end{matrix} \middle| \frac{1-z}{2}\right) \tag{5.31}$$

$$= \frac{\Gamma(\alpha+n+1)}{\Gamma(\alpha-n+1)n!} \left(\frac{z-1}{z+1}\right)^{n/2} F\left(\begin{matrix} -\alpha, & \alpha+1 \\ & 1+n \end{matrix} \middle| \frac{1-z}{2}\right), \qquad z \notin (-\infty, 1], \tag{5.32}$$

satisfies the recursion relation

$$(n-\alpha)(\alpha+n+1)y(n) + \frac{2(n+1)z}{\sqrt{(z^2-1)}} y(n+1) + y(n+2) = 0. \tag{5.33}$$

The branch of $(z^2-1)^{1/2}$ in (31) is the usual one, i.e.,

$$\begin{cases} (z^2-1)^{1/2} := |z^2-1|^{1/2} e^{i(\phi_1+\phi_2)/2}, \\ \phi_1 := \arg(z+1), \qquad \phi_2 := \arg(z-1), \qquad 0 \leqslant \phi_1, \phi_2 < 2\pi. \end{cases}$$

The representation (32) shows the computation of P_α^n is trivial when $\alpha \in Z$. I will assume henceforth that $\alpha \notin Z$.

The following asymptotic estimate holds:

$$P_\alpha^n(z) \sim -\sin(\pi\alpha)\sqrt{\left(\frac{2}{\pi}\right)}$$

$$\times\left[-e^{-1}\sqrt{\left(\frac{z-1}{z+1}\right)}\right]^n n^{n-1/2}\left[1+\frac{c_1}{n}+\frac{c_2}{n^2}+\cdots\right], \qquad n\to\infty.$$

Another solution of the recursion relation is $Q_\alpha^n(z)$, which has the behavior

$$Q_\alpha^n(z) \sim K\left[-e^{-1}\sqrt{\left(\frac{z+1}{z-1}\right)}\right]^n n^{n-1/2}\left[1+\frac{d_1}{n}+\frac{d_2}{n^2}+\cdots\right], \qquad n\to\infty.$$

(K is not required for our purposes.) It is easily seen that for $z\in\Omega$,

$$\Omega:=\{\text{Re } z>0\}-(0,1],$$

$w(n)=P_\alpha^n(z)$ is a minimal solution of the recursion. A useful normalization is (Erdélyi *et al.* (1953, vol. 1, p. 166 (4)))

$$(z+\sqrt{(z^2-1)})^\alpha = \sum_{k=0}^\infty \frac{\varepsilon_k P_\alpha^k(z)}{(\alpha+1)_k}.$$

Since $\sqrt{(z-1)}/\sqrt{(z+1)}$ maps the right-half plane into the unit disk the series converges for $z\in\Omega$. Rotenberg (1960) was, apparently, the first to notice that $P_\alpha^n(z)$ could be computed this way.

For the computation of Q_α (33) may be used in the forward but not the backward direction. However, there is another recursion relation which may be used to compute $Q_{\alpha+n}^m$ for all $z\notin[-1,1]$, see Erdélyi *et al.* (1953, vol. 1, p. 160 (2)),

$$(n+m+\alpha+1)y(n)-z(2n+2\alpha+3)y(n+1)$$

$$+(n-m+\alpha+2)y(n+2)=0. \quad \textbf{(5.34)}$$

The recursion is satisfied by $P_{\alpha+n}^m(z)$, $Q_{\alpha+n}^m(z)$. There is a fundamental set having the asymptotic behavior

$$y^{(h)}(n) \sim n^{\theta_h}(\xi_h(z))^n\left[1+\frac{c_h}{n}+\frac{d_h}{n^2}+\cdots\right].$$

$$\xi_1(z) = z+\sqrt{(z^2-1)}, \qquad \xi_2(z) = z-\sqrt{(z^2-1)}.$$

Since

$$Q_{\alpha+n}^m(z) = \sqrt{\left(\frac{\pi}{2}\right)}\frac{(-1)^m\Gamma(m+\alpha+n+1)[z-\sqrt{(z^2-1)}]^{\alpha+n+1/2}}{(z^2-1)^{1/4}\Gamma(\alpha+n+\frac{3}{2})}$$

$$\times F\left(\begin{matrix}\frac{1}{2}+m, & \frac{1}{2}-m \\ & \alpha+n+\frac{3}{2}\end{matrix}\middle|\frac{-z+\sqrt{(z^2-1)}}{2\sqrt{(z^2-1)}}\right),$$

we may identify $Q_{\alpha+n}^m$ with a constant (that is, independent of n) multiple of $y^{(2)}(n)$. $Q_{\alpha+n}^m(z)$ is thus the minimal solution of the recurrence (34) for

$z \notin [-1, 1]$. Unfortunately, as Gautschi (1967) points out, there does not seem to be a simple infinite series known for these functions which could serve as a normalization relation and thus if the Miller algorithm is to be applied to (34) the value $Q_\alpha^m(z)$ must be known. In the special case of the *toroidal functions* $Q_{-(1/2)+n}^m(z)$, however, the following relation *is* known (Erdélyi *et al.* (1953, vol. 1, p. 166 (3))):

$$Q_{-1/2}^m(z) + 2 \sum_{k=1}^{\infty} Q_{-(1/2)+k}^m(z)$$

$$= (-1)^m \sqrt{\left(\frac{\pi}{2}\right)} \Gamma(m + \tfrac{1}{2})(z+1)^{m/2}(z-1)^{-(m+1)/2}.$$

5.5 The Legendre function $Q_\nu^\mu(z)$

For the computation of the Legendre function $Q_\nu^\mu(z)$ when $|\tau| \leq 1$, where

$$\tau := \frac{-z + \sqrt{(z^2 - 1)}}{2\sqrt{(z^2 - 1)}}$$

(which is true, for instance, when $|z| > \sqrt{5}/2$), a method similar to the method of Chapter 9 for computing $_3F_2(1)$ can be used.

The recursion relation

$$(n + \mu + \nu + 1)y(n) - z(2n + 2\nu + 3)y(n+1)$$
$$+ (n + \nu - \mu + 2)y(n+2) = 0, \quad \textbf{(5.35)}$$

is stable for the computation of $Q_{\nu+n}^\mu(z)$ in the backward direction for all z off the cut $[-1, 1]$.

Consider the expansion

$$Q_{\nu+n}^\mu(z) = e^{\mu\pi i} \sqrt{\left(\frac{\pi}{2}\right)} \frac{\Gamma(n + \mu + \nu + 1)}{\Gamma(\nu + n + \tfrac{3}{2})} \frac{[z - \sqrt{(z^2 - 1)}]^{\nu+n+1/2}}{(z^2 - 1)^{1/4}}$$
$$\times F\left(\begin{matrix} \tfrac{1}{2} + \mu, & \tfrac{1}{2} - \mu \\ \nu + n + \tfrac{3}{2} \end{matrix} \middle| \tau\right).$$

The $(k+1)$th term in F above behaves like $O(\tau^k k^{-n-\nu-3/2})$. Thus a value of N can be found so that the Gaussian hypergeometric function can be computed to any desired degree of accuracy for $n = N+1$ and $n = N$. The recursion (35) can then be used in a stable manner to compute $Q_{\nu+N-1}^\mu, Q_{\nu+N-2}^\mu, \ldots, Q_\nu^\mu$.

5.6 The Jacobi polynomials $P_n^{(i\eta,-i\eta)}(-i\omega)$

In practical applications it is sometimes required to compute

$$z(n) := i^n P_n^{(i\eta,-i\eta)}(-i\omega), \qquad n \in Z^0,$$

where the parameters η, ω are real and $\eta > 0$, $\omega \geq 0$; see Gautschi (1969).

From the known relationship for Jacobi polynomials we see $z(n)$ satisfies

$$\frac{(n+1)^2+n^2}{(n+1)(n+2)}y(n)+\frac{(2n+3)\omega y(n+1)}{(n+2)}-y(n+2)=0, \qquad n\geq 0,$$

$$(5.36)$$

$$y(0)=1, \qquad y(1)=\omega-\eta. \tag{5.37}$$

Thus $z(n)$ is real.

The techniques of Section B.2 show this equation has two solutions with the property

$$y^{(h)}(n)\sim n^{-1/2}(-\omega+(-1)^h\sqrt{(\omega^2+1)})^n\left\{c_1^{(h)}+\frac{c_2^{(h)}}{n}+\cdots\right\},$$

$$n\to\infty, \quad h=1,2.$$

For $\omega\neq 0$ $y^{(1)}(n)$ is a dominant solution and $y^{(2)}(n)$ is a minimal solution, which we may identify with a multiple of

$$u(n)=i^{n+1}Q_n^{(i n,-i n)}(-i\omega),$$

by Szegö (1959, p. 223). By the same reference, p. 192, we see the desired solution $z(n)$ is nonminimal. I will identify $z(n)$ with $y^{(1)}(n)$. It would appear that (36), (37) could be used in the forward direction to generate $z(n)$. This is true as long as η is not too large. When η gets large something very curious happens. The results in Erdélyi *et al.* (1953, vol. 2, p. 172) show that

$$\frac{P_1^{(\alpha,-\alpha)}(x)}{P_0^{(\alpha,-\alpha)}(x)}-\frac{Q_1^{(\alpha,-\alpha)}(x)}{Q_0^{(\alpha,-\alpha)}(x)}=\frac{\Gamma(1+\alpha)\Gamma(1-\alpha)}{Q_0^{(\alpha,-\alpha)}(x)}\left(\frac{x+1}{x-1}\right)^{\alpha}. \tag{5.38}$$

Also

$$Q_0^{(\alpha,-\alpha)}=\Gamma(1+\alpha)\Gamma(1-\alpha)(x-1)^{-\alpha-1}(x+1)^{\alpha}F\left(\begin{matrix}\alpha+1, & 1\\ & 2\end{matrix}\middle|\frac{2}{1-x}\right).$$

Now

$$F\left(\begin{matrix}\alpha+1, & 1\\ & 2\end{matrix}\middle|\frac{2}{1-x}\right)=\sum_{k=0}^{\infty}\frac{(\alpha+1)_k}{(k+1)!}\left(\frac{2}{1-x}\right)^k$$

$$=-\left(\frac{1-x}{2\alpha}\right)+\sum_{k=0}^{\infty}\frac{(\alpha+1)_{k-1}}{k!}\left(\frac{2}{1-x}\right)^{k-1}$$

$$=\frac{(1-x)}{2\alpha}\left[\left(1-\frac{2}{1-x}\right)^{-\alpha}-1\right],$$

so

$$Q_0^{(\alpha,-\alpha)}(x)=-\frac{\Gamma(1+\alpha)\Gamma(1-\alpha)}{2\alpha}[1-(x+1)^{\alpha}(x-1)^{-\alpha}].$$

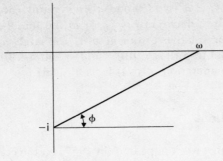

Fig. 5.1

Let

$$\frac{\omega + i}{\omega - i} := e^{2i\phi}, \qquad \phi = \arctan \frac{1}{\omega}.$$

Since $\omega > 0$, $0 < \phi \le \pi/2$, in accordance with Fig. 5.1.
Thus

$$u(0) = \frac{-\pi}{2 \sinh (\pi\eta)} (1 - e^{-2\eta\phi}),$$

and from (37) and (38) we find

$$\Delta := \frac{z(1)}{z(0)} - \frac{u(1)}{u(0)} = \frac{2\eta}{e^{2\eta\phi} - 1}. \tag{5.39}$$

Thus as η becomes large the ratio of the initial values of the minimal and the maximal solution become the same and so the two solutions will ultimately be indistinguishable (in finite arithmetic) even though they behave quite differently as $n \to \infty$. Thus one cannot determine $z(n)$ accurately from initial values when η is large unless multiple precision arithmetic is used.

For such cases Gautschi (1969) has devised the following technique. Let $v(n)$ be the solution of (36) defined by

$$v(0) := -\frac{u(1)}{u(0)}, \qquad v(1) := 1, \tag{5.40}$$

then

$$z(n) = \frac{u(n)}{u(0)} + \frac{\Delta[v(n)u(0)^2 + u(n)u(1)]}{u(1)^2 + u(0)^2}, \tag{5.41}$$

which shows that for small Δ the solution $z(n)$ initially follows $u(n)/u(0)$ closely until the dominance of $v(n)$ outweighs the smallness of Δ. All the quantities in (41) can be computed accurately, $u(n)$ by the simplified

Miller algorithm (4.7), Δ by (39) and $v(n)$ by using (36) with the initial conditions (40). The relative errors δ_0, δ_1 in the initial values $v(0)$, $v(1)$ give rise to similar relative errors in $v(n)$ for large n, namely, errors approximately equal to $\delta_0 u(1)^2/(u(0)^2 + u(1)^2)$ and $\delta_1 u(0)^2/(u(0)^2 + u(1)^2)$. This means the computation of $v(n)$ is well defined.

5.7 Bessel functions

Since Miller's original work appeared (in 1952) on the computation of the modified Bessel function $I_n(x)$, a large mathematical literature has accumulated on the computation of Bessel functions. It is probably safe to say that by now the computation of no other class of higher transcendental functions is better understood. Extremely effective algorithms based on the recurrence relations for these functions are now available and the behavior of the error involved has been well described, having been studied exhaustively by many writers. Miller-type algorithms, in fact, are the methods of choice for computing these functions on large-scale computers.

Important references on this topic are Stegun and Abramowitz (1957), Randels and Reeves (1958), Goldstein and Thaler (1959), Corbató and Uretsky (1959), Hitotumatu (1963), and Makinouchi (1966). In all of these papers the authors discuss what is essentially the Miller algorithm applied to the second-order recurrence satisfied by the functions. Many other references deal with related matters which are often of great interest, but I shall begin my discussion with a description of the Miller algorithm and treat other results later.

Consider the recurrence

$$y(n) - \frac{2(n+\nu+1)}{x} y(n+1) + y(n+2) = 0, \qquad x \neq 0. \tag{5.42}$$

As is well known, the equation has a fundamental set

$$y^{(1)}(n) = J_{n+\nu}(x), \qquad y^{(2)}(n) = Y_{n+\nu}(x). \tag{5.43}$$

I will henceforth assume that $0 \leqslant \nu < 1$.
We know that

$$J_{n+\nu}(x) = \frac{(x/2)^{n+\nu}}{\Gamma(n+\nu+1)} [1 + O(n^{-1})], \tag{5.44}$$

$$Y_{n+\nu}(x) = \frac{-(x/2)^{-n-\nu}\Gamma(n+\nu)}{\pi} [1 + O(n^{-1})], \tag{5.45}$$

and these estimates hold for all (complex) x and ν, $|\arg x| < \pi$. Obviously the Miller algorithm will converge when applied to the recurrence, provided a suitable normalization relation is available, and will yield $\{J_{n+\nu}(x)\}$. One of the questions I will address is the computation of $\{Y_{n+\nu}(x)\}$ for which the Miller algorithm yields a divergent process.

A convenient normalization is

$$\left(\frac{x}{2}\right)^{\nu} = \sum_{k=0}^{\infty} \frac{(2k+\nu)\Gamma(k+\nu)J_{2k+\nu}(x)}{k!}, \tag{5.46}$$

which converges for all (complex) x and ν.

Once $\{J_{\nu+n}(x)\}$ is computed for, say, $0 \leq n \leq N$ to the desired accuracy, Y_{ν} for ν nonintegral may be computed as follows (Goldstein and Thaler (1959)).

We have

$$Y_{\nu}(x) = \frac{J_{\nu}(x) \cos \nu\pi - J_{-\nu}(x)}{\sin \nu\pi}. \tag{5.47}$$

An expansion of $J_{-\nu}(x)$ in terms of $J_{2k+\nu}(x)$ is readily determined from a result in Erdélyi *et al.* (1953, vol. 2, p. 99, (2)) and we find

$$Y_{\nu}(x) = \sum_{k=0}^{\infty} d(k)J_{2k+\nu}(x), \tag{5.48}$$

where $\nu \neq 0$ and

$$d(0) = \cot \nu\pi - \frac{1}{\pi}\left(\frac{2}{x}\right)^{2\nu}\frac{\Gamma^2(1+\nu)}{\nu}, \tag{5.49}$$

$$d(1) = \left(\frac{2}{\pi}\right)\left(\frac{2}{x}\right)^{2\nu}\left(\frac{\nu+2}{1-\nu}\right)\Gamma^2(1+\nu), \tag{5.50}$$

$$d(n+1) = -\frac{(2n+\nu+2)(n+2\nu)(n+\nu)d(n)}{(n+1)(n+1-\nu)(2n+\nu)}, \qquad n \geq 1. \tag{5.51}$$

This expansion is easily modified to deal with the case where ν is an integer. For instance if $\nu = 0$, a limit gives

$$d(0) = \frac{2}{\pi}\left[\gamma + \ln\left(\frac{x}{2}\right)\right], \tag{5.52}$$

where $\gamma = 0.577\,21\ldots$ is Euler's constant and where the principal value of ln is used. $d(n)$ for $n > 0$ may then be computed from (50), (51).

Once Y_{ν} is found, $Y_{\nu+1}$ can be found from the Wronskian relation

$$Y_{\nu}(x)J_{\nu+1}(x) - Y_{\nu+1}(x)J_{\nu}(x) = \frac{2}{\pi x}. \tag{5.53}$$

Finally, $Y_{n+\nu}$ may be computed stably by using the recurrence in the forward direction.

If x is close to a zero of J_{ν}, then (52) does not provide a satisfactory way of computing $Y_{\nu+1}$. When this happens it is better to compute $Y_{\nu+1}$ from the series

$$Y_{\nu+1}(x) = \sum_{k=0}^{\infty} e(k)J_{k+\nu}(x). \tag{5.54}$$

This series is found by differentiating (48) and using the differential–difference relation satisfied by both Y_ν, J_ν (Erdélyi *et al.* (1954, vol. 2, p. 11 (54)). The result is

$$e(0) = -\frac{1}{\pi}\left(\frac{2}{x}\right)^{1+2\nu}\Gamma^2(1+\nu),$$

$$e(1) = d(0) - \tfrac{1}{2}d(1),$$

$$e(2n) = \frac{3\nu}{x}d(n), \qquad n \geqslant 1,$$

$$e(2n+1) = \tfrac{1}{2}(d(n) - d(n+1)), \qquad n \geqslant 1.$$

The Bessel functions $I_{n+\nu}$, $K_{n+\nu}$ are treated similarly. $I_{n+\nu}$ and $(-1)^n K_{n+\nu}$ satisfy

$$y(n) - \frac{2(n+\nu+1)}{x}y(n+1) - y(n+2) = 0.$$

The Miller algorithm will converge to $I_{n+\nu}$ and the appropriate normalization is

$$\left(\frac{x}{2}\right)^\nu = \sum_{k=0}^{\infty} \frac{(-1)^k (2k+\nu)\Gamma(k+\nu)}{k!} I_{k+\nu}(x), \tag{5.55}$$

and the appropriate expansion for K_ν is

$$K_\nu(x) = \sum_{k=0}^{\infty} f(k)I_{2k+\nu}(x), \qquad \nu \neq 0, \tag{5.56}$$

$$f(0) = -\frac{1}{2\nu}\left[\frac{\nu\pi}{\sin \nu\pi} - \left(\frac{2}{x}\right)^{2\nu}\Gamma^2(1+\nu)\right], \tag{5.57}$$

$$f(1) = \frac{\Gamma^2(1+\nu)(2+\nu)}{(1-\nu)}\left(\frac{2}{x}\right)^{2\nu}, \tag{5.58}$$

$$f(n+1) = \frac{(2n+\nu+2)(n+2\nu)(n+\nu)f(n)}{(n+1)(n+1-\nu)(2n+\nu)}, \qquad n \geqslant 1. \tag{5.59}$$

When $\nu = 0$ a limit must be taken in (56) and we find

$$f(0) = -\gamma - \ln\frac{x}{2}.$$

Subsequent values of $f(k)$ are then computed from (58), (59).

Certain numerical problems may be associated with the previous results when $|x|$ is large. For instance when x is large positive I_ν becomes large exponentially and there is a loss of significance in the evaluation of the expansion (54). In such cases it is preferable to evaluate K_ν from its integral representation

$$K_\nu(x) = \int_0^\infty e^{-x \cosh y} \cosh(\nu y)\, dy,$$

see Luke (1975, p. 397 ff) and the references given there for a technique for doing this.

Overflow problems can occur in these computations. Hitotumatu (1963) has given an estimate that may be useful in monitoring the overflow situation: let 10^p be the ratio between the largest and smallest positive numbers admissible in the computer. Overflow will be reached, roughly, when

$$(N+\tfrac{1}{3}) \log_{10}\left(\frac{2N}{ex}\right) = p-1.$$

Such overflow tends to occur when large values of N are required as is the case when I_ν, J_ν are to be computed for large values of x. The overflow problem can be circumvented by computing instead the ratio of successive quantities, e.g., $I_{n+\nu}/I_{n+\nu-1}$. See Section 4.5 for a discussion, as well as Amos (1974).

In such a situation it is preferable, when computing $I_{n+\nu}$, to normalize on $e^{-x}I_{n+\nu}$ by using the expansion

$$\frac{\Gamma(1+2\nu)}{\Gamma(1+\nu)}\frac{x^\nu}{2^{\nu+1}} = \sum_{k=0}^\infty \frac{(\nu+k)\Gamma(2\nu+k)}{k!}\, e^{-x}I_{k+\nu}(x),$$

rather than (54) since I_α grows exponentially with x.

The Olver error analysis of Section 4.4 gives a rough estimate of the error involved in computing $J_{n+\nu}$. To estimate the quantities ρ and τ one may use the asymptotic estimate for $J_{n+\nu}$, formula (44). To estimate a one needs an inequality for $J_{n+\nu}$ such as

$$|J_{n+\nu}(x)| \leq \frac{(x/2)^\nu}{\Gamma(n+\nu+1)}, \qquad \nu, x > 0.$$

$y^{(2)}(n)$ can be computed by using the difference equation in the forward direction with initial values 0, 1, and σ is determined by taking the maximum of the quantities $|y^{(2)}(n)/y^{(2)}(N+1)|^{1/N+1-n}$. The value of S_N is known in terms of computed quantities.

As an example, if $J_n(1)$ is to be computed taking $N = 5$, formula (4.40) gives an error bound for the value of $J_0(1)$ so computed of 4.00 (−5). The true error is 3.26 (−5) which is quite serviceable considering the crudeness of the estimates involved.

For the error involved in computing I_n, see Example 4.4.

Makinouchi (1966) discusses the error in more detail.

Obviously the recurrence satisfied by $I_{n+\nu}$, $K_{n+\nu}$ cannot be used in the backward direction to compute K_ν. But there are other recurrences which can be used to compute K_ν. These are discussed in Section 5.2 and Example 7.3.

5.8 Zeros of Bessel functions

The Miller algorithm serves as an ideal basis on which to construct a Newton-type interpolation formula for finding the zeros of $J_{n+\nu}(x)$; see Makinouchi (1966).

Let $y_N(n)$ (see (4.3)) be computed, as in the previous section, from the recurrence (42).

A better approximation, x^*, for a zero of $J_{n+\nu}(x)$ given an approximation x is

$$x^* = x - \frac{J_{n+\nu}(x)}{J'_{n+\nu}(x)}.$$

But

$$J'_{n+\nu}(x) = \tfrac{1}{2}[J_{n+\nu-1}(x) - J_{n+\nu+1}(x)],$$

and, by the recurrence, $J_{n+\nu-1}(x)$ is approximately $-J_{n+\nu+1}(x)$. Thus the new approximation, in terms of the computed quantities $y_N(n)$, is

$$x^* = x + \frac{y_N(n)}{y_N(n+1)}.$$

To begin the process one starts, say, with $N = 5$ and an initial guess for x and iterates on x till no change occurs. Then N is increased and the process repeated until subsequent values of x agree to the desired accuracy.

The process converges with enormous rapidity.

Starting values can be obtained from a number of different sources, such as McMahon's expansion, see Watson (1962, Ch. XV).

5.9 Eigenvalues of Mathieu's equation

Mathieu's differential equation may be written

$$y'' + (\lambda - 2q \cos 2x)y = 0. \tag{5.60}$$

The solutions I shall discuss here are periodic of period 2π; in general q is given and the eigenvalue λ is to be determined so the solution has this periodicity property. The Sturm–Liouville theory shows the equation (60) has an infinity of real simple eigenvalues. The literature on Mathieu functions is vast (the standard treatise is McLachlan (1947)) and much research has been devoted to the computation of the functions and their eigenvalues, e.g., Arscott *et al.* (1978), Arscott (1975), Delft University group (1973), Blanch and Clemm (1962), Kirkpatrick (1960), Blanch and Rhodes (1955), Blanch (1946), Goldstein (1927).

I shall present here a method for the computation of the eigenvalues which involves an interesting modification of the Miller algorithm.

The periodic solutions of the equation fall into four classes: even, odd, those having π as a period, those having π as an anti-period. The notation is as in Table 5.2.

I will discuss only the computation of eigenvalues for the solutions $ce_{2n}(x, q)$, the discussion for the other functions being very similar.

We can express the function as a trigonometric series

$$f(x) = \frac{\alpha(0)}{2} + \sum_{n=1}^{\infty} \alpha(n) \cos(2nx).$$

Substituting this expansion into the differential equation yields a recursion relation for the coefficients $\alpha(n)$,

$$y(n) - \left[\frac{\lambda - 4(n+1)^2}{q} \right] y(n+1) + y(n+2) = 0, \qquad n \geq 0, \tag{5.61}$$

$$\frac{\lambda}{2} y(0) - q y(1) = 0, \tag{5.62}$$

the last equation serving as a normalization relation.

First let's look at the behavior of the solutions of the recurrence. An appeal to the theory of Section B.2 shows the equation has a fundamental set with the property

$$y^{(1)}(n) = n^{-2n-1} \left(\frac{-e^2 q}{4} \right)^n [1 + O(n^{-1})], \tag{5.63}$$

$$y^{(2)}(n) = n^{2n-1} \left(\frac{-4}{e^2 q} \right)^n [1 + O(n^{-1})]. \tag{5.64}$$

Table 5.2

	Even/odd	π	Notation for eigenvalues
$ce_{2n}(x, q)$	even	period	a_{2n}
$ce_{2n+1}(x, q)$	even	anti-period	a_{2n+1}
$se_{2n+1}(x, q)$	odd	anti-period	b_{2n+1}
$se_{2n+2}(x, q)$	odd	period	b_{2n+2}

The first solution is obviously the one we are seeking. Since it is so strongly minimal the Miller algorithm applied to the equation converges with great rapidity.

One way of using the equations (61)–(62) is to establish an iterative procedure on λ. A (fixed) value of N is selected and a trial value of λ. One defines $y_N(N+1)=0$, $y_N(N)=1$, and then computes $y_N(n)$ for $n = N-1, N-2, \ldots, 0$, from the first equation. The second equation will not be satisfied but an improved value of λ, call it λ^*, may be computed from it by putting

$$\lambda^* = 2qy_N(1)/y_N(0).$$

The process is then repeated.

The problem with this procedure, as the reader probably appreciates, is that the convergence of what is actually a simple iteration process is only linear. It's a pity, too, since the N convergence is spectacular, roughly $O(K^N N^{-4N})$, as can be seen from (63), (64).

However, the Miller algorithm may be combined with a Newton–Raphson process to yield a rapidly convergent algorithm.

Consider $y(n)$ as a function of λ and differentiate the equation (61). We have

$$y'(n) - \left[\frac{\lambda - 4(n+1)^2}{q}\right]y'(n+1) + y'(n+2) = \frac{y(n+1)}{q}. \tag{5.65}$$

Pick N and define *two* sequences, $y_N(n)$, $y'_N(n)$, such that

$$y'_N(N+1) = y'_N(N) = y_N(N+1) = 0; \qquad y_N(N) = 1.$$

$y_N(n)$ and $y'_N(n)$ are then computed by using (61) and (65) alternately for $N = N-1, N-2, \ldots, 0$.

Now differentiate

$$f(\lambda) := \frac{\lambda}{2} y(0) - qy(1),$$

to get

$$f'(\lambda) = \frac{\lambda}{2} y'(0) - qy'(1) + \frac{y(0)}{2},$$

and compute the improved λ, call it λ^*, from

$$\lambda^* = \lambda - \frac{\left(\dfrac{\lambda}{2} y_N(0) - qy_N(1)\right)}{\left(\dfrac{\lambda}{2} y'_N(0) - qy'_N(1) + \dfrac{y_N(0)}{2}\right)}.$$

Table 5.3 Eigenvalue of
Mathieu's equation.

N	Limit of λ
10	264.388 236 658 93
15	263.190 710 673 55
20	263.190 710 670 37
25	263.190 710 670 37

The process is then repeated. Since the eigenvalues are simple the process
will converge (in λ) and convergence is quadratic.

Table 5.3 gives the results of the computation of the ninth eigenvalue
for $q = 60$ starting with the value $\lambda = 260$. No λ-computation (for a fixed
value of N) required more than 3 iterations.

Of course the procedure requires an estimate of λ. This can be
accomplished by using any of the standard search procedures. Also,
estimates are available both for small and for large q, see Arscott *et al.*
(1978) and the references given there.

6 Second-order nonhomogeneous equations: the Olver algorithm

6.1 Introduction: the algorithm

In 1967 F. W. J. Olver introduced a technique for generating solutions of the nonhomogeneous second-order equation

$$\mathcal{L}[y(n)] :\equiv y(n) + a(n)y(n+1) + b(n)y(n+2) = f(n), \qquad n \geqslant 0. \quad \textbf{(6.1)}$$

His algorithm, contained in the Olver papers (1967a, 1968), has several attractive features in common with the Miller algorithm discussed in Chapter 4, namely, that no initial values of the desired solution are required and that it is very amenable to computer implementation. The algorithm has engendered a vast amount of research, and extensions and analyses of the algorithm occupy a prominent place in the literature dealing with computational methods based on recurrence relations.

Since the second-order equation is of such great importance I will discuss in this chapter only that case and leave till Chapter 8 a discussion of generalizations of the method. I will assume at the start that the initial value $w(0)$ of the desired solution $w(n)$ is known, although the algorithm is readily adaptable to the case where only a normalization relation for $w(n)$ is available, as I shall show later. This minor specialization makes the features of the algorithm easier to observe.

The algorithm is defined by the following scheme. For an integer $N \geqslant 0$ let the sequence $w_N(n)$ be defined by

$$w_N(N+1) := 0, \qquad w_N(0) := w(0), \quad \textbf{(6.2)}$$

$$w_N(n) + a(n)w_N(n+1) + b(n)w_N(n+2) = f(n),$$
$$n = N-1, N-2, \ldots, 0. \quad \textbf{(6.3)}$$

We say the Olver algorithm converges to $w(n)$ if

$$\lim_{N \to \infty} w_N(n) = w(n), \qquad n \geqslant 0. \quad \textbf{(6.4)}$$

Sufficient conditions for convergence are easily formulated.

Theorem 6.1 *Let $w(n)$ satisfy the nonhomogeneous equation (1) and let* *the homogeneous equation*

$$y(n) + a(n)y(n+1) + b(n)y(n+2) = 0, \qquad \textbf{(6.5)}$$

have a fundamental set $\{y^{(h)}(n)\}$ with the property

$$\lim_{n \to \infty} y^{(1)}(n)/y^{(2)}(n) = 0, \qquad y^{(1)}(0) \neq 0. \qquad \textbf{(6.6)}$$

Let

$$\lim_{n \to \infty} w(n)/y^{(2)}(n) = 0. \qquad \textbf{(6.7)}$$

Then the Olver algorithm converges.

Proof Since $w(n)$ is a particular solution of (1) I can write

$$w_N(n) = \xi_N^{(1)} y^{(1)}(n) + \xi_N^{(2)} y^{(2)}(n) + w(n).$$

Putting $n = N+1$, $n = 0$ shows

$$\begin{aligned}
\xi_N^{(1)} &= \frac{y^{(2)}(0)w(N+1)}{y^{(1)}(0)y^{(2)}(N+1) - y^{(2)}(0)y^{(1)}(N+1)}, \\
\xi_N^{(2)} &= \frac{-y^{(1)}(0)w(N+1)}{y^{(1)}(0)y^{(2)}(N+1) - y^{(2)}(0)y^{(1)}(N+1)}.
\end{aligned} \qquad \textbf{(6.8)}$$

An application of (6) and (7) shows

$$\lim_{N \to \infty} \xi_N^{(1)} = \lim_{N \to \infty} \xi_N^{(2)} = 0,$$

and this proves the theorem. ∎

Example 6.1 Consider the modified moments

$$w(n) := \int_{-1}^{1} e^{-a(t+1)} T_{n+1}(t)\, dt, \qquad \textbf{(6.9)}$$

T_n being the nth Chebyshev polynomial, $w(n)$ satisfies the non-homogeneous equation

$$\left.\begin{aligned}
\mathscr{L}[y(n)] &:= y(n) + \frac{2}{a}(n+1)y(n+1) - \frac{(n+1)}{(n+3)} y(n+2) \\
&= f_1(n) + f_2(n), \\
f_1(n) &:= \frac{-2e^{-2a}}{a(n+3)}, \qquad f_2(n) := \frac{2(-1)^{n+1}}{a(n+3)};
\end{aligned}\right\} \qquad \textbf{(6.10)}$$

see Branders (1976).

Using the methods in Section B.2 we find there is a fundamental set for the homogeneous equation $y^{(1)}(n)$, $y^{(2)}(n)$ with the behavior

$$y^{(1)}(n) = n^{\theta_1} n^{-n} \left(\frac{-ae}{2}\right)^n [1 + O(n^{-1})],$$

$$y^{(2)}(n) = n^{\theta_2} n^n \left(\frac{2}{ae}\right)^n [1 + O(n^{-1})].$$

An asymptotic approximation for $w(n)$ may be obtained as follows. Write $\mathscr{L}[y(n)] = f_1(n)$, $\mathscr{L}[y(n)] = f_2(n)$. By dividing these equations by f_1 and f_2 respectively and applying the Δ operator we see that solutions of these equations satisfy third-order homogeneous equations of the type prescribed in Appendix B, and hence any solution of either original equation must possess a Birkhoff series. $w(n)$ will then be represented by the sum of these series. Finding the series by direct substitution is a simple matter and we obtain

$$w(n) = -n^{-2}[e^{-2a} + (-1)^{n+1}][1 + O(n^{-1})].$$

(The reason $w(n)$ decays so slowly is that it is essentially the $(n+1)$st coefficient for the expansion of an *odd* function in a Fourier cosine series.)

Computation of $w(n)$ in the forward direction using equation (10) will be unstable, since small errors will grow in accordance with the presence of the solution $y^{(2)}(n)$ in the computations. Also the use of the equation in the *backward* direction starting with values, say, $w(N)$, $w(N-1)$, will fail to converge for the same reason.

On the other hand, the Olver algorithm for the computation of $w(n)$ converges. I shall return to this example later. ∎

Example 6.2 Consider the recurrence

$$y(n) - \frac{2(n+v+1)}{z} y(n+1) + y(n+2) = \frac{-2}{\pi z}[1 + (-1)^n \cos v\pi],$$

$$z > 0, \quad n \geqslant 0.$$

Solutions of the homogeneous equation are

$$y^{(1)}(n) = J_{n+v}(z), \qquad y^{(2)}(n) = Y_{n+v}(z).$$

Olver's algorithm may be used to compute the solution

$$w(n) := \mathbf{E}_{n+v}(z) := \sin\frac{(n+v)\pi}{2}$$

$$\times \sum_{k=0}^{\infty} \frac{(-1)^k (z/2)^{2k}}{\Gamma(k+1+(n+v)/2)\Gamma(k+1-(n+v)/2)}$$

$$- \cos\frac{(n+v)\pi}{2} \sum_{k=0}^{\infty} \frac{(-1)^k (z/2)^{2k+1}}{\Gamma(k+\frac{3}{2}+(n+v)/2)\Gamma(k+\frac{3}{2}-(n+v)/2)}.$$

E_α is called Weber's function, see Erdélyi *et al.* (1953, 7.5.3), Watson (1962, Ch. X). $w(n)$ may be estimated for large n either by writing the right-hand side above in terms of hypergeometric functions and using the techniques of Luke (1969, vol. 1, Ch. 7) or by the method of the previous example. We find either way that

$$w(n) = \frac{2\sin^2[(n+v)\pi]/2}{\pi n}[1+O(n^{-1})], \qquad n \to \infty, \quad n+v \neq 2k.$$

(If $n+v$ is an even integer the above estimate must be modified. It then becomes $O(n^{-2})$.)

The conditions of Theorem 6.1 are clearly satisfied provided $J_v(z) \neq 0$. Thus the Olver algorithm may be used to compute $E_{v+n}(z)$.

For $v = 0$, $N = 13$, $z = 1$, Olver gives the data shown in Table 6.1. ■

Example 6.3 Consider the equation

$$y(n) - \frac{2(v+n+1)}{z}y(n+1) + y(n+2) = \frac{(z/2)^{n+v+1}}{(\sqrt\pi)\Gamma(n+v+5/2)},$$

$$z > 0, \quad n \geq 0,$$

satisfied by $y(n) = \mathbf{H}_{n+v}$, the Struve function (Erdélyi *et al.*, vol. 2, 7.5.4), Watson (1962, Ch. X). A fundamental set for the homogeneous equation is $\{y^{(h)}(n)\}$ of the previous example.

It is known that

$$\mathbf{H}_{n+v}(z) = \frac{2(z/2)^{n+v+1}}{(\sqrt\pi)\Gamma(n+v+5/2)}[1+O(n^{-1})], \qquad n \to \infty.$$

It is easy to see that the Olver algorithm for the computation of $\mathbf{H}_{n+v}(z)$ will converge, provided $J_v(z) \neq 0$. ■

Example 6.4 (Sadowski and Lozier (1972)). The integrals

$$S(n) := \int_{-1}^{1} T_{2n}(t)e^{-\alpha^2 t^2/2}\,\mathrm{d}t,$$

Table 6.1 Computation of $E_N(1)$.

n	Approx. $E_n = w_{14}(n)$
0	−0.568 656 627 (known)
3	0.248 805 382
6	0.018 919 443
9	0.071 668 637
12	0.004 479 865
14	0.000 000 000

The first digit in error has been underlined.

occur in plasma physics and satisfy

$$\alpha^2 \frac{(2n+3)}{8} y(n) + \frac{\alpha^2 + 2(2n+1)(2n+3)}{4} y(n+1)$$
$$- \frac{\alpha^2}{8}(2n+1)y(n+2) = -e^{-\alpha^2/2}.$$

Obviously $S(n)$ is bounded, while solutions of the homogeneous equation behave like $(2\alpha)^{-n}/n!$, $(-1)^n(2\alpha)^n\Gamma(n)$. Thus the Olver algorithm will converge. Concerning the computation of $S(0)$ and the construction of a normalization relation see the cited reference. ■

The system (3) is an $N \times N$ tridiagonal system in $w_N(1), w_N(2), \ldots, w_N(N)$ which can be written

$$\left.\begin{array}{l}
a(0)w_N(1) + b(0)w_N(2) = f(0) - w(0), \\
w_N(1) + a(1)w_N(2) + b(1)w_N(3) = f(1), \\
w_N(2) + a(2)w_N(3) + b(2)w_N(4) = f(2), \\
\qquad\qquad\vdots \\
w_N(N-2) + a(N-2)w_N(N-1) + b(N-2)w_N(N) = f(N-2), \\
w_N(N-1) + a(N-1)w_N(N) = f(N-1).
\end{array}\right\}$$

$$\textbf{(6.11)}$$

As such, the solution of the system can be accomplished by several different techniques, each of which has its own advantages. I shall discuss three approaches.

6.2 Solution by forward elimination (Method A)

Rewrite the first equation in (11) in the form

$$y(2)w_N(1) - y(1)w_N(2) = e(1), \qquad\qquad \textbf{(6.12)}$$

where

$$y(1) = 1, \qquad y(2) = \frac{-a(0)}{b(0)}, \qquad e(1) = \frac{w(0) - f(0)}{b(0)}.$$

The result of eliminating $w_N(1)$ from the previous equation (12) and the second member of (11) may be written

$$y(3)w_N(2) - y(2)w_N(3) = e(2),$$

where

$$y(3) = \frac{-a(1)y(2) - y(1)}{b(1)}, \qquad e(2) = \frac{e(1) - f(1)y(2)}{b(1)}.$$

Continuing, we obtain

$$y(n+1)w_N(n) - y(n)w_N(n+1) = e(n), \qquad n = 1, 2, \ldots, N, \qquad \textbf{(6.13)}$$

where $y(n)$, $e(n)$ satisfy the recurrences

$$y(n+1) = \frac{-a(n-1)y(n) - y(n-1)}{b(n-1)}, \qquad n = 1, 2, \ldots, N, \qquad \textbf{(6.14)}$$

$$e(n) = \frac{e(n-1) - f(n-1)y(n)}{b(n-1)}, \qquad n = 1, 2, \ldots, N. \qquad \textbf{(6.15)}$$

The equations (13)–(15) are used as follows:

(i) set

$$e(0) = w(0); \qquad y(0) = 0, \quad y(1) = 1;$$

(ii) use (15), (14) alternately in the forward direction to compute

$$e(n), \quad 1 \leq n \leq N, \qquad \text{and} \qquad y(n), \quad 2 \leq n \leq N+1;$$

(iii) set

$$w_N(N+1) = 0, \qquad w_N(N) = e(N)/y(N+1),$$

and use (13) in the backward direction to compute

$$w_N(N-1), w_N(N-2), \ldots, w_N(1).$$

This method of solution of the tridiagonal system will fail if and only if one of the numbers $y(2), y(3), \ldots, y(N+1)$ vanishes. In this case the system has either no solution or an infinity of solutions and the algorithm breaks down.

When some of the coefficients $b(n)$ vanish the equations become uncoupled. The algorithm is easily modified to take care of this situation. For example let $b(s) = 0$ but all other $b(n) \neq 0$. The first $s+1$ equations in (11) determine $w_n(1), w_N(2), \ldots, w_N(s+1)$ completely. They can be solved by application of the recurrence relations (14), (15) for $n = 1, 2, \ldots, s$ and the use of the back-substitution relation (13), beginning with

$$w_N(s+1) = \frac{f(s)y(s+1) - e(s)}{a(s+1)y(s+1) + y(s)}.$$

The remaining $N-s-1$ equations are solvable for $w_N(s+2)$, $w_N(s+3)$, $\ldots, w_N(N)$, by the method already described.

Ordinarily $w_N(n)$ is required for only a fixed range of values of n, say, $0 \leq n \leq m$. Olver has demonstrated a very useful technique for computing these recursively on N starting with $N = m$. His result also leads to a convenient error formulation for Method A.

First, note that the sequences $y(n)$, $e(n)$ do not depend on N. Thus we may replace N by $N+1$ in (13) and subtract this result from the original equation. We have

$$w_{N+1}(n) - w_N(n) = \frac{y(n)}{y(n+1)}[w_{N+1}(n+1) - w_N(n+1)], \tag{6.16}$$

and repeated application of this leads to

$$w_{N+1}(n) - w_N(n) = \frac{y(n)e(N+1)}{y(N+1)y(N+2)} := \eta_N(n), \qquad n = 0, 1, \dots, N+1. \tag{6.17}$$

Thus once the system (11) has been solved for $N = m$, N may be increased by using (17) and this requires tabulating only one additional member from each sequence, $\{y(n)\}$ and $\{e(n)\}$.

The formula also allows us to judge the effect of changing N into $N+1$. Suppose we want to compute $w(L)$ to D decimal places. Then the recurrence relations (14), (15) are applied from $n = 1$ past $n = L$ until an n is reached for which

$$\left| \frac{y(L)e(n+1)}{y(n+1)y(n+2)} \right| < \tfrac{1}{2} \times 10^{-D}.$$

For this value of n, $w_N(L)$ will agree with $w_{N+1}(L)$ to D places.

Replacing N by s in (17) and summing from $s = N$ to ∞ gives

$$E_N(n) := w_N(n) - w(n) = -\sum_{s=N}^{\infty} \eta_s(n). \tag{6.18}$$

Putting $N = n - 1$ shows

$$w(n) = \sum_{s=n-1}^{\infty} \eta_s(n), \tag{6.19}$$

and so

$$w_N(n) = \sum_{s=n-1}^{N-1} \eta_s(n), \qquad 1 \leqslant n \leqslant N. \tag{6.20}$$

Thus the Olver algorithm is equivalent to computing the partial sums (20) of the infinite series (19).

Let me discuss some numerical aspects of Method A.

When $w_N(n)$ is computed for $0 \leqslant n \leqslant N-1$ by means of equations (13)–(15), $6N-1$ multiplication–divisions are required, and $3N-1$ addition–subtractions.

Significant rounding errors may occasionally occur in the course of the computations. Since

$$y(n) = \frac{y^{(1)}(0)y^{(2)}(n) - y^{(2)}(0)y^{(1)}(n)}{D(0)},$$

we see that if the value of $y^{(1)}(0)$ is small compared with $y^{(1)}(1)y^{(2)}(0)/y^{(2)}(1)$, then initially $y(n)$ behaves like a multiple of $y^{(1)}(n)$ but soon is contaminated with rounding errors growing like $y^{(2)}(n)$ which causes a steady loss in significant figures. Olver (*ibid*) and Olver and Sookne (1972) indicate how this problem may be overcome.

Another possible loss of significance would be the computation of $e(n)$ from (15) due to cancellation in $e(n-1)$ and $f(n-1)e(n)$. Olver, however, remarks that this problem seems never to arise.

Formula (13) reveals that when a rounding error is introduced in $w_N(s)$ during back substitution it is multiplied by the factor $y(r)/y(s)$ when transmitted to $w_N(r)$, $r < s$. But this factor decays with diminishing r at a faster rate than $w_N(r)$ itself because $w(n)$ contains a component of $y^{(2)}(n)$ while $w(n)$ does not.

All in all, Method A is numerically a very safe scheme and difficulties will be encountered only in the most pathological of situations.

Note that it is also possible to eliminate the variables in the equations (11) in reverse order. The resulting set of equations can be written in the form

$$u_N(n+1)w_N(n) - u_N(n)w_N(n+1) = v_N(n),$$
$$n = N-1, N-2, \ldots, 0, \quad \textbf{(6.21)}$$

with initial conditions

$$u_N(N) = 1; \quad u_N(N-1) = -a(N-1); \quad v_N(N-1) = f(N-1),$$
$$\textbf{(6.22)}$$

and where

$$u_N(n-1) = -a(n-1)u_N(n) - b(n-1)u_N(n+1), \quad \textbf{(6.23)}$$
$$v_N(n-1) = b(n-1)v_N(n) + f(n-1)u_N(n),$$
$$n = N-1, N-2, \ldots, 1. \quad \textbf{(6.24)}$$

First the sequences $u_N(n)$, $v_N(n)$ are computed from (22), (23), (24) and the final equation of (21) is used to start the back substitution,

$$w_N(1) = (u_N(1)w(0) - v_N(0))/u_N(0).$$

$w_N(n)$, $2 \leq n \leq N$, is then computed from formula (21).

Notice that when $f(n) \equiv 0$ all the quantities $v_N(n)$ vanish and the result is the ordinary (simplified) Miller algorithm for the computation of a solution of a homogeneous equation.

However, in its application to nonhomogeneous equations this generalization of the Miller algorithm lacks the attractiveness of the original algorithm. Rather than simply multiplying the iterates $y_N(n)$ by a normalizing constant to obtain $w_N(n)$, the time-consuming process of back substitution must be used. Also there seems to be no simple algorithm such as the duality result (20) to increment the value of N.

6.3 The method of averaging (Method B)

This is essentially the method used by Wimp and Luke (1969).

First one generates a solution $z_N(n)$ of the *nonhomogeneous* equation by using the equation in the backward direction with starting values

$$z_N(N+1) = z_N(N) = 0.$$

Next, one generates a solution $y_N(n)$ of the *homogeneous* equation $\mathcal{L}[y(n)] = 0$ by using the equation in the backward direction with starting values

$$y_N(N+1) = 0; \qquad y_N(N) = 1. \tag{6.25}$$

Then we write

$$w_N(n) = \lambda(N)y_N(n) + z_N(n), \qquad 0 \leq n \leq N+1, \tag{6.26}$$

where

$$\lambda(N) := \frac{w(0) - z_N(0)}{y_N(0)}.$$

To see this, note that the right-hand side of (26) satisfies $\mathcal{L}[y(n)] = f(n)$ and has the correct values at $n = 0$, $n = N+1$. It must therefore be the desired solution $w_N(n)$.

This method has the advantage that increasing values of N in the computations is very easy since the intermediate sequences $y_N(n)$, $z_N(n)$ themselves satisfy difference equations (in N).

Theorem 6.2 *$y_N(n)$ satisfies the equation*

$$b(N)y_N(n) + a(N+1)y_{N+1}(n) + y_{N+2}(n) = 0,$$
$$0 \leq N = n, n+1, n+2, \ldots, \tag{6.27}$$

with the initial conditions

$$y_{n+1}(n) = 1, \qquad y_{n+2}(n) = -a(n). \tag{6.28}$$

Further, if $f(n) \neq 0$ $z_N(n)$ satisfies

$$z_N(n) - z_{N+1}(n)\left(1 - \frac{a(N+1)f(N)}{b(N)f(N+1)}\right) - \frac{f(N)}{b(N)} z_{N+2}(n)$$
$$\times \left(\frac{a(N+1)}{f(N+1)} - \frac{1}{f(N+2)}\right)$$
$$- z_{N+3}(n)\frac{f(N)}{b(N)f(N+2)} = 0, \qquad 0 \leq N = n, n+1, \ldots, \tag{6.29}$$

with the initial conditions

$$z_n(n) = 0, \qquad z_{n+1}(n) = f(n), \qquad z_{n+2}(n) = f(n) - a(n)f(n+1). \quad \textbf{(6.30)}$$

Proof The proof of (27), (28) is accomplished in the same way as that of Theorem 4.1. The proof of (29)–(30) is more difficult.

Write

$$z_N(n) = c_1(N)y^{(1)}(n) + c_2(N)y^{(2)}(n) + p(n), \qquad \textbf{(6.31)}$$

where $p(n)$ is any particular solution of $\mathscr{L}[y(n)] = f(n)$. Putting $n = N+1$, N, shows

$$c_1(N) = \frac{p(N+1)y^{(2)}(N) - p(N)y^{(2)}(N+1)}{D(N)}, \qquad \textbf{(6.32)}$$

$D(N)$ as in (A.2). $c_2(N)$ is the negative of (31) with $y^{(2)}$ replaced by $y^{(1)}$. A simple computation using the fact that $y^{(2)}$, p satisfy the homogeneous and nonhomogeneous equations respectively gives

$$c_1(N+1) = \frac{f(N)y^{(2)}(N+1)}{D(N)} + c_1(N),$$

and a similar formula holds for $c_2(N+1)$. Letting $N \to N+1$ in (31) and using these results shows

$$z_{N+1} = z_N + \frac{f(N)}{D(N)}[y^{(1)}(n)y^{(2)}(N+1) - y^{(2)}(n)y^{(1)}(N+1)].$$

Thus the quantity $(z_{N+1} - z_N)/f(N)$ satisfies the adjoint equation (with variable N). The theorem follows immediately. ∎

A similar algorithm can be constructed by using the difference equations in the forward rather than backward direction; see Olver (1967a). However this algorithm is unstable for the computation of the solution desired.

Despite the satisfying formal properties of Method B, care should be used in employing it. If the solutions of the difference equation are too dissimilar in their growth properties, this method of solving the tridiagonal system can become numerically unstable with increasing N due to the loss of significance involved in trying to determine a linear combination of two vectors with very large components, namely, $\{y_N(n)\}$ and $\{z_N(n)\}$, which is bounded. A sure indicator of this loss of significance is that the computed value of $w(0)$, namely $w_N(0)$, begins to differ from the exact value. This can usually be overcome by iteration, see Section 8.1, where the algorithm is given for higher-order equations.

The method requires $4N-3$ addition–subtractions and $5N-6$ multiplication–divisions, to compute $w_N(n)$, $0 \le n \le N-1$, better than

Method A and the same as Method C, to follow. The method has the advantage that only those $w_N(n)$ which are desired can be computed, starting with $w_N(0)$. Methods A and C require the computation of all the $w_N(n)$'s to obtain the earliest ones. If $w_N(n)$ is required only for $0 \leqslant n \leqslant s$, then only $4N - 3 + s$ multiplication–divisions are required.

Example 6.5 Take the functions of Example 6.1, $a = 2$. We get the sequence of approximations to $w(1)$ shown in Table 6.2. The last figure is off about 2 units in the ninth decimal place.

As N increases the accuracy of the iterates begins to decrease and very rapidly, too. The computed value of $w(0)$ (which is computed with exact data and should, therefore, be exact) is in error in the ninth, fifth and third decimal places for $N = 10$, 14 and 16, respectively. Although this problem can be overcome by iteration (see Section 8.1) the algorithm then loses any computational advantage over Method A. ■

6.4 The LU-decomposition (Method C)

Van der Cruyssen (1979b) has shown that the LU-decomposition can be applied to the tridiagonal system (11) with extremely advantageous results.

Let us write the system as

$$A\mathbf{y} = \mathbf{d}. \tag{6.33}$$

The basic idea is to solve (33) by using an LU-decomposition

$$A = LU = \begin{bmatrix} 1 & & & & (0) \\ -\beta(1) & 1 & & & \\ & \ddots & \ddots & & \\ (0) & -\beta(N-2) & \ddots & 1 & \\ & & & -\beta(N-1) & 1 \end{bmatrix}$$

$$\times \begin{bmatrix} -\alpha(0) & b(0) & & & \\ & -\alpha(1) & b(1) & & (0) \\ (0) & & \ddots & \ddots & \\ & & & -\alpha(N-2) & b(N-2) \\ & & & & -\alpha(N-1) \end{bmatrix} \tag{6.34}$$

Table 6.2   Values of
$\int_{-1}^{1} e^{-2(t+1)} T_2(t)\, dt.$

N	$w_N(1)$
4	0.036 552 065 6
6	0.036 630 625 7
8	0.036 631 273 5
10	0.036 631 279 6

The αs and βs are computed from the recurrences

$$\beta(n) = \alpha(n-1)^{-1}, \quad \alpha(n) = -a(n) - \beta(n)b(n-1),$$

$$n = 1, 2, \ldots, N-1, \quad \textbf{(6.35)}$$

$$\alpha(0) = -a(0).$$

(33) is then seen to be equivalent to the system

$$L\mathbf{x} = \mathbf{d}; \quad U\mathbf{y} = \mathbf{x},$$

$\mathbf{x} := [x(0), x(1), \ldots, x(N-1)]^T$, which can be solved by the formulas

$$x(n) = f(n) + \beta(n)x(n-1), \quad n = 1, 2, \ldots, N-1,$$

$$x(0) = f(0) - w(0), \quad \textbf{(6.36)}$$

and

$$w_N(n) = \frac{b(n-1)w_N(n+1) - x(n-1)}{\alpha(n-1)}, \quad n = N-1, N-2, \ldots, 1,$$

$$w_N(N) = \frac{-x(N-1)}{\alpha(N-1)}. \quad \textbf{(6.37)}$$

The composite of the previous formulas (35)–(37) will be called Method C.

The above variables and the variables $y(n)$, $e(n)$ in the Olver identities of Method A are related by

$$y(n+1) = \frac{\alpha(0)\alpha(1)\cdots\alpha(n-1)}{b(0)b(1)\cdots b(n-1)},$$

$$e(n) = -\frac{\alpha(0)\alpha(1)\cdots\alpha(n-2)x(n-1)}{b(0)b(1)\cdots b(n-1)},$$

and in terms of the new variables we have

$$\eta_N(n) := \phi_N(n)x(N),$$

(see (17)), with $\phi_N(n)$ given by

$$\phi_{N+1}(n) = -\frac{\phi_N(n)b(N)}{\alpha(N+1)}, \qquad N = n, n+1, n+2, \ldots,$$

$$\phi_N(N+1) = -\alpha(N)^{-1}.$$

Method C requires $5N-2$ multiplication–divisions and $3N-1$ addition–subtractions, slightly superior to Method A.

As van der Cruyssen points out, the classes of problems for which Method C and Method A can be applied are the same. Method C breaks down iff there is an index $n < N+1$ such that $\alpha(0), \alpha(1), \ldots, \alpha(n-1) \neq 0$ while $\alpha(n) = 0$. But this implies that $y(n+1) = 0$, so Method A breaks down, and vice versa. For Method A the vanishing of a $b(n)$ necessitates an adaptation. Method C does not require modification in this case. The stability properties of Methods A and C are the same, cancellation of significant figures in one algorithm implying the same for the other algorithm.

6.5 Adaptation to general normalizing conditions

Let us assume that instead of the value $w(0)$ we have the normalization relation

$$S := \sum_{k=0}^{\infty} c(k)w(k) \neq 0.$$

The obvious approach is to solve the following system of equations:

$$w_N(N+1) = 0, \qquad \sum_{k=0}^{N+1} c(k)w_N(k) = S,$$

$$w_N(n) + a(n)w_N(n+1) + b(n)w_N(n+2) = f(n),$$

$$n = N-1, N-2, \ldots, 0. \quad (\mathbf{6.38})$$

We have the convergence

Theorem 6.3 *Let $w(n)$ satisfy the nonhomogeneous equation (1) and let the homogeneous equation (5) have a fundamental set $\{y^{(h)}(n)\}$ with the property*

(i) $\quad S_N^{(1)}, S_N^{(3)}, \dfrac{y^{(1)}(N+1)}{y^{(2)}(N+1)} S_N^{(2)}, \dfrac{w(N+1)}{y^{(2)}(N+1)} S_N^{(2)}$

converge, the latter two to 0, *the first to other than* 0, *where*

$$S_N^{(h)} := \sum_{k=0}^{N+1} c(k)y^{(h)}(k), \qquad h = 1, 2, 3,$$

$$y^{(3)}(n) := w(n).$$

(ii) $\quad |S_N^{(2)}| \rightarrow \infty.$

Then the modified Olver algorithm converges and

$$\lim_{N \to \infty} w_N(n) = w(n), \qquad n \geqslant 0.$$

Proof I have

$$w_N(n) = \xi_N^{(1)} y^{(1)}(n) + \xi_N^{(2)} y^{(2)}(n) + w(n). \tag{6.39}$$

Let

$$S^{(h)} := \lim_{N \to \infty} S_N^{(h)}.$$

Plugging in the conditions (38) and solving (39) gives

$$\xi_N^{(1)} = \frac{w(N+1)S_N^{(2)} - y^{(2)}(N+1)(S_N^{(3)} - S)}{y^{(2)}(N+1)S_N^{(1)} - y^{(1)}(N+1)S_N^{(2)}}. \tag{6.40}$$

It is easily seen that $\xi_N^{(1)} \to (S - S^{(3)})/S^{(1)} = 0$. Likewise, $\xi_N^{(2)} \to 0$. ∎

Let us first consider the adaptation of Method A to this algorithm. When the forward elimination process is applied, we have the following pivotal equations:

$$y(n+1)w_N(n) - y(n)w_N(n+1) + q(n) \sum_{s=n+1}^{N} c(s)w_N(s) = e(n),$$

$$n = 0, 1, \ldots, N, \quad (6.41)$$

where

$$q(0) = 1, \qquad q(n) = [b(0)b(1) \cdots b(n-1)]^{-1}, \qquad n \geqslant 1,$$
$$y(0) = 0, \qquad y(1) = c(0), \qquad e(0) = S.$$

First we compute

$$y(n+1) = -\frac{a(n-1)y(n) - y(n-1)}{b(n-1)} + q(n)c(n), \qquad n = 1, 2, \ldots, N,$$

$$e(n) = \frac{e(n-1) - f(n-1)y(n)}{b(n-1)}, \qquad n = 1, 2, \ldots, N.$$

The final equation of (41) is

$$y(N+1)w_N(N) = e(N), \tag{6.42}$$

and this yields the value of $w_N(N)$. After this $w_N(N-1)$, $w_N(N-2)$, $\ldots, w_N(0)$ are determined by use of the pivotal equations in the backward direction.

To analyze the truncation error of this algorithm write

$$\eta(n) := w_{N+1}(n) - w_N(n). \tag{6.43}$$

From the pivotal equation (41) we obtain

$$y(n+1)\eta(n) = y(n)\eta(n+1) - q(n)(c(n+1)\eta(n+1)$$
$$+ \cdots + c(N)\eta(N)), \qquad n < N,$$

so

$$|\eta(n)| \le \rho(n)[|\eta(n+1)| + |\eta(n+2)| + \cdots + |\eta(N)|], \qquad n < N, \qquad \textbf{(6.44)}$$

where $\rho(n)$ is the greater of

$$\left|\frac{y(n) - q(n)c(n+1)}{y(n+1)}\right|, \qquad \left|\frac{q(n)}{y(n+1)}\right| \sup_{2 \le s < \infty} |c(n+s)|.$$

The equations (38), (14) and (43) yield $\eta(N+1) = e(N+1)/y_{N+1}(N+2)$. Using this and (44) shows

$$|\eta(N)| \le \rho(N) |e(N+1)/y(N+2)|,$$

and thus by induction we have

$$|\eta(n)| \le \rho(n)(1 + \rho(n+1))(1 + \rho(n+2))$$
$$\cdots (1 + \rho(N)) |e(N+1)|/|y(N+2)|, \qquad n \le N-1.$$

The left-hand side of this relation is $|w_{N+1}(n) - w_N(n)|$. Replacing N by $N+1, N+2, \ldots$ and summing gives, after an application of the previous theorem,

$$|E_N(n)| = |w_N(n) - w(n)|$$
$$\le \rho(n)(1 + \rho(n+1)) \cdots (1 + \rho(N))K_N, \qquad n \le N-1,$$

where

$$K_N = \left|\frac{e(N+1)}{y(N+2)}\right| + (1 + \rho(N+1)) \left|\frac{e(N+2)}{y(N+3)}\right|$$
$$+ (1 + \rho(N+1))(1 + \rho(N+2)) \left|\frac{e(N+3)}{y(N+4)}\right| + \cdots,$$

provided this series converges. Also

$$|E_N(N)| \le \rho(N)K_N, \qquad |w(N+1)| \le K_N.$$

Example 6.6 Compute to 5D the solution $w(n)$ of the homogeneous equation

$$(2n+1)y(n) - 12(n+1)y(n+1) + (2n+3)y(n+2) = 0,$$

with

$$\tfrac{1}{2}w(0) + w(1) + w(2) + \cdots = 1.$$

A quick asymptotic analysis reveals the equation has solutions behaving like $n^\theta(3\pm\sqrt{2})^n$. Thus there is a minimal solution and the algorithm will converge. One finds

$$e(n) = q(n) = \frac{1}{2n+1}, \quad S = 1; \quad c(r) = \varepsilon_r/2.$$

Olver (1967a) has tabulated $y(n)$, $e(n)$, $w_N(n)$ for $N = 6, 13$. For $N = 6$, one has $\{\rho(n)\} = \{1, 0.166\,666, \quad 0.204\,301, \quad 0.198\,892, \quad 0.192\,914, 0.188\,768, 0.185\,913, \ldots\}$. $e(n)/y(n+1)$ decays so rapidly that

$$K_N \approx |e(N+1)/y(N+2)|,$$

and, for $N = 6$, this is $8.955\,(-7)$. Thus an approximate bound for the error in approximating $w(0)$ by $w_N(0)$ is $2.6\,(-6)$, a figure confirmed by numerical computations. ∎

Method B is adaptable to an arbitrary normalization in a straightforward way. Let $\{y_N(n)\}$, $\{z_N(n)\}$ be computed as in Section 6.3. Then

$$w_N(n) := \frac{(S - T_N)}{S_N} y_N(n) + z_N(n), \quad T_N := \sum_{k=0}^{N+1} c(k) z_N(k),$$

$$S_N := \sum_{k=0}^{N+1} c(k) y_N(k).$$

This process is subject to the same numerical hazards as Method B.

Olver has shown (1967a) that if two sequences are generated from the homogeneous and nonhomogeneous equations respectively by using these equations in the *forward* direction and an appropriate linear combination of them is formed to give $w_N(n)$, then an error bound can be derived in terms of already computed quantities which is simpler than the formulation here.

Example 6.7 The integrals

$$w(n) := W_n(u, z) := \int_u^\infty e^{-y - z/y} y^{-n} \, dy, \quad z > 0, \quad n \geqslant 0,$$

are the quaintly titled 'leaky aquifer functions' (see Thomas *et al.* (1978)). Integration by parts shows $w(n)$ satisfies

$$y(n) + ny(n+1) - zy(n+2) = e^{-u - z/u} u^{-n},$$

where the right-hand side is interpreted as 0 if $u = 0$. Letting $y = ut$ and integrating by parts shows

$$W_n(u, z) = \frac{u^{1-n} e^{-u - z/u}}{n} + O(n^{-2}), \quad n \to \infty.$$

Solutions of the homogeneous equation are

$$y^{(1)}(n) = (-1)^n z^{-n/2} I_{n-1}[2\sqrt{z}], \qquad y^{(2)}(n) = z^{-n/2} K_{n-1}[2\sqrt{z}].$$

A brief computation shows that the Olver algorithm of Section 6.1 will converge; the convergence is quite rapid, in fact. For instance, $w_{10}(2)$ is accurate to about 12 S.F. Of course these computations require a knowledge of $w(0)$. Is it possible to do better, one wonders, by using the algorithm of this section with an appropriate normalization? A normalization is easily found by multiplying $w(n)$ by $z^n/n! := c(n)$ and summing under the integral sign. We find

$$S := \sum_{k=0}^{\infty} c(k)w(k) = e^{-u}. \tag{6.45}$$

The sums in Theorem 6.3 converge and the solutions involved display the appropriate behavior, so it seems the computation of $w(n)$ by the general method based on the normalization (45) should present no problems.

However, if the reader attempts such a computation he will find the algorithm does not converge. What is the explanation of this?

The cause for the divergence of the algorithm is a surprising consequence of a known series (Erdélyi *et al.* (1953, vol. 2, p. 99 (9))),

$$\frac{z^{\nu/2}}{\Gamma(1+\nu)} = \sum_{k=0}^{\infty} \frac{z^{k/2}(-1)^k}{k!} I_{k+\nu}(2\sqrt{z}), \qquad z > 0,$$

which holds for all ν. Setting $\nu = -1$ shows that $S^{(1)} = \sum c(k)y^{(1)}(k) = 0$! Obviously $S^{(1)}$ must be bounded away from zero if convergence is to be demonstrated in any straightforward fashion from equation (40). In fact the algorithm can be shown to diverge. A more detailed analysis reveals that while $\xi_N^{(2)} \to 0$, we have $|\xi_N^{(1)}| \to \infty$, and the consequences of this are obvious. Of course this does not preclude the computation of $w(n)$ based on other sums, for instance sums of the form

$$S(m) = \sum_{k=0}^{\infty} c(k)(-1)^k w(k+m) = \int_u^{\infty} \frac{e^{-y-2z/y}}{y^m}\, dy.$$

One finds that

$$2zS(2) - S(0) = -e^{-u-2z/u},$$

and this can serve to normalize $w_N(n)$.

For a discussion of the history and computational techniques associated with these integrals see the cited reference. ∎

Amos and Burgmeier (1973) give explicit algorithms for the application of the Olver method to many commonly occurring integrals including incomplete Laplace transforms, repeated integrals, and moment integrals. The integrals considered are

$$w(n) = \int_0^c e^{-\beta t} t^\mu f_n(t)\, \mathrm{d}t,$$

with $f_n = J_{n+\alpha}(t)$, $I_{n+\alpha}(t)$, $J_{n+\alpha}(\sqrt{t})$, $I_{n+\alpha}(\sqrt{t})$, $\mathbf{H}_{n+\alpha}(t)$, $\mathbf{L}_{n+\alpha}(t)$ (Struve functions), $t^{-\mu} i^n \operatorname{erfc}(t)$, $t^{-\mu} i^n \operatorname{erfc}(\sqrt{t})$, $t^{-\mu-1/2} i^n \operatorname{erfc}(\sqrt{t})$, and incomplete moment integrals involving Struve functions and the exponential function.

6.6   Conclusions

There are as many variants of Olver's algorithm as there are methods of solving systems of equations and when the defining set of equations is found to be ill-conditioned, methods appropriate to such systems should be used. Method A, for instance, is essentially Gaussian elimination without partial pivoting. In some cases it may be advisable to use Gaussian elimination *with* partial pivoting to prevent the loss of significant figures. Cash (1980a,b) discusses an approach which incorporates partial pivoting when it is needed. The equations may also be solved iteratively, and sometimes this is desirable. Cash (1977a,b) discusses this approach and gives a generalization to third-order difference equations.

A number of writers have treated the generalization of the algorithm to higher-order difference equations. This topic is discussed in Chapter 7.

7 Higher-order systems: homogeneous equations

7.1 Observations on higher-order equations

Generally speaking the solutions of a higher-order homogeneous linear difference equation may be classified according to their growth properties. There may be minimal or dominant solutions (see Definitions 2.7, 2.8) and intermediate solutions, i.e., solutions for which two other solutions can be found such that one dominates and the other is dominated by the intermediate solution. Straightforward recursion in the forward direction is effective for computing dominant solutions. A generalization of Miller's algorithm given in Section 2 which follows can be used to compute minimal solutions. However, neither forward recurrence nor the Miller algorithm is numerically stable for the computation of an intermediate solution. Faced with the task of computing an intermediate solution for an equation of higher order, the analyst may have several options. If the desired solution is a solution of a nonhomogeneous equation and inter-mediate in behavior among the solutions of a fundamental set for $\mathscr{L}[y] = 0$ it may be computable by a method due to Wimp and Luke which utilizes the equation in the backward direction and requires only one normalization relation for the desired solution. Although the solution need not be minimal it must satisfy certain growth requirements with respect to a fundamental set for the homogeneous equation. Regrettably the method tends to become unstable for large N, the point at which the backward computations begin. Nevertheless the method, presented in Section 8.2, can be very effective for certain equations, generally equations of order not greater than three or four.

Another possibility for computing intermediate solutions of the *homogeneous* equation is to use the method of averaging presented in Section 4. This method suffers from two drawbacks: it requires several normalizations for the solution and these are not usually easy to come by. Also no convergence estimates are presently available. The manner in which round-off error propagates is in general unknown and there is the possibility that using the method on unsuitable equations can lead to a numerical disaster. Nevertheless, used on an appropriate equation the technique can yield striking results and several examples of this are given in Section 4. The reader should bear in mind that the functions so computed are important and are difficult to compute by any other method.

Now, for the second-order homogeneous equation Olver's algorithm can be used to compute both minimal and intermediate solutions. The extension of Olver's method to higher-order equations has been considered by several authors, e.g., Oliver (1968a, 1968b), Cash (1978, 1981), Lozier (1980). Lozier's method, the one I shall present in Chapter 8, has features in common with the algorithms given by these other writers and when applied to appropriate equations may be used to compute any solution whatsoever. One of its attractive features is its formulation as the problem of finding the solution of an operator equation $\mathscr{L}y = f$, where $\mathscr{L}$ is an infinite matrix operator acting on the space of complex sequences. It thus suggests a future research tool for analyzing the very puzzling class of algorithms discussed in the final section of this chapter.

7.2 The Miller algorithm

In this section I shall discuss computation by backward recursion based on the linear homogeneous difference equation of order $\sigma \geq 2$,

$$\mathscr{L}[y(n)] = \sum_{\nu=0}^{\sigma} A_\nu(n)y(n+\nu) = 0,$$

$$A_0 = 1, \quad A_\sigma(n) \neq 0, \quad n \geq 0. \quad \textbf{(7.1)}$$

As we shall see, the convergence of the algorithm, which is a straightforward generalization of the algorithm discussed in Chapter 4, depends on the behavior of a fundamental set of solutions for the equation *adjoint* to (1). For $\sigma = 2$ the solutions of the adjoint equation are simply related to the solutions of the equation itself and this enables one to formulate simple and effective criteria to decide the convergence of the algorithm and to derive error estimates. For higher-order equations on the other hand the solutions of the adjoint involve determinants of the solutions of the original equation and this makes the convergence of the method harder to assess. Obviously explicit error bounds are even more difficult to come by and usually the accuracy of the method is judged by comparing common figures of successive iterates.

The algorithm proceeds as follows. Let N be an integer ≥ 0 and define $y_N(n)$, $0 \leq n \leq N+\sigma-1$ by

$$y_N(N+j) = \delta_{0j}, \quad 0 \leq j \leq \sigma-1,$$
$$\mathscr{L}[y_N(n)] = 0, \quad n = N-1, N-2, \ldots, 0. \quad \textbf{(7.2)}$$

Suppose we are given the normalization relationship

$$S := \sum_{k=0}^{\infty} c(k)w(k), \qquad S \neq 0, \tag{7.3}$$

where $w(n)$ is the desired solution of (1).

Take

$$w_N(n) := Sy_N(n)/S_N, \tag{7.4}$$

$$S_N := \sum_{k=0}^{N} c(k)y_N(k). \tag{7.5}$$

If

$$\lim_{N \to \infty} w_N(n) = w(n), \qquad n \geq 0, \tag{7.6}$$

we say the generalized Miller algorithm defined by (1)–(4) converges (to $w(n)$).

Let us analyze the above algorithm. By the theory in Appendix A we know that $w(n)$ may be completed to a fundamental set, $\{y^{(h)}(n)\}$, $y^{(1)}(n) := w(n)$. Since $y_N(n)$ satisfies the equation I have

$$y_N(n) = \sum_{h=1}^{\sigma} \xi_N^{(h)} y^{(h)}(n), \tag{7.7}$$

where $\xi_N^{(h)}$ is independent of n. Setting $n = N, N+1, \ldots, N+\sigma-1$ we find that

$$\xi_N^{(h)} = T^{(h)}(N)/D(N) = \hat{y}^{(h)}(N), \tag{7.8}$$

(see A-V, A-VI). Thus $\xi_N^{(h)}$ is a solution of the adjoint equation (with variable N). Note that D cannot be 0. I can therefore write

$$w_N(n) = \frac{S \sum_{h=1}^{\sigma} T^{(h)}(N) y^{(h)}(n)}{\sum_{h=1}^{\sigma} T^{(h)}(N) S_N^{(h)}}, \qquad S_N^{(h)} := \sum_{k=0}^{N} c(k) y^{(h)}(k). \tag{7.9}$$

Let

$$R_N^{(h)} := T^{(h)}(N)/T^{(1)}(N). \tag{7.10}$$

The following theorem is immediate.

Theorem 7.1 *Let $T^{(1)}(N) \neq 0$ for N sufficiently large and*

$$\lim_{N \to \infty} R_N^{(h)} = \lim_{N \to \infty} R_N^{(h)} S_N^{(h)} = 0, \qquad 2 \leq h \leq \sigma. \tag{7.11}$$

Then the computation of $w(n)$ by the generalized Miller algorithm converges. ∎

Remark It is clear that the convergence of the algorithm is independent of the choice of functions used to complete the fundamental set.

As with the original Miller algorithm, if the value of $w(0)$ is known the algorithm and the conditions for convergence simplify considerably. This means we taken $c(k) = \delta_{k0}$.

Corollary *Let $w(0) \neq 0$ and*

$$\lim_{N \to \infty} R_N^{(h)} = 0, \qquad 2 \leq h \leq \sigma. \tag{7.12}$$

Then

$$\lim_{N \to \infty} \frac{w(0) y_N(n)}{y_N(0)} = w(n), \qquad n \geq 0. \quad \blacksquare \tag{7.13}$$

Note that convergence of the simplified Miller algorithm is equivalent to the statement that the adjoint equation possesses a dominant solution.

Example 7.1 Let

$$w(n) := \int_0^\infty e^{-t^\sigma - P(t)} t^n \, dt, \qquad n \geq 0,$$

$$P(t) := \sum_{r=1}^{\sigma-1} a_r t^r, \qquad \sigma \geq 2, \tag{7.14}$$

where the a_r are complex constants.

A single integration by parts shows that $w(n)$ satisfies the equation

$$y(n) - (n+1)^{-1} \sum_{\nu=1}^{\sigma} \nu a_\nu y(n+\nu) = 0, \qquad a_\sigma = 1. \tag{7.15}$$

De Bruijn's analysis (1961, p. 119) gives the leading term for the Birkhoff series for $w(n)$. We have

$$w(n) \sim \frac{\Gamma\!\left(\dfrac{n+1}{\sigma}\right)}{\sigma} \exp\{\mu_2 n^{1-1/\sigma} + \mu_3 n^{1-2/\sigma} + \cdots + \mu_\sigma n^{1/\sigma}\} q(\sigma, n),$$

$$\tag{7.16}$$

where $q(\sigma, n)$ is an asymptotic series of the form (B.13) and

$$\mu_2 = -a_{\sigma-1} \sigma^{(1/\sigma)-1},$$

$$\mu_3 = \left[\frac{a_{\sigma-1}^2}{2} \left(1 - \frac{1}{\sigma}\right)^2 - a_{\sigma-2} \right] \sigma^{(2/\sigma)-1},$$

etc.

Note that when $\sigma = 2$ $w(n)$ is essentially the parabolic cylinder function $U(n + \frac{1}{2}, a_1/\sqrt{2})$ the theory and computation of which are discussed by Miller in Abramowitz and Stegun (1964, Ch. 19).

By the theory in Appendix B we know there exists a Birkhoff set for (15), $\{y^{(h)}(n)\}$, where $y^{(h)}$ has as an asymptotic expansion the Birkhoff series on the right with n replaced by $ne^{2\pi i(h-1)}$, $1 \leq h \leq \sigma$. This gives σ formal series solutions so there are no more. We identify $w(n)$ with $y^{(1)}(n)$.

The behavior of $R_N^{(h)}$ can be inferred from Theorem B.8. We have

$$\frac{T^{(h)}(N)}{T^{(1)}(N)} = \frac{\hat{y}^{(h)}(N)}{\hat{y}^{(1)}(N)} = N^{\theta_h} e^{V^{(h)}(N)}[1 + o(1)],$$

where

$$V^{(h)}(N) = \mu_2 N^{1-1/\sigma}[1 - e^{-2\pi i(h-1)/\sigma}] + \cdots.$$

Now Re $V^{(h)}(n)$ is positive or negative according to whether

$$f^{(h)}(\gamma) := \sin \frac{\pi(h-1)}{\sigma} \sin \left(\gamma - \frac{\pi(h-1)}{\sigma} \right),$$

is positive or negative, where $\gamma := \arg a_{\sigma-1}$. For $|\gamma| < \pi$, $f^{(h)}$ has zeros only at the points $(h-1)\pi/\sigma$ and $(h-1)\pi/\sigma - \pi$, so $f^{(h)}$ is of one sign for γ between these points. Since

$$f^{(h)}(0) = -\sin^2 ((h-1)\pi/\sigma) < 0,$$

we have

$$f^{(h)}(\gamma) < 0, \qquad \frac{(h-1)\pi}{\sigma} - \pi < \gamma < \frac{(h-1)\pi}{\sigma},$$

or

$$f^{(h)}(\gamma) < 0, \qquad -\frac{\pi}{\sigma} < \gamma < \frac{\pi}{\sigma},$$

and for this range of values of $\arg a_{\sigma-1}$ the first part of (9) is fulfilled.

To construct an appropriate normalization relationship, let

$$e^{P(t)} = \Gamma\left(1 + \frac{1}{\sigma}\right) \sum_{k=0}^{\infty} c(k)t^k. \tag{7.17}$$

Since the left-hand side is an entire function of order $\sigma - 1$ we have, by Boas (1954, p. 10),

$$|c(k)| < Ck^{-k/\sigma}, \qquad k \geq 1,$$

so

$$k^{\mu} |c(k)| |w(k)| \leq C'k^{\theta} \exp\left\{ -\frac{k}{\sigma} (1 + \ln \sigma)[1 + C''k^{-1/\sigma}] \right\} = o(1),$$

for all μ.

An easy computation shows

$$\sum_{k=0}^{\infty} c(k)w(k)=1. \tag{7.18}$$

The results on exponential sums (Theorems B.5, 6) can now be invoked to show the second condition in (9) is also satisfied.

Thus the Miller algorithm based on (15), (18) will converge to $w(n)$ for $|\arg a_{\sigma-1}| < \pi/\sigma$. ∎

Example 7.2 This example treats an important class of hypergeometric functions. Let

$$w(n):=\frac{(-1)^n z^n (a_{p+2})_n}{(\gamma)_{2n}(b_p)_n}{}_{p+2}F_{p+1}\left(\begin{matrix} n+a_{p+2} \\ n+b_p, \quad 2n+\gamma+1 \end{matrix}\bigg| z\right),$$

where $\beta+1$, γ, a_i, b_j are complex parameters, $a_{p+2}=\beta+1$. (For the product notation see Appendix C.) We assume none of these parameters is a negative integer or zero and that $z\neq0$, $|\arg(1-z)|<\pi$.

The formulas (C.12, 13) show $w(n)$ satisfies the recurrence

$$\sum_{\nu=0}^{p+2}\left[A_\nu(n)+\frac{B_\nu(n)}{z}\right]y(n+\nu)=0, \tag{7.19}$$

where

$$A_\nu(n)=\frac{(-1)^p(2n+\gamma)_{p+3}(n+\gamma+\nu-a_{p+2})}{\Gamma(p+3-\nu)(2n+\gamma+\nu)_{\nu+1}(n+a_{p+2})}$$
$$\times {}_{p+4}F_{p+3}\left(\begin{matrix} \nu-p-2, \quad 2n+\gamma+\nu, \quad n+\gamma+\nu+1-a_{p+2} \\ 2n+\gamma+2\nu+1, \quad n+\gamma+\nu-a_{p+2} \end{matrix}\bigg| 1\right),$$
$$1\leq\nu\leq p+2,$$

$$B_\nu(n)=\frac{(-1)^p(2n+\gamma)_{p+3}(n+\gamma+\nu+1-b_p)}{\Gamma(p+2-\nu)(2n+\gamma+\nu+1)_\nu(n+a_{p+2})}$$
$$\times {}_{p+2}F_{p+1}\left(\begin{matrix} \nu-p-1, \quad 2n+\gamma+\nu+1, \quad n+\gamma+\nu+2-b_p \\ 2n+\gamma+2\nu+1, \quad n+\gamma+\nu+1-b_p \end{matrix}\bigg| 1\right),$$
$$1\leq\nu\leq p+1,$$

with $A_0:=1$, $B_0=B_{p+2}:=0$. Note that the above functions are terminating series and thus for any value of ν they are rational functions of n. Therefore the equation will possess a Birkhoff set.

A result of Wimp and of Luke (1969, vol. 2, p. 7, with $m=0$ and then $w=0$) produces the normalization relation

$$1=\sum_{k=0}^{\infty}\frac{(\gamma)_k}{k!}w(k). \tag{7.20}$$

Since

$$A_\nu(n) = \frac{(-p-2)_\nu}{\nu!} + O(n^{-1}),$$

$$B_\nu(n) = \frac{4(-p)_{\nu-1}}{\Gamma(\nu)} + O(n^{-1}),$$

there are two linearly independent solutions of the recurrence, $y^{(1)}(n)$, $y^{(2)}(n)$, with the behavior

$$y^{(1)}(n) \sim (-1)^n \lambda_1^n q_1(n), \qquad n \to \infty,$$
$$y^{(2)}(n) \sim (-1)^n \lambda_2^n q_2(n), \qquad n \to \infty,$$

where q_i is a Birkhoff series of the form (B.13) with $\rho = 1$ and

$$\lambda_1 = (\tau - \surd(\tau^2 - 1))^2, \qquad \lambda_2(\tau + \surd(\tau^2 - 1))^2, \qquad \tau = z^{-1/2},$$

cuts for $\surd(\tau^2 - 1)$ being established as in Example 5.3. Thus $|\lambda_1| < 1$, $|\lambda_2| > 1$. It can be shown (see Wimp (1969)) that the remaining members of any canonical set $\{y^{(h)}(n)\}$ are purely algebraic or algebraic–logarithmic in behavior as $n \to \infty$.

A simple computation shows that the conditions (9) are satisfied so the generalized Miller algorithm based on (19) and (20) will converge (exponentially) to $w(n)$. ∎

Example 7.3 In this section I develop an algorithm for computing the confluent hypergeometric function $\Psi(a, c; x)$. The algorithm is more rapidly convergent than the one given in Section 5.2, the error being roughly on the order of $\exp(-\alpha x^{1/3} N^{2/3})$ rather than $\exp(-2\surd(Nx))$, the penalty that must be paid being that the computations must be done with a four-term rather than a three-term recurrence.

In Wimp (1974) it is shown that the function

$$w(n) := G_{23}^{31}\left(x \,\middle|\, \begin{matrix} 1-n, & n+1 \\ 1, & a, & a+1-c \end{matrix}\right) \middle/ \Gamma(a)\Gamma(a+1-c),$$

satisfies the recurrence

$$y(n) + A_1(n)y(n+1) + A_2(n)y(n+2) + A_3(n)y(n+3) = 0, \qquad (7.21)$$

where

$$A_1 := \frac{1}{\gamma(n)}\{(2n+2)\gamma(n) - (2n+1)[\gamma(n+1) + 2x]\},$$

$$A_2 := \frac{(2n+1)}{(2n+5)\gamma(n)}\{[(2n+4)(n+c+2-a)(n+3-a)$$
$$- (n+2-a)(n+c+1-a)(2n+5)] - 2x(2n+5)\},$$

$$A_3 := \frac{-(2n+1)(n+3-a)(n+2+c-a)}{(2n+5)\gamma(n)},$$

$$\gamma(n) := (n+a)(n+a+1-c).$$

The function of interest is

$$w(0) = x^a \Psi(a, c; x).$$

A normalization is given by

$$1 = \sum_{k=0}^{\infty} \varepsilon_k w(k). \tag{7.22}$$

In the cited reference, the following result is shown:

Theorem 7.2   *Let neither a nor $a + 1 - c$ be a negative integer nor 0, $x \neq 0$, $|\arg x| < \pi$.*

Then $w(n)$ may be computed by applying the Miller algorithm to (21), (22); further,

$$E_N(n) = O(N^\theta \exp(3\kappa(N^2 |x|)^{1/3})), \qquad N \to \infty, \tag{7.23}$$

for some θ where

$$\kappa = \max_{h=1,2} \cos\left[\frac{\arg x}{3} + \frac{4\pi h}{3}\right] < 0, \qquad |\arg x| < \pi. \quad \blacksquare$$

Equation (21) can be used to compute $K_\nu(x)$. Let $c = 2a$, $a = \nu + \frac{1}{2}$, $x \to 2x$. The recurrence is

$$y(n) + B_1(n)y(n+1) + B_2(n)y(n+2) + B_3(n)y(n+3) = 0, \tag{7.24}$$

where

$$B_1 := -\frac{1}{\gamma(n)}[(n+\tfrac{1}{2})(3n + 8x + \tfrac{7}{2}) + \nu^2],$$

$$B_2 := \frac{(n+\tfrac{1}{2})}{(n+\tfrac{5}{2})\gamma(n)}[(n+\tfrac{5}{2})(3n - 8x + \tfrac{11}{2}) + \nu^2],$$

$$B_3 := \frac{-(n+\tfrac{1}{2})[(n+\tfrac{5}{2})^2 - \nu^2]}{(n+\tfrac{5}{2})\gamma(n)},$$

$$\gamma(n) := (n+\tfrac{1}{2})^2 - \nu^2.$$

We may use the normalization (22). The quantity required is

$$w(0) = \sqrt{\left(\frac{2x}{\pi}\right)} e^x K_\nu(x).$$

The Miller algorithm converges for all (complex) ν and $|\arg x| < \pi$ in accordance with the error estimate (23).

Explicit formulas for $w(n)$ may be obtained by using the same methods as in Example 5.2. $\blacksquare$

Example 7.4 The so-called Bickley functions are defined by

$$w(n) := Ki_n(x) := \int_x^\infty Ki_{n-1}(t)\, dt, \qquad n \geqslant 1, \quad x > 0,$$

$$Ki_0(x) := K_0(x),$$

see Blair *et al.* (1978).

It is easy to show that Ki_n satisfies

$$y(n) + \frac{(n+1)}{x} y(n+1) - y(n+2) - \frac{(n+2)}{x} y(n+3) = 0. \tag{7.25}$$

The usual techniques reveal the equation has a fundamental set of solutions $\{y^{(h)}(n)\}$ with the behavior

$$y^{(1)}(n) = n^{-n-1/2}(-xe)^n[1 + O(n^{-1})], \qquad y^{(2)}(n) = n^{-1/2}[1 + O(n^{-1})],$$
$$y^{(3)}(n) = (-1)^n n^{-1/2}[1 + O(n^{-1})].$$

An explicit representation is

$$Ki_n(x) = \int_1^\infty e^{-xu} u^{-n}(u^2 - 1)^{-1/2}\, du. \tag{7.26}$$

Note that Ki_n is monotone decreasing in n. Also

$$Ki_n(x) > \int_1^\infty e^{-xu} u^{-n-1} = \frac{e^{-x}}{n} + O(n^{-2}),$$

as an integration by parts shows. Thus the required function is a constant multiple of $y^{(2)}(n)$.

An interesting additional fact (which is easily verified) is that $y^{(1)}(n)$ is a constant multiple of

$$Ii_n(x) := (-x)^n \sum_{r=0}^\infty \frac{(\tfrac{1}{2})_r x^{2r}}{r!\,(n+2r)!}, \tag{7.27}$$

which is the nth repeated integral of $I_0(x)$ (times $(-1)^n$).

Thus the Miller algorithm applied to (25) converges for (27) but not for (26). Forward recursion is (weakly) stable for the latter. ∎

Example 7.5 Consider the integrals

$$T(n; x) = \int_0^\infty t^n e^{-t^2 - x/t}\, dt, \qquad x > 0,$$

(see Cole and Pescatore (1979)). T satisfies

$$xy(n) + (n+2)y(n+1) - 2y(n+3) = 0.$$

This recurrence is analyzed in Appendix B. There is a fundamental set $\{y^{(h)}(n)\}$ with the behavior

$$y^{(1)}(n) = n^{n/2}(2e)^{-n/2}(1 + O(n^{-1/2})),$$

$$y^{(2)}(n) = n^{n/2}(-2e)^{-n/2}(1 + O(n^{-1/2})),$$

$$y^{(3)}(n) = n^{-n-3/2}(-xe)^n(1 + O(n^{-1})).$$

It is clear that $T(n; x) \to \infty$. In fact Laplace's method shows that $T(n; x) \sim \sqrt{(\pi/2)}y^{(1)}(n)$. Since the minimal solution $y^{(3)}(n)$ is not the one being sought, the Miller algorithm will diverge.

However, the equation can be used in the *forward* direction in a stable fashion to compute $T(n; x)$. This requires three initial values. ■

As in the algorithm for the second-order equation the quantities $y_N(n)$, $S_N(n)$ (see (2), (5)) themselves satisfy difference equations (in N).

Theorem 7.3   (Duality theorem) $y_N(n)$ *satisfies*

$$\sum_{\nu=0}^{\sigma} A_{\sigma-\nu}(N+\nu)y_{N+\nu}(n) = 0, \qquad 0 \leqslant N = n, n+1, n+2, \ldots, \qquad \textbf{(7.28)}$$

with the initial conditions

$$y_n(n) = 1, \qquad y_{n+1}(n) = -A_1(n),$$

$$y_{n+2}(n) = A_1(n)A_1(n+1) - A_2(n), \ldots, \qquad \textbf{(7.29)}$$

and $S_N(n)$ *(see (4.15)) satisfies*

$$\sum_{\nu=0}^{\sigma} A_{\sigma-\nu}(N+\nu)S_{N+\nu}(n) = c(N+\sigma),$$

$$0 \leqslant N = n, n+1, n+2, \ldots, \qquad \textbf{(7.30)}$$

with the initial conditions

$$S_n(n) = c(n), \qquad S_{n+1}(n) = -A_1(n)c(n) + c(n+1),$$

$$S_{n+2}(n) = c(n)[A_1(n)A_1(n+1) - A_2(n)] \qquad \textbf{(7.31)}$$

$$-c(n+1)A_1(n+1) + c(n+2), \ldots.$$

Proof   (28) follows immediately from the representation (7) and the theory in A-VI. Now let $n \to k$ in (28), multiply the equation by $c(k)$ and sum from $k = n$ to $N + \sigma$:

$$\sum_{k=n}^{N+\sigma} \sum_{\nu=0}^{\sigma} A_{\sigma-\nu}(N+\nu)c(k)y_{N+\nu}(k) = 0.$$

Now $y_{N+\nu}(n) = 0$ for $N + \nu + 1 \leqslant n \leqslant N + \nu + \sigma - 1$, so if empty sums are interpreted as 0 we can write this as

$$\sum_{\nu=0}^{\sigma} A_{\sigma-\nu}(N+\nu)\left[\sum_{k=n}^{N+\nu} c(k)y_{N+\nu}(k) + \sum_{k=N+\nu+\sigma}^{N+\sigma} c(k)y_{N+\nu}(k)\right]$$

$$= c(N+\sigma)y_N(N+\sigma)A_\sigma(N) + \sum_{\nu=0}^{\sigma} A_{\sigma-\nu}(N+\nu)S_{N+\nu}(n)$$

$$= -c(N+\sigma) + \sum_{\nu=0}^{\sigma} A_{\sigma-\nu}(N+\nu)S_{N+\nu}(n) = 0. \quad\blacksquare$$

7.3 The matrix formulation: stability and weak stability

Zahar (1977) has given a matrix formulation for the Miller algorithm which, although it is only easily applied to the simplified algorithm, is interesting for the light it sheds on convergence and the fact that it enables one to define a concept of stability analogous to that introduced for forward computation in matrix systems in Chapter 2.

In what follows $\|\cdot\|$ will stand for the supremum vector norm (although the results obtained will hold for any other norm by the topological equivalence of vector norms on $\mathcal{R}^\sigma$).

I will write the system as

$$\mathbf{y}(n) + A(n)\mathbf{y}(n+1) = \mathbf{0}, \tag{7.32}$$

where $A(n)$ is a $\sigma \times \sigma$ matrix, $A(n)$ nonsingular, $\mathbf{y}$ a $\sigma \times 1$ vector.

Put $\mathbf{y}_N(N) := \mathbf{e}_N = (1, 0, 0, \ldots, 0)^{\mathrm{T}}$ and use the equation (32) in the backward direction to compute $\mathbf{y}_N(n)$, $n = N-1, N-2, \ldots, 0$.

To effect a normalization we take $\|\mathbf{y}_N(0)\|$ as a scale factor and define

$$\mathbf{w}_N(n) := \mathbf{y}_N(n)/\|\mathbf{y}_N(0)\|, \qquad 0 \leq n \leq N. \tag{7.33}$$

Since $A(n)$ is nonsingular, all the vectors above exist and are nonzero for $0 \leq n \leq N$.

If the limit

$$\lim_{N\to\infty} \mathbf{w}_N(n) = \mathbf{w}(n), \tag{7.34}$$

exists and is nonzero for a given sequence $\mathbf{e}_N$, we say that this matrix version of the simplified Miller algorithm converges (to $\mathbf{w}(n)$).

To analyze the convergence of the process let $Y(n)$ be a fundamental matrix of solutions $\{\mathbf{y}^{(h)}(n)\}$,

$$Y(n) = [\mathbf{y}^{(1)}(n), \mathbf{y}^{(2)}(n), \ldots, \mathbf{y}^{(\sigma)}(n)].$$

It is easily seen that

$$\mathbf{w}_N(n) = \frac{Y(n)Y^{-1}(N)\mathbf{e}_N}{\|Y(0)Y^{-1}(N)\mathbf{e}_N\|}. \tag{7.35}$$

Thus convergence of $\mathbf{w}_N(0)$ is equivalent to convergence of $\mathbf{w}_N(n)$, $0 \leq n \leq N$.

The matrix $Y(n)Y^{-1}(N)$ is independent of the choice of solutions $\{\mathbf{y}^{(h)}(n)\}$ for

$$Y(n)Y^{-1}(N) = (-1)^{N+n}A(n)A(n+1)\cdots A(N-1).$$

We define the adjoint of (32) to be the equation

$$\hat{\mathbf{y}}(n) + [A^{-1}(n)]^*\hat{\mathbf{y}}(n+1) = 0, \qquad n \geq 0. \tag{7.36}$$

(Note this differs from the notation used elsewhere in this book insofar as the *conjugate* transpose is employed.) Each column of $[Y^{-1}(n)]^*$ satisfies (36).

With each choice of $\{\mathbf{y}^{(h)}(n)\}$ we associate a fundamental set of solutions of the adjoint equation $\{\hat{\mathbf{y}}^{(h)}(n)\}$ such that $\mathbf{y}^{(h)}(n)$ is the hth column of $Y(n)$ and $\hat{\mathbf{y}}^{(h)}(n)$ is the hth column of $[Y^{-1}(n)]^*$. Thus

$$Y(n) = [\mathbf{y}^{(1)}(n), \mathbf{y}^{(2)}(n), \ldots, \mathbf{y}^{(\sigma)}(n)], \tag{7.37}$$

$$[Y^{-1}(n)]^* = [\hat{\mathbf{y}}^{(1)}(n), \hat{\mathbf{y}}^{(2)}(n), \ldots, \hat{\mathbf{y}}^{(\sigma)}(n)]. \tag{7.38}$$

Obviously, the $\hat{\mathbf{y}}^{(h)}(n)$ can be computed from the minors of $Y(n)$. In fact, for the σth-order scalar equation, this is essentially the relationship given in A-VI.

Denote by $V(n) = [\boldsymbol{\alpha}^{(h)}(n)]$, $[V^{-1}(n)]^* = [\hat{\boldsymbol{\alpha}}^{(h)}(n)]$ those solutions of the equation and its adjoint respectively having the property $V(0) = [V^{-1}(0)]^* = I$, where I is the $\sigma \times \sigma$ identity matrix.

We have

$$\mathbf{w}_N(0) = \frac{V^{-1}(N)\mathbf{e}_N}{\|V^{-1}(N)\mathbf{e}_N\|}. \tag{7.39}$$

But

$$V^{-1}(N)\mathbf{e}_N = [(\hat{\boldsymbol{\alpha}}^{(1)}(N), \mathbf{e}_N), (\hat{\boldsymbol{\alpha}}^{(2)}(N), \mathbf{e}_N), \ldots, (\hat{\boldsymbol{\alpha}}^{(\sigma)}(N), \mathbf{e}_N)]^T, \tag{7.40}$$

$(\cdot, \cdot)$ denoting the standard inner product.

This equation makes clear the role of the solutions of the adjoint equation in the convergence of the modified Miller algorithm.

Theorem 7.4 (Zahar) *The following statements are equivalent:*

(i)   *the modified Miller algorithm converges;*
(ii) *for the fundamental matrix $V(n)$ the vector $V^{-1}(N)\mathbf{e}_N/\|V^{-1}(N)\mathbf{e}_N\|$ converges to a nonzero limit as $N \to \infty$;*
(iii) *the adjoint equation possesses a fundamental matrix $[\hat{\mathbf{y}}^{(h)}(n)]$ satisfying*

$$\lim_{N\to\infty} \frac{(\hat{\mathbf{y}}^{(h)}(n), \mathbf{e}_N)}{(\hat{\mathbf{y}}^{(1)}(n), \mathbf{e}_N)} = 0, \qquad 2 \leq h \leq \sigma. \tag{7.41}$$

Proof Since convergence is equivalent to convergence of $\mathbf{w}_N(0)$, by the previous remarks (i) and (ii) are equivalent. I now show that (ii) $\Rightarrow$ (iii) $\Rightarrow$ (i).

Let (ii) hold. Equation (40) shows that

$$\lim_{N \to \infty} \frac{(\hat{\boldsymbol{\alpha}}^{(h)}(N), \mathbf{e}_N)}{\|V^{-1}(N)\mathbf{e}_N\|} := c^{(h)}, \qquad 1 \leqslant h \leqslant \sigma, \tag{7.42}$$

for some constants $c^{(h)}$ not all zero. Suppose $c^{(k)} \neq 0$. Let $\mathbf{y}^{(1)}(n)$ be the solution of the equation satisfying $\mathbf{y}^{(1)}(0) = [c^{(1)}, \ldots, c^{(\sigma)}]^{\mathrm{T}}$ and let $Y(0)$ be the matrix

$$Y(0) = [\mathbf{y}^{(1)}(0), \boldsymbol{\alpha}^{(1)}(0), \ldots, \boldsymbol{\alpha}^{(k-2)}(0), \boldsymbol{\alpha}^{(k)}(0), \ldots, \boldsymbol{\alpha}^{(\sigma)}(0)].$$

A computation gives

$$Y^{-1}(0) = [\boldsymbol{\alpha}^{(2)}(0), \ldots, \boldsymbol{\alpha}^{(k)}(0), \mathbf{f}^{(k)}(0), \boldsymbol{\alpha}^{(k+1)}(0), \ldots, \boldsymbol{\alpha}^{(\sigma)}(0)],$$

where

$$\mathbf{f}^{(k)}(0) = \left[\frac{1}{c^{(k)}}, \frac{-c^{(1)}}{c^{(k)}}, \ldots, \frac{-c^{(k-1)}}{c^{(k)}}, \frac{-c^{(k+1)}}{c^{(k)}}, \ldots, \frac{-c^{(\sigma)}}{c^{(k)}} \right]^{\mathrm{T}}.$$

Thus if $\hat{\mathbf{y}}^{(h)}(n)$ are the corresponding adjoint solutions we have, by superposition of columns of $[Y^{-1}(0)]^*$,

$$\hat{\mathbf{y}}^{(1)}(n) = \frac{1}{\bar{c}^{(k)}} \hat{\boldsymbol{\alpha}}^{(k)}(n),$$

$$\hat{\mathbf{y}}^{(i)}(n) = \hat{\boldsymbol{\alpha}}^{(i-1)}(n) - \frac{\bar{c}^{(i-1)}}{\bar{c}^{(k)}} \hat{\boldsymbol{\alpha}}^{(k)}(n), \qquad i = 2, 3, \ldots, k,$$

$$\hat{\mathbf{y}}^{(i)}(n) = \hat{\boldsymbol{\alpha}}^{(i+1)}(n) - \frac{\bar{c}^{(i+1)}}{\bar{c}^{(k)}} \hat{\boldsymbol{\alpha}}^{(k)}(n), \qquad i = k+1, k+2, \ldots, \sigma,$$

bars denoting complex conjugates. Thus for $i \neq 1$,

$$\frac{(\hat{\mathbf{y}}^{(i)}(N), \mathbf{e}_N)}{(\hat{\mathbf{y}}^{(1)}(N), \mathbf{e}_N)} = \frac{c^{(k)}(\hat{\boldsymbol{\alpha}}^{(i \mp 1)}(N), \mathbf{e}_N)}{(\hat{\boldsymbol{\alpha}}^{(k)}(N), \mathbf{e}_N)} - c^{(i \mp 1)},$$

and from equation (42) it follows that

$$\lim_{N \to \infty} \frac{(\hat{\mathbf{y}}^{(i)}(N), \mathbf{e}_N)}{(\hat{\mathbf{y}}^{(1)}(N), \mathbf{e}_N)} = c^{(k)} \frac{c^{(i \mp 1)}}{c^{(k)}} - c^{(i \mp 1)} = 0,$$

proving (iii).

Finally suppose (iii) holds and let

$$Y(n) = [\mathbf{y}^{(1)}(n), \ldots, \mathbf{y}^{(\sigma)}(n)],$$

be the fundamental matrix for (32) corresponding to $\{\hat{\mathbf{y}}^{(h)}(n)\}$. Then

$$Y^{-1}(N)\mathbf{e}_N = [(\hat{\mathbf{y}}^{(1)}(N), \mathbf{e}_N), \ldots, (\hat{\mathbf{y}}^{(\sigma)}(N), \mathbf{e}_N)]^{\mathrm{T}},$$

and

$$Y(n) Y^{-1}(N) e_N = \sum_{i=1}^{\sigma} \mathbf{y}^{(i)}(n) (\hat{\mathbf{y}}^{(i)}(N), e_N).$$

Because of (41) and the fact that $\mathbf{y}^{(1)}(n) \neq \mathbf{0}$ we have

$$\| Y(0) Y^{-1}(N) e_N \| = \| \mathbf{y}^{(1)}(0) \| |(\hat{\mathbf{y}}^{(1)}(N), e_N)| (1 + o(1)), \qquad N \to \infty.$$

Thus it follows by again using (41) that

$$\begin{aligned}
\lim_{N \to \infty} w_N(n) &= \lim_{N \to \infty} \frac{Y(n) Y^{-1}(N) e_N}{\| Y(0) Y^{-1}(N) e_N \|} \\
&= \lim_{N \to \infty} \frac{\sum_{i=1}^{\sigma} \mathbf{y}^{(i)}(n) (\hat{\mathbf{y}}^{(i)}(N), e_N)}{\| \mathbf{y}^{(1)}(0) \| |(\hat{\mathbf{y}}^{(1)}(N), e_N)|} \\
&= \mathbf{y}^{(1)}(n) / \| \mathbf{y}^{(1)}(0) \|.
\end{aligned} \qquad (7.43)$$

This finishes the proof. ■

As opposed to studying convergence we could ask how errors introduced at one point in the computation grow or decrease as the computation proceeds. This is the problem of stability, analyzed for forward computation in Chapter 2. Let us recall some of the main results. Define

$$\alpha(k, n) := \frac{\| \mathbf{w}(k) \|}{\| \mathbf{w}(n) \|} \| Y(n) Y^{-1}(k) \|,$$

where $\mathbf{w}(n)$ is the desired solution of the recurrence. It was shown that if $\sigma \geq 2$ and the recursion has a solution $\mathbf{y}^*(n)$ such that $\underline{\lim} \| \mathbf{w}(n) \| / \| \mathbf{y}^*(n) \| = 0$, then forward recurrence is unstable for $\mathbf{w}(n)$. Conversely if forward recurrence is unstable for $\mathbf{w}(n)$ in the sense that $\overline{\lim}_{n \to \infty} \alpha(k, n) = \infty$ for some fixed $k \geq 0$, then a solution $\mathbf{y}^*(n)$ of (32) exists having this property.

Let us similarly examine the problem of stability when Miller's algorithm is applied to the equation.

We fix N and assume an error is introduced at a point $n = k \leq N$ in the following fashion,

$$\begin{aligned}
\bar{\mathbf{y}}_N(k) &= \mathbf{y}_N(k) + \| \mathbf{y}_N(k) \| \, \mathbf{e}, \qquad \| \mathbf{e} \| = \varepsilon, \\
&= Y(k) Y^{-1}(N) e_N + \| Y(k) Y^{-1}(N) e_N \| \, \mathbf{e}.
\end{aligned}$$

Then for $0 \leq n \leq k$,

$$\begin{aligned}
\bar{\mathbf{y}}_N(n) &= Y(n) Y^{-1}(k) (Y(k) Y^{-1}(N) e_N + \| Y(k) Y^{-1}(N) e_N \| \, \mathbf{e}), \\
&= \mathbf{y}_N(n) + Y(n) Y^{-1}(k) \| Y(k) Y^{-1}(N) e_N \| \, \mathbf{e}.
\end{aligned}$$

Using the formula (35) for $\mathbf{w}_N(n)$ gives

$$\bar{\mathbf{w}}_N(n) = \mathbf{w}_N(n) + \frac{Y(n) Y^{-1}(k) \| Y(k) Y^{-1}(N) e_N \| \, \mathbf{e}}{\| Y(0) Y^{-1}(N) e_N \|},$$

where $\bar{\mathbf{w}}_N(n)$ is the computed value of $\mathbf{w}_N(n)$. Using (35) again gives

$$\frac{\|\bar{\mathbf{w}}_N(n) - \mathbf{w}_N(n)\|}{\|\mathbf{w}_N(n)\|} = \frac{\|Y(k)Y^{-1}(N)\mathbf{e}_N\|}{\|Y(n)Y^{-1}(N)\mathbf{e}_N\|} \|Y(n)Y^{-1}(k)\mathbf{e}\|. \tag{7.44}$$

Since the only situation of practical interest is where convergence occurs I will assume for the remainder of this section that condition (iii) of Theorem 4 holds. I take $Y(n)$ to be the fundamental matrix defined when $[Y^{-1}(n)]^* = [\hat{\mathbf{y}}^{(h)}(n)]$, where the $\{\hat{\mathbf{y}}^{(h)}(n)\}$ satisfy (41).

From (44) and (43) we have

$$\lim_{N \to \infty} \frac{\|\bar{\mathbf{w}}_N(n) - \mathbf{w}_N(n)\|}{\|\mathbf{w}_N(n)\|} = \frac{\|\mathbf{y}^{(1)}(k)\|}{\|\mathbf{y}^{(1)}(n)\|} \|Y(n)Y^{-1}(k)\mathbf{e}\|.$$

We are interested in the maximum error of the term on the right which leads us to define the amplification factor for backward recurrence as

$$\beta(n, k) := \frac{\|\mathbf{w}(n)\|}{\|\mathbf{w}(k)\|} \|Y(k)Y^{-1}(n)\|, \tag{7.45}$$

where $0 \leq k \leq n$ and $\mathbf{w}(n) := \mathbf{y}^{(1)}(n)$ is the solution to be computed. Then the right-hand side above is bounded by $\varepsilon\beta(n, k)$.

Definition 7.1 *The convergence for the Miller algorithm is said to be stable for* $\mathbf{w}(n)$ *if there is a constant* $C > 0$ *such that*

$$\sup_{\substack{k,n \geq 0 \\ n < k}} \beta(n, k) = C < \infty. \quad \blacksquare \tag{7.46}$$

Note that $\alpha(k, n)$, $\beta(n, k)$ are not antisymmetric in n, k since the factor $Y(k)Y^{-1}(n)$ in each is the same. This difference has the effect of introducing solutions of the adjoint equation into the stability analysis.

Theorem 7.5 *Let the Miller algorithm converge to* $\mathbf{w}(n)$. *If the convergence is stable, then for all solutions* $\hat{\mathbf{y}}(n)$ *of the adjoint,*

$$\overline{\lim_{k \to \infty}} \|\mathbf{w}(k)\| \|\hat{\mathbf{y}}(k)\| \leq C < \infty, \tag{7.47}$$

for some constant $C > 0$. *Conversely if convergence is unstable in the fashion*

$$\overline{\lim_{k \to \infty}} \beta(n, k) = \infty \quad \text{for some fixed } n \geq 0, \tag{7.48}$$

there exists a solution $\hat{\mathbf{y}}(n)$ *of the adjoint equation with the property*

$$\overline{\lim_{k \to \infty}} \|\mathbf{w}(k)\| \|\hat{\mathbf{y}}(k)\| = \infty. \tag{7.49}$$

Proof First suppose that n is fixed. Since $\|\mathbf{w}(n)\| \neq 0$ (by definition of convergence),

$$\overline{\lim_{k \to \infty}} \frac{\|\mathbf{w}(k)\|}{\|\mathbf{w}(n)\|} \|Y(n) Y^{-1}(k)\| = \infty \Leftrightarrow$$

$$\overline{\lim_{k \to \infty}} \|\mathbf{w}(k)\| \|Y(n) Y^{-1}(k)\| = \infty \Leftrightarrow$$

$$\overline{\lim_{k \to \infty}} \|\mathbf{w}(k)\| \|Y^{-1}(k)\| = \infty,$$

which follows from the inequalities

$$\|Y(n) Y^{-1}(k)\| \le \|Y(n)\| \|Y^{-1}(k)\|,$$

$$\|Y^{-1}(k)\| = \|Y^{-1}(n) Y(n) Y^{-1}(k)\| \le \|Y^{-1}(n)\| \|Y(n) Y^{-1}(k)\|.$$

Furthermore the rows of $Y^{-1}(k)$ are the vectors $\hat{\mathbf{y}}^{(h)}(k)^*$ so for some h, $1 \le h \le \sigma$,

$$\overline{\lim_{k \to \infty}} \frac{\|\mathbf{w}(k)\|}{\|\mathbf{w}(n)\|} \|Y(n) Y^{-1}(k)\| = \infty \Leftrightarrow \tag{7.50}$$

$$\overline{\lim_{k \to \infty}} \|\mathbf{w}(k)\| \|\hat{\mathbf{y}}^{(h)}(k)\| = \infty.$$

Now suppose the convergence is stable. Then the right-hand side of (50) cannot hold, for if it did, the left-hand side would hold for any n, contradicting stability. This implies that (47) must hold for $\hat{\mathbf{y}}^{(1)}(n)$, $\hat{\mathbf{y}}^{(2)}(n), \dots, \hat{\mathbf{y}}^{(\sigma)}(n)$ and consequently for any adjoint solution $\hat{\mathbf{y}}(n)$. This proves the theorem. ■

As in the case for computation in the forward direction we are, practically speaking, concerned only with the calculation of $\mathbf{w}(n)$ for $n \le K$. The above criterion is then too stringent and the following weaker form of stability is of more value.

Definition 7.2 *Convergence in the Miller algorithm is said to be* weakly stable if *for each $K > 0$ there exists a constant $C \equiv C(K)$ such that*

$$\sup_{\substack{n < k \\ k \le K}} \beta(n, k) = C < \infty \quad ■ \tag{7.51}$$

Compare this with the Definitions 2.2, 2.3 of *stable at k* and *weakly stable* for computation in the forward direction.

Equation (50) shows that convergence is weakly stable if for a fixed $n \ge 0$

$$\overline{\lim_{k \to \infty}} \frac{\|\mathbf{w}(k)\|}{\|\mathbf{w}(n)\|} \|Y(n) Y^{-1}(k)\| = C < \infty \Leftrightarrow$$

$$\overline{\lim_{k \to \infty}} \|\mathbf{w}(k)\| \|\hat{\mathbf{y}}^{(h)}(k)\| = C_1 < \infty,$$

for all h, $1 \leq h \leq \sigma$. Since the set $0 \leq n \leq K$ is finite (51) will hold if the left-hand side of the previous relation is true for $n \leq K$. We thus have the

Corollary *Let the Miller algorithm converge to* $\mathbf{w}(n)$. *Then the convergence is weakly stable iff*

$$\varlimsup_{k \to \infty} \|\mathbf{w}(k)\| \, \|\hat{\mathbf{y}}^{(h)}(k)\| = C_1 < \infty, \qquad 1 \leq h \leq \sigma. \quad \blacksquare \tag{7.52}$$

The theorem and corollary describe convergence of the Miller algorithm in terms of a fundamental set for the adjoint equation rather than for the equation itself. Unfortunately it is usually the solutions themselves rather than the adjoint solutions whose asymptotic character is known. Nevertheless since each $\hat{\mathbf{y}}^{(h)}(n)$ is a column of $[Y^{-1}(n)]^*$ each component of that solution is a $\sigma - 1$ order minor of $Y(n)$ divided by $|Y(n)|$ and $\mathbf{y}^{(h)}(n)$ is absent from the minor. If the solutions $\{\mathbf{y}^{(h)}(n)\}$ are suitably differentiated asymptotically so that no adverse cancellation occurs one would expect $\|\hat{\mathbf{y}}^{(h)}(n)\|$ to behave roughly like $\|\mathbf{y}^{(h)}(n)\|^{-1}$, and if this is indeed the case, then (52) shows that for stable convergence one requires something like

$$\varlimsup_{k \to \infty} \frac{\|\mathbf{w}(k)\|}{\|\mathbf{y}^{(h)}(k)\|} = C_1 < \infty,$$

along with condition (iii) of Theorem 5.

Wimp (1969) has shown that when the matrix $A(n)$ possesses an asymptotic expansion in inverse powers of $n^{1/\omega}$, ω an integer ≥ 1, then $\|\hat{\mathbf{y}}^{(h)}(n)\|$ usually *does* behave like $\|\mathbf{y}^{(h)}(n)\|^{-1}$. Note that in this case the members of a fundamental set possess Birkhoff expansions (see Appendix B).

I will state the result for the case of the scalar equation $\mathcal{L}[y(n)] = 0$. It is simply a reformulation of a result in Theorem B.8.

Theorem 7.6 *Let* $\{y^{(h)}(n)\}$ *be a Birkhoff set for* $\mathcal{L}[y(n)] = 0$. *Then if the corresponding asymptotic series are free from logarithms, there are constants* c_h, θ_h, $1 \leq h \leq \sigma$, $c_h \neq 0$, *such that*

$$\hat{y}^{(h)}(n) = \frac{c_h n^{\theta_h}}{y^{(h)}(n)} [1 + o(1)], \qquad n \to \infty. \quad \blacksquare \tag{7.53}$$

Remark The Theorem B.6 actually states more, namely that there is a complete asymptotic expansion for $\hat{y}^{(h)}(n)$ for which the first term on the right above is the leading term.

Also, as the following shows, one really needs powers of n in formula (53).

Example 7.6 Consider the equation in Example 7.8 (Section 7.4). There $y^{(1)}(n) = 1$, $y^{(2)}(n) = (-1)^n$, $y^{(3)}(n) = n!$ and

$$\hat{y}^{(1)}(n) = \frac{\frac{1}{2}n^{-3}}{y^{(1)}(n)}[1 + o(1)], \qquad \hat{y}^{(2)}(n) = \frac{\frac{1}{2}n^{-3}}{y^{(2)}(n)}[1 + o(1)],$$

$$\hat{y}^{(3)}(n) = -\frac{n^{-5}}{y^{(3)}(n)}[1 + o(1)]. \qquad \blacksquare$$

It is clear that no ingenious alternative choice for the initial vector $\mathbf{e}_N$ can improve the convergence *rate* or make an otherwise unstable algorithm stable. To see this note that, from the definition of $\hat{\mathbf{y}}^{(h)}(n)$,

$$(\hat{\mathbf{y}}^{(h)}(N), \mathbf{e}_N)$$
$$= \|[\mathbf{y}^{(1)}(N), \mathbf{y}^{(2)}(N), \dots, \mathbf{y}^{(h-1)}(N), \mathbf{e}_N, \mathbf{y}^{(h+1)}(N), \dots, \mathbf{y}^{(\sigma)}(N)]\|/|Y(n)|.$$

Thus, unless $\mathbf{e}_N$ is a constant multiple of $\mathbf{w}(N) = \mathbf{y}^{(1)}(N)$ (in which case the algorithm will be exact) one of the above inner products must be $\neq 0$ for an infinite number of N. Thus if $\mathbf{e}_N$ makes one of the inner products small, i.e., is nearly in the span of a $\sigma - 1$ subset of the solutions of the equation, other of the products will be large. This does not mean of course that temporary computational benefits cannot accrue by choosing the components of $\mathbf{e}_N$ to mimic the asymptotic behavior of $\mathbf{w}(N)$, but merely that this can't be done in general on a systematic basis.

When the preceding theory is applied to the scalar equation $\mathscr{L}[y(n)] = 0$ the result is

Corollary Let $\mathbf{e}_N = (1, 0, 0, \dots, 0)^{\mathrm{T}}$. *The Miller algorithm converges for the scalar equation if and only if the adjoint equation possesses a dominant solution.* $\blacksquare$

It is sometimes assumed that a sufficient condition for convergence of Miller's algorithm is that $\mathscr{L}[y(n)] = 0$ possesses a minimal solution $y^{(1)}(n)$. In fact when $\sigma = 2$, the existence of such a solution is necessary and sufficient for convergence and convergence is to $y^{(1)}(n)$ (Theorem 4.2, Corollary). But when $\sigma > 2$ convergence, when it occurs, need not be to a minimal solution, as the following example shows.

Example 7.7 (Zahar (1977)) Consider the third-order equation

$$A_0(n)y(n) + A_1(n)y(n+1) + A_2(n)y(n+2) + A_3(n)y(n+3) = 0,$$

$$A_3(n) = 1,$$

whose solutions are

$$y^{(1)}(n) = 1, \qquad y^{(2)}(n) = \tfrac{1}{2}[1 + (-1)^n]n^{-1/2}, \qquad y^{(3)}(n) = n.$$

(Expressions for A_i, which are not important for my purposes, are given in the cited reference.)

The corresponding solutions of the adjoint are

$$\hat{y}^{(1)}(n) = \frac{(n+2)(n+1)^{-1/2}[1-(-1)^n]-(n+1)(n+2)^{-1/2}}{2D(n+1)},$$

$$\hat{y}^{(2)}(n) = \frac{-1}{D(n+1)},$$

$$\hat{y}^{(3)}(n) = \frac{[1+(-1)^n](n+2)^{-1/2}-[1-(-1)^n](n+1)^{-1/2}}{D(n+1)}.$$

Thus $\hat{y}^{(1)}(n)$ is a maximal solution of the adjoint and $\hat{y}^{(2)}(n)/\hat{y}^{(1)}(n) \to 0$, $\hat{y}^{(3)}(n)/\hat{y}^{(1)}(n) \to 0$. Thus from the formula (43) we see that the Miller algorithm converges to $y^{(1)}(n)$. But $y^{(1)}(n)$ is not the minimal solution of the original equation. $y^{(2)}(n)$ is, both in the ordinary and the vector sense, since $\mathbf{y}^{(2)}(n) = [y^{(1)}(n), y^{(1)}(n+1)]^T$.

This phenomenon occurs because the zeros of the solution $y^{(2)}(n)$ cause cancellation in the calculation of the adjoint solutions $\hat{y}^{(h)}(n)$. This cannot happen when the equation is of Birkhoff type for then the solutions will be sufficiently differentiated asymptotically.

The convergence of Miller's algorithm is unstable, however, since

$$\lim_{n \to \infty} \|\mathbf{y}^{(1)}(n)\| \, \|\hat{\mathbf{y}}^{(h)}(n)\| = \infty, \qquad h = 1, 2.$$

The recurrence is weakly stable for $y^{(3)}(n)$ when used in the forward direction.

Zahar (1977) also discusses the situation where general initial conditions are given with results similar to ours of Theorem 1. ■

7.4 The Clenshaw averaging process

When the equation has no minimal solution (or, to be more precise, when the adjoint equation has no maximal solution) and u of the functions

$$R^{(h)}(n) := T^{(h)}(n)/T^{(1)}(n),$$

behave similarly as $n \to \infty$ but the ratio of any one of these to each of the other $\sigma - u$ values of $R^{(h)}$ approaches zero as $n \to \infty$, a generalization of the method in Section 4.7 which is due to Luke can often be used to obtain any one of the first u solutions corresponding to the u values of $R^{(h)}$.

Example 7.8 The equation

$$(n+1)(n^2+5n+5)y(n)-(n^2+3n+1)y(n+1)$$
$$-(n+1)(n^2+5n+5)y(n+2)+(n^2+3n+1)y(n+2)=0,$$

has a fundamental set

$$y^{(1)}(n) = 1, \qquad y^{(2)}(n) = (-1)^n, \qquad y^{(3)}(n) = n!.$$

The corresponding solutions of the adjoint are

$$\hat{y}^{(1)}(n) = \frac{(n+3)}{\tau(n)}, \qquad \hat{y}^{(2)}(n) = \frac{(-1)^n(n+1)}{\tau(n)},$$

$$\hat{y}^{(3)}(n) = \frac{-2}{(n+1)!\,\tau(n)},$$

$$\tau(n) = 2[(n+1)(n+2)-1][(n+2)(n+3)-1].$$

Thus

$$R^{(2)}(n) = \frac{(-1)^n(n+1)}{(n+3)}, \qquad R^{(3)}(n) = \frac{-2}{(n+3)(n+1)!}.$$

$R^{(1)}$, $R^{(2)}$ behave similarly and $R^{(1)}/R^{(3)}$, $R^{(2)}/R^{(3)} \to 0$. The Miller algorithm for the computation of either $y^{(1)}$, $y^{(2)}$ will not converge, but the Clenshaw averaging process based on suitable normalization relations will, as we shall see. ■

In what follows $S_N^{(h,j)}$, $R^{(h)}$, $c^{(j)}$ will be as defined in Section 4.7 with $1 \le h, j \le \sigma$.

Theorem 7.7   *Let u normalization relationships be known for the solution* $w(n) = y^{(1)}(n)$ *with coefficients* $c^{(j)}(k)$, $k \ge 0$, $1 \le j \le u$, *and*

$$\lim_{N \to \infty} S_N^{(h,j)} = S^{(h,j)} < \infty, \qquad 2 \le h \le u, \tag{7.54}$$

with $|S^{(h,j)}|_1^u \ne 0$, $S^{(1,1)} \ne 0$.

Let $y_N(n)$ *be as defined in* (4.3) *with* $c(k) = c^{(1)}(k)$ *nonzero for N sufficiently large,* $0 \le n \le N$. *Let*

$$w_N(n) := S^{(1,1)} y_N(n) \Big/ \sum_{k=0}^{N} c^{(1)}(k) y_N(k), \tag{7.55}$$

and let $R^{(h)}(n)$ *be bounded and bounded away from 0 as* $n \to \infty$, $2 \le h \le u$, *while*

$$\lim_{N \to \infty} R^{(h)}(N) = \lim_{N \to \infty} S_N^{(h,j)} R^{(h)}(N) = 0,$$

$$u+1 \le h \le \sigma, \quad 1 \le j \le u. \tag{7.56}$$

Then for N_1 *sufficiently large we can determine* N_r, $2 \le r \le u$, $N_1 < N_2 < \cdots < N_u$, *such that the system of equations*

$$\sum_{\nu=1}^{u} \pi_\nu \sum_{k=0}^{N_\nu} c^{(j)}(k) w_{N_\nu}(k) = S^{(1,j)}, \qquad \sum_{\nu=1}^{u} \pi_\nu = S^{(1,1)}, \tag{7.57}$$

has a unique solution $\{\pi_h\}$.

Let $|R^{(h)}(N_j)|_1^u$ be bounded away from 0 as $N_1 \to \infty$. Then

$$\lim_{N_1 \to \infty} \sum_{\nu=1}^{u} \pi_\nu w_{N_\nu}(n) = y^{(1)}(n), \qquad n \geqslant 0.$$

Proof Equation (57), by the formula (55) for $w_N(n)$, may be written

$$\sum_{\nu=1}^{u} \pi_\nu^* \sum_{h=1}^{\sigma} R^{(h)}(N_\nu) S_{N_\nu}^{(h,j)} = S^{(1,j)}, \qquad 1 \leqslant j \leqslant u,$$
$$\pi_\nu^* := \pi_\nu T^{(1)}(N_\nu)/S_{N_\nu}, \tag{7.58}$$
$$S_N := \sum_{k=0}^{N} c^{(1)}(k) y_N(k).$$

The determinant of the system is

$$|R^{(h)}(N_j)|_1^u |S^{(h,j)}|_1^u (1 + o(1)), \qquad N_1 \to \infty. \tag{7.59}$$

The N_r can be chosen so that (59) is not zero or else the $R^{(h)}$ are linearly dependent. Further, by the hypothesis on $|R^{(h)}(N_j)|$, π_ν^*, and hence π_ν is bounded as $N_1 \to \infty$.
Write

$$\sum_{\nu=1}^{u} \pi_\nu w_{N_\nu}(n) = \sum_{r=1}^{\sigma} d^{(r)} y^{(r)}(n), \qquad d^{(r)} \equiv d^{(r)}(N_1, N_2, \ldots, N_u). \tag{7.60}$$

Then

$$d^{(s)} = \sum_{\nu=1}^{u} \frac{\pi_\nu R^{(s)}(N_\nu)}{\sum_{h=1}^{\sigma} R^{(h)}(N_\nu) S_{N_\nu}^{(h,1)}}.$$

Clearly

$$\lim_{N_1 \to \infty} d^{(r)} = 0, \qquad u+1 \leqslant s \leqslant \sigma.$$

Also from the π_ν^* equation (58) and the boundedness of the π_ν we have

$$\sum_{h=1}^{u} S^{(h,j)} d^{(h)} = S^{(1,j)} + o(1), \qquad N_1 \to \infty, \quad 1 \leqslant j \leqslant u.$$

But since this system has a unique solution, we find

$$\lim_{N_1 \to \infty} d^{(h)} = \begin{cases} 1, & h = 1, \\ 0, & h \neq 1. \end{cases}$$

This fact used in equation (60) completes the proof. ∎

Example 7.9 In the previous example, if the normalization relations satisfy

$$\sum_{k=0}^{\sigma} c^{(j)}(k)y^{(h)}(k) = S^{(h,j)},$$

with $S^{(1,1)}S^{(2,2)} - S^{(2,1)}S^{(1,2)} \neq 0$ the Miller algorithm converges to 1 provided N_1, N_2 are chosen appropriately. This has to be done so that $|R^{(h)}(N_j)|$ is bounded away from 0 as $N_1 \to \infty$. But

$$|R^{(h)}(N_j)| = \frac{(-1)^{N_2}(N_2+1)}{(N_2+3)} - \frac{(-1)^{N_1}(N_1+1)}{(N_1+3)},$$

and if N_2 is odd, N_1 is even, for example, the above quantity satisfies the prescribed condition. ■

Example 7.10 Let $\lambda > 0$ and consider the expansion

$$e^{-\lambda/x} = \sum_{n=0}^{\infty} \varepsilon_n C_n(\lambda)T_n^*(x), \qquad 0 \leq x \leq 1, \tag{7.61}$$

where $T_n^*(x) = T_n(2x-1)$ is the shifted Chebyshev polynomial. Despite the rather strange character of the expansion (we are expanding a function in orthogonal polynomials over a segment issuing from an essential singularity) the fact that the expansion converges is clear from the theory of Fourier series since $e^{-\lambda/x} \in C^1[0, 1]$.

The coefficients $C_n(\lambda)$ can be found from a result given by Luke and Wimp (1963). They are G-functions:

$$C_n(\lambda) = \pi^{-1/2}G_{23}^{30}\left(\lambda \left| \begin{matrix} 1-n, & n+1 \\ 1, & \frac{1}{2}, & 0 \end{matrix} \right.\right)$$

$$= (-1)^n\pi^{-1/2}G_{23}^{21}\left(\lambda \left| \begin{matrix} 1-n, & n+1 \\ \frac{1}{2}, & 0, & 1 \end{matrix} \right.\right). \tag{7.62}$$

We also have

$$C_n(\lambda) = (-1)^n\left\{ -2\lambda^{1/2}{}_2F_2\left(\begin{matrix} \frac{1}{2}+n, & \frac{1}{2}-n \\ \frac{1}{2}, & \frac{3}{2} \end{matrix} \right| -\lambda \right)\right.$$

$$\left. + 2(\sqrt{\pi})n\lambda {}_2F_2\left(\begin{matrix} -n, & n \\ \frac{3}{2}, & 2 \end{matrix} \right| -\lambda \right)\right\}, \qquad n > 0, \tag{7.63}$$

which follows on using the expansion formula in Erdélyi et al. (1953, vol. 1, p. 208 (5)).

Putting $n = 0$ in (62) gives the interesting result

$$C_0(\lambda) = \frac{2}{\sqrt{\pi}} \text{Erfc}(\lambda),$$

in the notation of Erdélyi et al. (1953, 9.9). Using the methods in Wimp

(1967) on the Barnes integral representation of the G-function in (62) it may be verified that $C_n(\lambda)$ satisfies the four-term recurrence

$$ny(n)+[3(n+1)-4\lambda]y(n+1)+[3(n+2)+4\lambda]y(n+2)$$
$$+(n+3)y(n+3)=0, \qquad n\geqslant 0. \quad \textbf{(7.64)}$$

The techniques of Section B.2 show that this equation has a Birkhoff set $\{y^{(h)}(n)\}$ behaving like

$$y^{(h)}(n)=n^{\theta}(-1)^n \exp(3\lambda^{1/3}n^{2/3}\omega_h)[1+O(n^{-1/3})], \qquad n \to \infty,$$

$$\omega_1:=1, \qquad \omega_2:=-\tfrac{1}{2}+\frac{\sqrt{3}\,i}{2}, \qquad \omega_3:=-\tfrac{1}{2}-\frac{\sqrt{3}\,i}{2}. \qquad \textbf{(7.65)}$$

Since $|y^{(1)}(n)| \to \infty$, $C_n(\lambda)$ is represented by a linear combination of the solutions $y^{(2)}(n)$, $y^{(3)}(n)$. These solutions have the same growth, so the Miller algorithm applied to the equation (64) will fail.

Let

$$\boldsymbol{\rho}:=\{1,0,-1,0,1,0,\ldots\}.$$

Then the relations

$$T_n^*(0)=(-1)^n, \qquad T_n^*(\tfrac{1}{2})=\rho_n, \qquad T_n^*(1)=1,$$

can be used to determine three normalizations

$$\sum_{n=0}^{\infty} \varepsilon_N C_n(\lambda)(-1)^n = 0; \qquad \qquad \textbf{(7.66)}$$

$$\sum_{n=0}^{\infty} \varepsilon_n C_n(\lambda)\rho_n = e^{-2\lambda}; \qquad \qquad \textbf{(7.67)}$$

$$\sum_{n=0}^{\infty} \varepsilon_n C_n(\lambda) = e^{-\lambda}. \qquad \qquad \textbf{(7.68)}$$

Table 7.1 Some coefficients for $e^{-1/x} = \sum_{n=0}^{\infty} D_n T_n^*(x), 0 \leqslant x \leqslant 1$.

n	D_n
0	0.157 299 207 1
1	0.201 018 166 5
2	0.026 142 328 3
3	−0.020 135 038 7
4	0.002 113 170 0
5	0.002 915 963 3

The first is needed because the presence of $ny(n)$ presents the determination of $y_N(0)$ from the difference equation. Thus (66) is used to compute $y_N(0)$. Then (67) and (68) are used to compute π_1, π_2 in accordance with (57). For $N_1 = 54$, $N_2 = 55$, $\lambda = 1$, the coefficients so computed are accurate to 10 places and a partial list is given in Table 7.1. As (65) shows, the coefficients decay slowly; $D_{10} \sim 1.6$ (-4); $D_{15} \sim 6.9$ (-6); $D_{20} \sim 4.5$ (-7). ■

Example 7.11 (The Chebyshev series for the exponential integral)   We start with another expansion found in Luke and Wimp, that for the hypergeometric function $(\lambda x)^a \Psi(a, c; \lambda x)$ in series of Chebyshev polynomials $T_n^*(1/x)$. Putting $a = c = 1$, $\lambda \to -\lambda$, $\lambda > 0$, taking real parts and using the fact that

$$\text{Re}\,\{-\lambda x \Psi(1, 1; -\lambda x)\} = \lambda x e^{-\lambda x} E^*(\lambda x),$$

we have the expansion

$$\lambda x e^{-\lambda x} E^*(\lambda x) = \sum_{n=0}^{\infty} \varepsilon_n C_n(\lambda) T_n^*(1/x), \qquad \lambda > 0, \quad 1 \leqslant x \leqslant \infty. \tag{7.69}$$

An actual expression for $C_n(\lambda)$ (in terms of G-functions) is possible but not very interesting. However, we have

$$C_0(\lambda) = 2\lambda^{1/2} e^{-\lambda} \,\text{Erfi}\,(\lambda^{1/2}), \tag{7.70}$$

in terms of the modified error function.

The recursion relation satisfied by C_n is

$$(n+1)y(n) + [3n+4-4\lambda]y(n+1)$$
$$+ [3n+5+4\lambda]y(n+2) + (n+2)y(n+3) = 0. \tag{7.71}$$

The usual analysis shows there is a basis of solutions to the equation having the behavior

$$O(\exp(3\lambda^{1/3} n^{2/3} \omega_h)), \qquad \omega_h \text{ as in (65)},$$

and thus the standard Miller algorithm fails for the computation of $C_n(\lambda)$ while the Clenshaw averaging process succeeds.

One normalization is given by the known series

$$1 = \sum_{n=0}^{\infty} \varepsilon_n (-1)^n C_n(\lambda), \tag{7.72}$$

and the second will be based on the value of $C_0(\lambda)$ in (70). It is interesting that attempting to use the normalization

$$\frac{1}{4\lambda} = \sum_{n=1}^{\infty} (-1)^{n+1} n^2 C_n(\lambda),$$

obtained by differentiating (69) with respect to x and letting $x \to \infty$ yields apparently a divergent algorithm. The reasons for this are not known.

For $\lambda = 2$ and $N_1 = 50$, $N_2 = 51$, the sum $\sum \varepsilon_n C_n(\lambda)$ of the coefficients obtained by the Clenshaw averaging procedure yields

$$\frac{2}{e^2} E^*(2) \approx 1.340\,965\,419\,577\,2.$$

The true value has instead 799 in the last three places. Some of the coefficients are

$C_1 = 0.102\,386\,406\,756,$

$C_2 = -0.064\,031\,023\,661,$

$C_3 = -0.019\,013\,607\,256,$

$C_4 = 0.011\,804\,626\,306.$ ∎

7.5 Topics for future research: infinite systems

The Miller algorithm can be extended in a straightforward way to difference equations of infinite order. Such equations are not a technical oddity but rather occur frequently and are satisfied by functions which are both important and difficult to compute.

Since none of the convergence problems have been resolved for these algorithms I simply present examples that will demonstrate the variety of functions to be computed this way and the sort of convergence one may expect.

7.5.1 Basic series for functions satisfying functional equations

Denote by H the class of functions analytic at 0. (Since I am interested only in establishing the algorithm I will disregard convergence considerations in what is to follow.) Let $p_k(z) \in H$, $k = 0, 1, 2, \ldots$ and let $f \in H$ while $\mathscr{L}[f] \equiv 0$ for some linear operator $\mathscr{L}: H \to H$.

The problem is to find the coefficients $c(k)$ in the expansion

$$f(z) = \sum_{k=0}^{\infty} c(k) p_k(z), \qquad c(0) := 1. \tag{7.73}$$

Let $\mathscr{L}(p_k) = p_k^*$ and

$$p_k^*(z) := \sum_{s=0}^{\infty} a(k, s) z^s, \tag{7.74}$$

(functions other than z^s can be used above for added generality). I assume the vectors $[a(0, j), a(1, j), \ldots]$ are linearly independent.

Taking $\mathscr{L}$ of (73) gives

$$0 \equiv \sum_{k=0}^{\infty} c(k)p_k^*(z), \qquad (\textbf{7.75})$$

and using (74) gives

$$0 \equiv \sum_{k=0}^{\infty} c(k) \sum_{s=0}^{\infty} a(k, s)z^s, \qquad (\textbf{7.76})$$

which is satisfied if

$$\sum_{k=0}^{\infty} c(k)a(k, n) = 0, \qquad n \geqslant 0. \qquad (\textbf{7.77})$$

It is this system of equations of infinite order on which our Miller algorithm is to be based.

Let $N \gg 0$ and define

$$c_N(n) := \begin{cases} 0, & n > N; \\ 1, & n = N. \end{cases} \qquad (\textbf{7.78})$$

Compute $c_N(n)$, $0 \leqslant n \leqslant N-1$, by

$$\sum_{k=0}^{N} c_N(k)a(k, n) = 0, \qquad n = N-1, N-2, \ldots, 0, \qquad (\textbf{7.79})$$

and form the ratio $c_N(n)/c_N(0)$. By *convergence* of the algorithm I mean the property

$$c(n) \approx c(0)c_N(n)/c_N(0), \qquad N \to \infty. \qquad (\textbf{7.80})$$

All the algorithms in this section will be of the form (77)–(80).

Note it is not necessary that $c(0) = 1$ or even to know $c(0)$. I have used this assumption only for convenience. Any series relationship involving the $c(k)$ can be used for normalization.

When the infinite matrix $[a(k, s)]$ has zeros below a diagonal it is not necessary to solve equations to implement the algorithm. This is almost always the case. In particular it is the case when $\mathscr{L}$ maps polynomials into polynomials in such a way that

$$L(z^k) = O(z^{k+\tau}), \qquad z \to 0, \quad \tau \geqslant 0.$$

What one then does is to generate $\tau + 1$ different solutions of the system, call them $c_N^{(j)}(n)$, $1 \leqslant j \leqslant \tau + 1$, corresponding to linearly independent initial conditions

$$[c_N^{(j)}(N), c_N^{(j)}(N+1), \ldots, c_N^{(j)}(N+\tau)]^{\mathrm{T}} = \mathbf{e}^{(j)}, \qquad (\textbf{7.81})$$

where $\mathbf{e}^{(j)}$ is a unit vector in $R^{\tau+1}$. Then for each j, $c_N^{(j)}(n)$, $0 \leqslant n \leqslant N-1$, is generated from

$$\sum_{k=n-\tau}^{N+\tau} c_N^{(i)}(k)a(k, n) = 0, \qquad n = N+\tau-1, N+\tau-2, \ldots, \tau. \tag{7.82}$$

(This does not require solving equations.) Finally for $c_N(n)$ the linear combination

$$A_1 c_N^{(1)}(n) + A_2 c_N^{(2)}(n) + \cdots + A_{\tau+1} c_N^{(\tau+1)}(n) \tag{7.83}$$

is taken, where the A_j are chosen to satisfy

$$A_1 c_N^{(1)}(0) + A_2 c_N^{(2)}(0) + \cdots + A_{\tau+1} c_N^{(\tau+1)}(0) = 1,$$
$$\sum_{k=0}^{N+\tau} a(k, n) \sum_{j=1}^{\tau+1} A_j c_N^{(j)}(k) = 0, \qquad n = \tau-1, \tau-2, \ldots, 0. \tag{7.84}$$

Example 7.12 Determine the coefficients in the Chebyshev polynomial expansion

$$\frac{1}{\Gamma(z+1)} = \sum_{n=0}^{\infty} C(n)T_n^*(z). \tag{7.85}$$

Here $\mathscr{L}[f(z)] = (z+1)f(z+1) - f(z)$,

$$a(k, n) = \frac{(k)_n}{(1/2)_n} \binom{k}{n}[1+(-1)^{k+n+1}] + \frac{(k)_{n-1}}{(\frac{1}{2})_{n-1}} \binom{k}{n-1},$$

with an obvious interpretation for $n = 0$. For normalization either the exact (from the integral) value of $C(0)$ may be used or a particular value of the sum (85), e.g., $z = 1$, $1 = \sum_{k=0}^{\infty} C(k)$. The former and $N = 20$ results in Table 7.2.

The first digit in error has been underlined, see Luke (1969, vol. 2, p. 300). The same level of accuracy is maintained in all the coefficients, about 11 S.F. In this example the equations (79) were solved although the algorithm (81) ff. with $\tau = 1$ could have been used. ∎

Table 7.2 Chebyshev coefficients for $\Gamma(z+1)^{-1}$.

n	Approx. $C(n)$	
0	1.063 773 007 8	(exact)
1	−4.985 587 29$\underline{5}$ 6	(−3)
2	−6.419 254 36$\underline{2}$ 9	(−2)
3	5.065 798 6$\underline{5}$0 9	(−3)
4	4.166 091 $\underline{4}$50 9	(−4)

Table 7.3 Approximate derivatives of $\Gamma(z+1)^{-1}$, $N=30$.

n	Approx. $A(n)$
0	1
1	0.577 215 664 901 536 (Euler's constant)
2	−0.655 878 071 520 245
3	−0.042 002 635 034 117
4	0.166 538 611 382 250
5	−0.042 197 734 555 495

Example 7.13 Compute the derivatives

$$A(k)=\frac{d^k}{dz^k}\frac{1}{\Gamma(z+1)}\bigg|_{z=0}.$$

L is as above, $p_k(z)=z^k$, and

$$a(k,n)=\binom{k+1}{n}-\delta_{kn}.$$

The aforementioned computational scheme is used with $\tau=1$: two sequences $c_N^{(1)}(n)$, $c_N^{(1)}(2)$, are generated and then combined in accordance with (83), (84). For $N=45$ all $A(n)$ are accurate to 16 S.F. ∎

It is interesting that a requirement for the success of the method seems to be that the functional equation $L(f)\equiv0$ in some sense must *characterize* $f(z)$. How the functional equation for $1/\Gamma(z+1)$ characterizes that function can be formulated precisely and is a famous result in classical analysis. On the other hand an attempt to compute the Taylor series coefficients for the number-theoretic function $\xi(z)$ based on the functional equation $\xi(1-z)-\xi(z)=0$ fails. The equation hardly characterizes $\xi(z)$; in fact it is satisfied by any function even about $\tfrac{1}{2}$. In this case finite blocks of the coefficient matrix $[a(k,n)]$ are singular. These observations do not preclude successful computations based on more descriptive functional equations, or on functional equations for closely related functions.

7.5.2 Stieltjes moment integrals

Integrals of the type

$$\sigma(n):=\int_0^\infty t^n e^{-P(t)}\,dt,\qquad n\geq0,$$

occur frequently. I will assume $e^{-P(t)} = O(e^{-\alpha t})$ for some $\alpha > 0$. Let P be entire with

$$P(t) = a(1)t + a(2)t^2 + \cdots.$$

An integration by parts shows $\sigma(n)$ satisfies

$$\sigma(n) = \frac{1}{n+1} \sum_{k=0}^{\infty} (k+1)a(k+1)\sigma(n+k+1),$$

a difference equation of infinite order.

Let

$$\sigma_N(n) = \begin{cases} 0, & n > N; \\ 1, & n = N; \end{cases}$$

and compute $\sigma_N(n)$, $0 \leq n \leq N-1$, from (79) in the usual way. A convenient normalization is found from

$$\int_0^\infty e^{-P(t)} P'(t)\, dt = \sum_{k=0}^{\infty} (k+1)a(k+1) \int_0^\infty e^{-P(t)} t^k\, dt$$

$$= \sum_{k=0}^{\infty} (k+1)a(k+1)\sigma(k) = -e^{-P(t)}\big|_0^\infty = 1.$$

Example 7.14 Compute

$$\sigma(n) = e \int_0^\infty t^n e^{-e^t}\, dt, \qquad n \geq 0.$$

'Explicit' check values can be found from

$$\sigma(n) = e \int_1^\infty (\ln u)^n e^{-u} \frac{du}{u}$$

$$= e \frac{\partial^n}{\partial \alpha^n} \int_1^\infty u^{\alpha-1} e^{-u}\, du\Big|_{\alpha=0}$$

$$= e \frac{\partial^n}{\partial \alpha^n} \left[\Gamma(\alpha) - \sum_{s=0}^{\infty} \frac{(-1)^s}{(s+\alpha)s!} \right]_{\alpha=0}$$

$$= n!\, e\left\{ S(n+1) + (-1)^{n+1} \sum_{s=1}^{\infty} \frac{(-1)^s}{s!\, s^{n+1}} \right\},$$

where $S(n)$ are the well-tabulated coefficients in the series

$$\Gamma(z+1) = 1 + S(1)z + S(2)z^2 + \cdots.$$

Taking $N = 40$ in (78)–(79) provides the results in Table 7.4.

The same level of significance is maintained through $\sigma(40)$, as seems to be typical of the Miller algorithm when it is successful. What makes the

Table 7.4   Stieltjes moments.

n	Approx. $\sigma(n)$	
0	0.596 347 362 323 194	⎫
1	0.265 965 385 032 409	Accurate to
2	0.193 560 650 277 724	all places
3	0.180 562 883 493 93$\underline{5}$	
4	0.197 517 611 937 41$\underline{2}$	

method useful in this case is that numerical quadrature schemes to compute $\sigma(n)$ become numerically hazardous with increasing n. ∎

Example 7.15   Compute the modified Bessel function $K_0(x)$, $x>0$.
I will use the representation

$$K_0(x) = \int_0^\infty e^{-x\cosh t}\, dt.$$

Let

$$\sigma(n):= \int_0^\infty e^{-x\cosh t} t^{2n}\, dt.$$

Then $\sigma(0) = K_0(x)$ and a normalization is furnished by

$$K_{1/2}(x) = \sqrt{\left(\frac{\pi}{2x}\right)}e^{-x} = \int_0^\infty e^{-x\cosh t}\cosh\frac{t}{2}\, dt$$

$$= \int_0^\infty e^{-x\cosh t} \sum_{k=0}^\infty \frac{(t/2)^{2k}}{(2k)!}\, dt = \sum_{k=0}^\infty \frac{\sigma(k)}{2^{2k}(2k)!}.$$

Integration by parts in the integral for $K_0(x)$ gives the difference equation of infinite order

$$\sigma(n) = \frac{x}{(2n+1)} \sum_{k=0}^\infty \frac{\sigma(n+k+1)}{(2k+1)!}.$$

Table 7.5   Approximations to $K_0(x)$.

N	$x=1$	$x=2$	$x=4.5$
8	0.421 024 $\underline{3}$04 686 283	0.034 739 504 $\underline{6}$70 305	0.006 399 857 2$\underline{5}$0 455
12	0.421 024 438 $\underline{7}$00 743	0.034 739 504 386 $\underline{3}$88	0.006 399 857 243 23$\underline{8}$
16	0.421 024 438 $\underline{2}3\underline{7}$ 208	0.034 739 504 386 279	0.006 399 857 243 234
20	0.421 024 438 240 7$\underline{4}$2	0.034 739 504 386 279	0.006 399 857 243 234
24	0.421 024 438 240 708	0.034 739 504 386 279	0.006 399 857 243 234

As in the previous example, this equation provides the desired Miller algorithm.

Table 7.5 shows approximate values of $K_0(x)$ for a range of values of N and x.

Obviously the accuracy increases with increasing x but even for x rather small it is still good. For $N = 8$, $x = 1$ the computed value of $K_0(x)$ is good to 6 S.F. ■

It is interesting that when $P(t)$ is *not* entire convergence apparently can be selective; in other words, certain of the $\sigma_N(n)$ may converge as $N \to \infty$ and others not. For instance, let

$$g(s) = \int_0^\infty e^{-P(t)} t^s \, dt, \qquad P(t) = \ln \left(\frac{e^t - 1}{t} \right),$$

and suppose the computation of the Riemann ζ-function is attempted using this integral. The connecting formula is

$$g(s) = \Gamma(s+2)\zeta(s+2).$$

Taking $\sigma(n) = g(s+n)$, I find the appropriate difference equation to be

$$\sigma(n) = \frac{1}{(n+s+1)} \sum_{k=0}^\infty \alpha(k)\sigma(n+k+1),$$
$$\alpha(k) = B_{k+1}(1)/(k+1)!.$$

Several normalizations are possible. To simplify the discussion I assume $\sigma(0) = 1$. In Table 7.6 I have taken $s = 2.5$ but the behavior displayed seems typical of all s. Apparently the $\sigma_N(n)$ with even n converge (and rather rapidly) while those with n odd diverge. The reasons for this strange phenomenon remain to be explored.

Table 7.6 ζ-Function computations. $s = 2.5$

n/N	10	15	$\sigma_N(n)$ 20	25	30
0	1	1	1	1	1
1	−0.819 095 11	17.589 134	−1.742 327 4	27.608 493	−1.211 808 1
2	−20.483 743	−38.715 562	−39.479 229	−39.478 427	−39.778 418
3	131.244 6	−690.525 46	68.781 262	−1 089.939 8	47.840 267
4	1 445.676 2	1 554.598 7	1 558.568 5	1 558.546 7	1 558.545 5
5	−538.793 64	27 468.045	−2 715.656 1	43 029.125	−1 888.658 1

8 Higher-order systems (continued): the nonhomogeneous case

8.1 Introduction

Suppose we wish to compute a solution of the equation

$$\mathcal{L}[y(n)]:= \sum_{\nu=0}^{\sigma} A_\nu(n)y(n+\nu) = f(n), \tag{8.1}$$

by, say, using the equation in the backward direction. One approach would be to convert the equation into a $(\sigma+1)$th order *homogeneous equation* by writing

$$\frac{\mathcal{L}[y(n+1)]}{f(n+1)} - \frac{\mathcal{L}[y(n)]}{f(n)} = 0.$$

The σ linearly independent solutions of the homogeneous equation $\mathcal{L}[y(n)]=0$ and any constant multiple of a particular solution of (1) are all solutions of this equation. Thus the Miller algorithm succeeds when applied to (1) only when the adjoint solution corresponding to the desired solution dominates all the other solutions in some fundamental set for the adjoint. Informally (but sometimes inaccurately, as explained in Section 7.2) we may say the Miller algorithm succeeds when the desired solution of (1) is 'small' compared to the members of any fundamental set $\mathcal{S}$ for the homogeneous equation.

However, there may be a distinct advantage in computing with the nonhomogeneous equation since the solution so computed need not be small compared to the members of $\mathcal{S}$. Usually in fact the required particular solution of a nonhomogeneous equation has asymptotic properties intermediate to those of the members of $\mathcal{S}$.

Example 8.1 Here $F(x)$ is the incomplete Laplace transform of $I_\nu(x)$,

$$F(x):= 2^\nu(\mu+\nu+1)\Gamma(\nu+1)x^{-(\mu+\nu+1)}e^{(a-1)x}\int_0^x e^{-at}t^\mu I_\nu(t)\,\mathrm{d}t. \tag{8.2}$$

We wish to determine the coefficients $E(n)$ for the expansion of this function in shifted Chebyshev polynomials,

$$F(x) = \sum_{n=0}^{\infty} E(n) T_n^* \left(\frac{x}{z} \right), \qquad z \neq 0, \tag{8.3}$$

where z is a range parameter and x lies on the segment $[0, z]$.

We will need the related Chebyshev expansion of I_ν,

$$I_\nu(x) = \frac{(x/2)^\nu e^x}{\Gamma(\nu+1)} \sum_{n=0}^{\infty} C(n) T_n^* \left(\frac{x}{z} \right). \tag{8.4}$$

The $C(n)$ may be determined explicitly in terms of hypergeometric functions of the form $_2F_2$; see Wimp and Luke (1969).

$E(n)$ satisfies the nonhomogeneous equation

$$\frac{2}{\varepsilon_n} y(n) + \frac{1}{y}(n+\mu+\nu+2+y) y(n+1) + \frac{1}{y}(n-\mu-\nu+1-y) y(n+2)$$

$$- y(n+3)$$

$$= \frac{(\mu+\nu+1)}{y} [C(n+1) - C(n+2)], \qquad n \geq 0, \tag{8.5}$$

where

$$y := \frac{z(1-a)}{4}.$$

There is a fundamental set for the related homogeneous equation having the behavior

$$y^{(h)}(n) = \lambda_h^n n^{(\mu_h n + \theta_h)} [1 + O(n^{-1})], \tag{8.6}$$

where

$$\lambda_1 = 4/ez(1-a), \qquad \mu_1 = 1; \qquad \theta_1 = -\mu - \nu - \tfrac{3}{2};$$
$$\lambda_2 = -1, \qquad \mu_2 = 0; \qquad \theta_2 = 2\mu + 2\nu + 1;$$
$$\lambda_3 = -ez(1-a)/4, \qquad \mu_3 = -1; \qquad \theta_3 = -\mu - \nu - \tfrac{3}{2}.$$

However, the required solution $E(n) := y^{(0)}(n)$ has the behavior (6) multiplied by a constant c_0 with

$$\lambda_0 = -ez/2, \quad \mu_0 = -1, \quad \theta_0 = -\nu - 2,$$

$$c_0 = \frac{e^{-z} 2^{2\nu+3/2} \Gamma(\nu+1)(\mu+\nu+1)}{\pi(a+1)}.$$

Depending on the value of a, $E(n)$ and $y^{(3)}(n)$ will compete for being the smallest solution of the fourth-order homogeneous equation formed from (5) and thus the convergence of the Miller algorithm will be lackadaisical at best and nonexistent at worst. ∎

The methods to be discussed in this chapter, however, can be used to compute intermediate solutions such as $E(n)$ very efficiently.

8.2   The Wimp–Luke method

This algorithm is as follows. Let $w(n)$ be the desired solution of the equation and let the normalization condition

$$S := \sum_{k=0}^{\infty} c(k)w(k),$$

be given.

For $N \gg 1$ generate *two* sequences, $\{y_N(n)\}$, $\{z_N(n)\}$ for $0 \leqslant n \leqslant N + \sigma - 1$ by putting

$$
\begin{aligned}
y_N(N+j) &:= \delta_{0j}, & 0 \leqslant j \leqslant \sigma - 1; \\
\mathcal{L}\{y_N(n)\} &= 0, & n = N-1, N-2, \ldots, 0;
\end{aligned}
\tag{8.7}
$$

$$
\begin{aligned}
z_N(N+j) &:= 0, & 0 \leqslant j \leqslant \sigma - 1; \\
\mathcal{L}\{z_N(n)\} &= f(n), & n = N-1, N-2, \ldots, 0.
\end{aligned}
\tag{8.8}
$$

Define

$$w_N(n) := \left(\frac{S - T_N}{S_N}\right) y_N(n) + z_N(n), \qquad 0 \leqslant n \leqslant N + \sigma - 1, \tag{8.9}$$

where

$$T_N := \sum_{k=0}^{N} c(k) z_N(k), \qquad S_N := \sum_{k=0}^{N} c(k) y_N(k).$$

Then, hopefully,

$$\lim_{N \to \infty} w_N(n) = w(n), \qquad n \geqslant 0. \tag{8.10}$$

If this is so we say the Wimp–Luke algorithm converges (to $w(n)$).

Necessary and sufficient conditions for the convergence of the algorithm can be described simply. Let $\{y^{(h)}(n)\}$ be a fundamental set for the homogeneous equation and

$$S_N^{(h)} := \sum_{k=0}^{N} c(k) y^{(h)}(k),$$

$$R_N := S - \sum_{k=0}^{N} c(k) w(k).$$

Let $c_i \equiv c_i(N)$ be defined by the following system of equations:

$$-w(N+j) = c_1 y^{(1)}(N+j) + \cdots + c_\sigma y^{(\sigma)}(N+j), \qquad 1 \leq j \leq \sigma - 1,$$
$$R_N = c_1 S_N^{(1)} + \cdots + c_\sigma S_N^{(\sigma)}. \tag{8.11}$$

Theorem 8.1 *Let c_i be defined as in equation (11).*
Then the Wimp–Luke algorithm converges to $w(n)$ if and only if

$$\lim_{N \to \infty} c_i(N) = 0, \tag{8.12}$$

for some basis $\{y^{(h)}(n)\}$ of $\mathcal{L}[y] = 0$.

Proof We may write

$$w_N(n) = c_1 y^{(1)}(n) + \cdots + c_\sigma y^{(\sigma)}(n) + w(n).$$

Putting in the boundary conditions and normalization relationship gives the system of equations (11), and the necessity and sufficiency of the condition (12) is clear. ∎

Given enough asymptotic information about $y^{(h)}(n)$, $w(n)$, the conditions (12) are usually easy enough to verify in any particular case, although the computations may be tedious. Alternative conditions can be given which involve the growth of solutions of a nonhomogeneous form of the adjoint equation, but it does not seem useful to do this.

It can be shown for the previous example that

$$w_N(n) = w(n) + O\left(\left(\frac{ez}{2}\right)^N N^{\theta - N}\right),$$

for some θ. Although the algorithm possesses excellent theoretical properties, its application can give rise to rather serious numerical instabilities. For the case in point the nature of this is made clear by the easily derived formulas,

$$z_N(n) = w(n) + K_1 \left(\frac{2}{1-a}\right)^{N+1} N^{\mu - 1/2} [1 + O(N^{-1})], \qquad N \to \infty,$$

$$y_N(n) = K_2 \left(\frac{-4N}{ez(1-a)}\right)^N N^{\mu + \nu + 3/2} [1 + O(N^{-1})], \qquad N \to \infty,$$

where K_1, K_2 depend on n.

For certain ranges of values of a ($-1 < a < 3$) we are in the position of having to determine a linear combination of two large quantities which will be small. It is clear that for these values of a there will be an appreciable loss of significant figures.

To illustrate this phenomenon in general, let us define as a measure of

Table 8.1

a	ω	ω^*
0	4.05×10^{-8}	$\leqslant 1.00 \times 10^{-15}$ (1 iteration)
0.1	3.92×10^{-7}	$\leqslant 1.00 \times 10^{-15}$ (1 iteration)
0.2	4.16×10^{-5}	$\leqslant 1.00 \times 10^{-15}$ (1 iteration)
0.3	1.74×10^{-3}	$\leqslant 1.00 \times 10^{-15}$ (1 iteration)
0.4	1.45×10^{-1}	$\leqslant 1.00 \times 10^{-15}$ (1 iteration)
0.5	4.90×10	$\leqslant 1.00 \times 10^{-15}$ (2 iterations)
0.6	2.45×10^{4}	$\leqslant 4.00 \times 10^{-14}$ (2 iterations)
0.7	7.54×10^{7}	$\leqslant 1.09 \times 10^{-10}$ (2 iterations)
0.8	7.78×10^{14}	$\leqslant 4.22 \times 10^{-5}$ (3 iterations)

the loss of significance the quantity

$$\omega := \left| S - \sum_{k=0}^{N} c(k) w_N(k) \right|.$$

If all computations were done with infinite precision ω would be 0.

In the example, when 16 places are carried in the computations and $N = 32$, even in the most favorable case it is likely that none of the $w(n)$ computed by the algorithm is accurate to more than 8 places.

Usually a tremendous improvement in the efficiency of the algorithm can be made by *iterating*. One uses the values

$$w_N(N), w_N(N-1), \ldots, w_N(N-\sigma+1)$$

just computed as starting values

$$z_{N'}(N'+\sigma-1), z_{N'}(N'+\sigma-2), \ldots, z_{N'}(N')$$

rather than 0s for a new sequence $z_{N'}(n), N' = N - \sigma$. This procedure often serves to reduce considerably the size of the elements in the sequence $z_N(n)$ and hence to mitigate the loss of significance caused by subtracting large numbers. This procedure can be repeated until no further reduction in ω occurs.

Let ω^* be the smallest value of ω so obtained. Table 8.1 illustrates the improvement possible in this example and the number of iterations needed. In all of the data, $z = 8$, $\mu = \nu = 0$, $N = 32$.

8.3 The Lozier algorithm

Before I state the algorithm I will formulate the problem in terms of the solution of an equation involving a linear operator $\mathcal{D}$ acting on the space $\mathscr{C}_S$ of complex sequences.

Let $\mathcal{D}_\sigma$ denote the set of all infinite upper triangular band matrices of the form

$$\begin{bmatrix} d_0(0) & \cdot & \cdot & \cdot & d_\sigma(0) & & \\ & d_0(1) & \cdot & & \cdot & d_\sigma(1) & (0) \\ & & \cdot & & & & \cdot \\ & & & \cdot & & & \\ & & & & \cdot & & \\ & & d_0(n) & & \cdot & & \cdot & d_\sigma(n) \\ & (0) & & & & \cdot & \\ & & & & & & \cdot \end{bmatrix},$$

where $d_j \in \mathcal{C}_S$ and $d_0(n), d_\sigma(n) \neq 0$ for infinitely many values of n. It is convenient to use the concise notation $\langle d_0, d_1, \ldots, d_\sigma \rangle := D$ for elements of $\mathcal{D}_\sigma$. Each $D \in \mathcal{D}_\sigma$ is a linear operator $D : \mathcal{C}_S \to \mathcal{C}_S$ when we define Dx by ordinary matrix multiplication. In fact $\mathcal{D}_\sigma$ is the class of linear difference operators of order σ on $\mathcal{C}_S$.

If d_0, d_σ have no terms equal to 0, we shall say the associated operator D is *nonsingular*. d_0 and d_σ are called the *leading* and *trailing coefficients* of D respectively.

The notation x^i will indicate a sequence whose first term is $x(i)$,

$$x^i := (x(i), x(i+1), \ldots),$$

and $x^{i,r}$ will denote the finite subsequence of x^i whose final term is $x(r)$,

$$x^{i,r} := (x(i), x(i+1), \ldots, x(r)).$$

The point i is called the *initial point*, r the *terminal point* of the sequence. Similarly, D^i will indicate that i is the initial point of each of the $\sigma + 1$ sequences used to define the operator $D^i := \langle d_0^i, d_1^i, \ldots, d_\sigma^i \rangle$. When the superscript is omitted it is understood that the initial points are 0 unless the context dictates otherwise. D_j^i denotes the matrix obtained from D^i by deleting the first j columns, $0 \leqslant j \leqslant \sigma$, and $D_j^{i,r}$ the leading principal matrix of order $r - i + 1$ of D_j^i as below:

$$D_j^{i,r} := \begin{bmatrix} d_j(i) & \cdot & \cdot & \cdot & d_\sigma(i) & & & \\ \vdots & & & & & \cdot & & \\ d_0(i+j) & & & d_j(i+j) & & & d_\sigma(i+j) & \\ & \cdot & & \vdots & & & & \cdot \\ & & d_0(m) & & & d_j(m) & & d_\sigma(m) \\ & & & \cdot & & & \cdot & \vdots \\ & & & d_0(r-k) & & & d_j(r-k) & d_\sigma(r-k) \\ & & & & \cdot & & & \vdots \\ & & & & & d_0(r) & \cdot & \cdot & \cdot & d_j(r) \end{bmatrix}$$

The general nonhomogeneous linear difference equation

$$\sum_{k=0}^{\sigma} d_k(n)y(n+k) = f(n). \tag{8.13}$$

can now be written $Dy = f$ (where $f \in \mathscr{C}_s$, $D \in \mathscr{D}_\sigma$ are given) and is equivalent to the infinite algebraic system $Dy(n) = f(n)$, $n \geq 0$. Of course auxiliary conditions are required to uniquely specify y. In this treatment those conditions will consist of specifying two finite subsequences

$$y^{i,i+j-1}, \qquad y^{r,r+k-1},$$

where $j + k = \sigma$, $j, k \geq 0$, or possibly only one of these. When $r \geq i + j$ these are called *initial* and *terminal conditions* respectively.

To get an idea of what sort of computations will be involved in establishing an algorithm, let us consider the following finite boundary-value problem:

Find $y^{i+j,r-1}$ such that

$$Dy(n) = f(n), \qquad n = i, i+1, \ldots, r-j-1, \tag{8.14}$$

with initial and terminal conditions respectively

$$y(i+n) = \alpha(n), \qquad n = 0, 1, \ldots, j-1, \tag{8.15}$$

$$y(r+n) = \beta(n), \qquad n = 0, 1, \ldots, k-1, \tag{8.16}$$

$$k = \sigma - j, \qquad i \geq 0, \qquad r \geq i + j.$$

This means the vectors

$$y^{i,i+j-1}, \qquad y^{r,r+k-1}, \tag{8.17}$$

are known. Note that if $j = 0$, then the first of these is empty, if $k = 0$ the second. The first corresponds to the absence of (15), the second to the absence of (16).

The desired subsequence $y^{i+j,r-1}$ satisfies a system of linear equations which may be expressed in matrix form using our previous notation as

$$D_j^{i,r-i-1}y^{i+j,r-1} = f^{i,r-j-1} - \begin{bmatrix} D_0^{i,i+j-1}y^{i,i+j-1} \\ \cdots\cdots\cdots\cdots\cdots \\ 0 \end{bmatrix}$$

$$- \begin{bmatrix} 0 \\ \cdots\cdots\cdots\cdots\cdots \\ D_\sigma^{r-\sigma,r-j-1}y^{r,r+k-1} \end{bmatrix}, \tag{8.18}$$

$$k := \sigma - j, \qquad r \geq i + j, \qquad i \geq 0,$$

provided

$$r - i - j \geq \max(j, k). \tag{8.19}$$

The partitioned column vector comprising the first matrix on the right-hand side has j entries in the upper part and that of the second matrix k entries in the lower part. The length of the other column vectors in the equation is $r-i-j$ and this gives rise to the restriction (19). (This is really irrelevant for the algorithm since we shall be considering the case $r \to \infty$.)

The solution of the original boundary-value problem is thus equivalent to the solution of the linear system (18) which has a unique solution iff $D_j^{i,r-i-1}$ is nonsingular. If in addition D itself is nonsingular, then the required solution $y^{i,r+k-1}$ extends uniquely to $y \in \mathscr{C}_S$ by forward and backward recurrence. (Nonsingularity, however, is not a necessary condition for this extension to exist. For example, some of the leading and trailing coefficients of D appearing in the submatrix $D_j^{i,r-i-1}$ could be zero.)

First let's look at the nonsingular triangular cases of (18). One occurs when $j = \sigma$ and then (18) assumes the form

$$D_\sigma^{i,r-\sigma-1} y^{i+\sigma,r-1} = f^{i,r-\sigma-1} - \begin{bmatrix} D_0^{i,i+\sigma-1} y^{i,i+\sigma-1} \\ \cdots\cdots\cdots\cdots\cdots \\ 0 \end{bmatrix}. \tag{8.20}$$

Writing these equations out gives

$$d_\sigma(i)y(i+\sigma) = f(i) - \sum_{s=0}^{\sigma-1} d_s(i)y(i+s),$$

$$d_{\sigma-1}(i+1)y(i+\sigma) + d_\sigma(i+1)y(i+\sigma+1) = f(i+1)$$

$$\begin{matrix} \cdot \\ \cdot \\ \cdot \end{matrix} \qquad - \sum_{s=0}^{\sigma-2} d_s(i+1)y(i+s+1),$$

$$\sum_{s=0}^{\sigma} d_s(i+\sigma)y(i+\sigma+s) = g(i+\sigma),$$

$$\begin{matrix} \cdot \\ \cdot \\ \cdot \end{matrix}$$

$$\sum_{s=0}^{\sigma} d_s(r-\sigma-1)y(r-\sigma-1+s) = g(r-\sigma-1).$$

$$\tag{8.21}$$

This system is lower triangular and thus can be solved by determining $y(i+\sigma)$ from the first equation (recall, $y^{i,i+\sigma-1}$ is known), $y(i+\sigma+1)$ from the second, etc. This is obviously equivalent to finding $y(i+\sigma)$, $\ldots, y(r-1)$ from (14) by forward recurrence starting with the initial values (15).

The second triangular case occurs when $j = 0$, and then

$$D_\sigma^{i,r-1} y^{i,r-1} = f^{i,r-1} - \begin{bmatrix} 0 \\ \cdots\cdots\cdots\cdots\cdots \\ D_\sigma^{r-\sigma,r-1} y^{r,r+\sigma-1} \end{bmatrix}. \tag{8.22}$$

Writing these equations out gives

$$\sum_{s=0}^{\sigma} d_s(i)y(i+s) = f(i),$$

$$\vdots$$

$$\sum_{s=0}^{\sigma} d_s(r-\sigma-1)y(r-\sigma-1+s) = f(r-\sigma-1),$$

$$\vdots$$

$$d_0(r-2)y(r-2) + d_1(r-2)y(r-1) = f(r-2) - \sum_{s=2}^{\sigma} d_s(r-2)y(r-2+s),$$

$$d_0(r-1)y(r-1) = f(r-1) - \sum_{s=1}^{\sigma} d_s(r-1)y(r-1+s).$$

$$(8.23)$$

This system is upper triangular and can be solved by determining $y(r-1)$ from the last equation, $y(r-2)$ from the next to the last equation, etc. This is equivalent to computing $y(r-1), y(r-2), \ldots, y(i)$ from (14) by backward recurrence starting with the terminal values (16).

Of course the intent is to solve the nontriangular case of (18) where $0 < j < \sigma$. Our goal is to determine a method to solve the system by performing, in effect, a forward recurrence of order j followed by a backward recurrence of order $\sigma - j$.

To explore the possibility of doing this will take a number of preliminary definitions and theorems.

Definition 8.1 *Let the boundary-value problem* (14)–(16), *or equivalently,* (15), (16), (18), *be nonsingular. The problem is said to be* factorizable *if there exist difference operators* $A^i := \langle a_0^i, a_1^i, \ldots, a_j^i \rangle$ *of order j and* $B^{i+j} := \langle b_0^{i+j}, b_1^{i+j}, \ldots, b_k^{i+j} \rangle$ *of order $k := \sigma - j$ such that*

$$D_j^{i,r-i-1} = A_j^{i,r-i-1} B_0^{i+j,r-1}. \tag{8.24}$$

If in addition the sequences a_j^i and b_0^{i+j} are free from zeros and the infinite matrix factorization,

$$D_j^i = A_j^i B^{i+j}, \tag{8.25}$$

is valid we say the difference operator D is (i, j)-factorizable. ∎

Note that $D_j^{i,r-i-1}$ and D_j^i are band matrices each having a total bandwidth $\sigma + 1$, lower bandwidth j, and upper bandwidth k. A nonsingular factorizable boundary-value problem is one for which a finite matrix factorization of $D_j^{i,r-i-1}$ exists with the left factor lower triangular with total bandwidth $j+1$ and the right factor upper triangular with total bandwidth $k+1 := \sigma + 1 - j$. Obviously the operators A^i, B^{i+j} may be arbitrary, provided only that they satisfy (24) as far as the original problem is concerned.

Lozier (1980) establishes the three following results:

Theorem 8.2 *Let D be (i, j)-factorizable. Then for every $r \geq i + j + \max(j, \sigma - j)$ the boundary-value problem (14)–(16) is nonsingular and factorizable.* ∎

Theorem 8.3 *Let (14)–(16) be nonsingular and factorizable. If A^i, B^{i+j} are difference operators of order j and $k = \sigma - j$ respectively such that (24) is valid, then the solution of (14)–(16) is identical with the solution of*

$$B_0^{i+j,r-1} y^{i+j,r-1} = z^{i+j,r-1} - \begin{bmatrix} 0 \\ \cdots\cdots\cdots\cdots\cdots\cdots\cdots\cdots\cdots \\ (A_j^{r-\sigma,r-i-1})^{-1} D_\sigma^{r-\sigma,r-i-1} y^{r,r+k-1} \end{bmatrix}, \quad (8.26)$$

where $z^{i+j,r-1}$ is the solution of

$$A_j^{i,r-i-1} z^{i+j,r-1} = f^{i,r-i-1} - \begin{bmatrix} D_0^{i,i+j-1} y^{i,i+j-1} \\ \cdots\cdots\cdots\cdots\cdots\cdots \\ 0 \end{bmatrix}. \quad ∎ \qquad (8.27)$$

It is important to note that if in addition D is (i, j)-factorizable in the form (25), then

$$D_\sigma^{r-\sigma,r-i-1} = A_j^{r-\sigma,r-i-1} B_k^{r-k,r-1}, \qquad (8.28)$$

and the algorithm simplifies; i.e., $B_k^{r-k,r-1}$ may be used for the product of the two matrices on the right-hand side of (26).

When is an operator factorizable? The following provides one answer.

Theorem 8.4

(i) *Let (14)–(16) be nonsingular. Then it is factorizable if every leading principal submatrix of $D_j^{i,r-i-1}$ is nonsingular.*

(ii) *D is (i, j)-factorizable if every leading principal submatrix of D_j^i is nonsingular. Furthermore, this condition is necessary if D is nonsingular.*

Proofs Of these theorems I shall prove only the last since the proof shows how (i, j)-factorizations may be constructed by Gaussian elimination in natural order, i.e., without row or column interchanges. The product (25) of course is nothing more than the LU decomposition of D_j^i.

Since the proof of (i) is contained in the first $r - i - j$ stages of the proof of (ii) we prove only (ii), starting with sufficiency.

The first row of D_j^i is used to annihilate every nonzero element except the first in the first column. This affects only the j rows immediately following the first. Also the trailing coefficient in each row remains unchanged. The multipliers used in these annihilations go into positions 2 through $j+1$ of the first column of the lower triangular factor A_j^i. The first element is 1 and every other element of the first column of A_j^i is 0. This completes the first stage.

Now suppose we have completed the $(n-1)$st stage. Let $\bar{D}_j^i$ denote the transformation of the original D_j^i at this point. The leading principal submatrix $\bar{D}_j^i$ of order n is upper triangular. Therefore the pivotal element is $\neq 0$, and the nth row of $\bar{D}_j^i$ may be used to annihilate every nonzero element below the nth element in the nth column. Only the j rows immediately following the nth row are affected. The trailing coefficient in each of these rows is not changed. The multipliers at this stage go into positions $n+1$ through $n+j$ of the nth column of A_j^i; 1 goes into the nth position, 0 into every other position. This completes the nth stage.

Let B^{i+j} be the infinite upper triangular matrix which results by performing this sequence of operations on D_j^i. By construction, both A_j^i and B^{i+j} have the bandwidths required for (i, j)-factorizability. Also, formally multiplying A_j^i on the right by B^{i+j} gives D_j^i (see (25)).

To prove the necessity condition in part (ii) note that the factorization has the property that leading and trailing coefficients of both A_j^i and B^{i+j} are nonzero if D is nonsingular. Therefore no row nor column interchanges could be made without widening the bandwidth. Thus Gaussian elimination in natural order is the only possibility for arriving at the factorization, which is possible only if every leading principal minor is nonzero. This completes the proof. ∎

Note that a unique (i, j)-factorization does not exist since the rows may be scaled differently. This would result in the appearance of entries on the main diagonal of A_j^i that are other than 1. In fact the scaling used in the (0, 1)-factorization for the second-order equation (see (25)) apparently gives an algorithm different from Olver's. However, a computation gives $\sigma = 2$, $j = k = 1$, $i = 0$, and

$$A_1 = \begin{bmatrix} 1 & 0 & 0 & \cdots \\ 1/a(0) & 1 & 0 & \cdots \\ 0 & 1/\alpha(1) & 1 & \cdots \\ 0 & 0 & 1/\alpha(2) & \cdots \end{bmatrix},$$

$$B^1 = \begin{bmatrix} a(0) & b(0) & 0 & 0 & \cdots \\ 0 & \alpha(1) & b(1) & 0 & \cdots \\ 0 & 0 & \alpha(2) & b(2) & \cdots \\ 0 & 0 & 0 & \alpha(3) & \cdots \end{bmatrix},$$

where $\alpha(k)$ satisfies

$$\alpha(k) = a(k) - \frac{b(k-1)}{\alpha(k-1)}, \qquad k = 1, 2, \dots, \quad \alpha(0) = a(0),$$

and the corresponding recurrences for z, y are given by (26), (27). The identification

$$z(n) = -\frac{b(n-1)}{y(n)} e(n), \qquad \alpha(n) = -b(n)\frac{y(n+2)}{y(n+1)},$$

$e(n), y(n), b(n)$ as defined in Section 6.2 reveals that the algorithm is nothing more than Olver's algorithm. In fact choosing a different scaling may confer certain computational advantages, but this has not yet been thoroughly investigated.

Thus Olver's algorithm can be generalized to the system $Dy = f$ by posing the problem as a boundary-value problem of the form (14) with initial conditions (15) and terminal conditions of the form

$$y_r(r) = y_r(r+1) = \cdots = y_r(r+k-1) = 0, \qquad k := \sigma - j.$$

Its solution, y_r, will be an approximation to y over some finite subsequence $y^{i,m}$ and the algorithm will consist of the following computations:

(i)   the factorization of D_j^i;
(ii) the solution of the system (27) which is accomplished by solving the system in the forward direction;
(iii) the solution of the system (26) which is done in the backward direction.

One question is what sorts of conditions can be placed on the desired solution and on the system which will assure convergence of the algorithm as $r \to \infty$? To explore this it is necessary to establish a classification for the growth properties of the solutions of such systems. Recall that a sequence $\{y(n)\}$ *dominates* a sequence $\{z(n)\}$ if $z(n)/y(n) \to 0$.

Let $D \in \mathcal{D}_\sigma$. The kernel of D is $\mathcal{K}(D) := \{x \in \mathcal{C}_S \mid Dx = 0\}$. Assume that $\mathcal{K}(D)$ has dimension σ so that there is a fundamental set $\langle y^{(1)}, y^{(2)}, \dots, y^{(\sigma)} \rangle$ for $\mathcal{K}$. A sufficient condition for this, as follows from the classical theory of difference equations, is that D be nonsingular. We will call $\{y^{(h)}\}$ a *basis for* D.

Definition 8.2 D *is totally separable (as $n \to \infty$) if there is a basis $\{y^{(h)}\}$ for D such that $y^{(s+1)}$ dominates $y^{(s)}$, $1 \leqslant s \leqslant \sigma$.*
   If

$$|y^{(i)}(n+1)/y^{(i)}(n)| > |y^{(j)}(n+1)/y^{(j)}(n)|, \qquad n > n_0, \quad i > j, \tag{8.29}$$

then $\{y^{(h)}\}$ is said to be totally ranked. ∎

Remarks   These bases are not in general unique and neither property implies the other. The concept (29), due to Oliver (1968a), is equivalent to requiring that ultimately $|y^{(i)}/y^{(j)}|$, $j > i$, be monotone decreasing, but the limit may be > 0.

It can be shown (Lozier, 1980) that every totally ranked basis generates the same subspaces of $\mathcal{K}(D)$ and these satisfy $\mathcal{K}_1 \subset \mathcal{K}_2 \subset \cdots \subset \mathcal{K}_\sigma = \mathcal{K}(D)$ with proper set inclusion at each stage.

We are interested in classifying solutions not only of $Dy = 0$ but also of $Dy = f$. The following allows us to do this.

Definition 8.3   *Let D be totally separable with totally ranked basis $\{y^{(h)}\}$. Then y is said to be of type s, where $0 < s < \sigma$, if*

$$\lim_{n \to \infty} y(n)/y^{(s+1)}(n) = 0, \tag{8.30}$$

and

$$0 < \overline{\lim_{n \to \infty}} \, |y(n)/y^{(s)}(n)| \le \infty. \tag{8.31}$$

If (30) is true for $s = 0$, type $(y) = 0$, and if (31) is true for $s = \sigma$, then type $(y) = \sigma$. ∎

Example 8.2   Consider the solutions in Example 8.1, $a \ne \pm 1$. D is totally ranked and totally separable and

$$\text{type } (E(n)) = \begin{cases} 1, & \left| \dfrac{1-a}{2} \right| < 1; \\[2mm] 0, & \left| \dfrac{1-a}{2} \right| > 1. \end{cases}$$

Type (E) is not defined when $a = 1$, the equation then reducing to one of the second order. ∎

Obviously type (y) is independent of the basis used. Thus without ambiguity we can define type (y) with respect to D.

Not every operator with constant coefficients is totally separable and it is sometimes desirable to introduce a classification system which includes all constant-coefficient operators. Lozier calls such operators *separable* but I shall not develop that idea here.

Theorem 8.5   *Let y be a solution of $Dy = f$ where D has a totally ranked basis $\{y^{(h)}\}$. Let $j = $ type (y). Then y is uniquely determined by its values $y(i), y(i+1), \ldots, y(i+j-1)$ provided that $i \ge 0$ is a point at which the leading principal minor of the Casorati determinant $D(i)$ (see (A.2)) is nonzero.*

Proof I omit the proof which is straightforward but tedious. See the cited reference. ∎

This theorem clarifies the number of initial values needed in the approximating boundary-value problem to specify y, namely, j values. I will now derive the algorithm for such a case.

Let $j \geqslant \text{type } w$, where w is the desired solution of $Dy = f$. Since $y_r^{i,i+j-1} = w^{i,i+j-1}$, $y_r^{r,r+k-1} = 0$, the linear system to be solved is

$$D_j^{i,r-j-1} y_r^{i+j,r-1} = f^{i,r-j-1} - \left[\begin{matrix} D_0^{i,i+j-1} y^{i,i+j-1} \\ \cdots\cdots\cdots\cdots \\ 0 \end{matrix} \right]. \tag{8.32}$$

The value of r is to be determined during the course of the computations. (We saw in Chapter 6 that this was one of the strengths of the Olver algorithm.)

In view of the previous theorem we find w is the unique solution of the infinite system

$$A_j^i B^{i+j} y^{i+j} = f^i - \left[\begin{matrix} D_0^{i,i+j-1} y^{i,i+j-1} \\ \cdots\cdots\cdots\cdots \\ 0 \end{matrix} \right]. \tag{8.33}$$

Now assume that D is (i, j)-factorizable. (This is not a crippling assumption; in practical situations it is nearly always the case, or the difference equation can be modified so that it is true.) Then the system may be triangularized by Gaussian elimination, so the solution of (32) may be obtained as the solution of

$$B_0^{i+j,r-1} y_r^{i+j,r-1} = z^{i+j,r-1}, \tag{8.34}$$

where $z^{i+j,r-1}$ is a finite subsequence of the solution of

$$A_j^i z^{i+j} = f^i - \left[\begin{matrix} D_0^{i,i+j-1} y^{i,i+j-1} \\ \cdots\cdots\cdots\cdots \\ 0 \end{matrix} \right]. \tag{8.35}$$

Comparison of (35) and (33) with respect to the identity $D_j^i = A_j^i B^{i+j}$ shows that

$$B^{i+j} y^{i+j} = z^{i+j}.$$

Thus (34) merely represents the finite system that results from truncating (33) at the $(r-i-j)$ row and column. As mentioned before, the algorithm proceeds in three stages:

(i)   factorization of D_j^i to get A_j^i and B^{i+j};
(ii) *forward elimination* using (35);
(iii) *back substitution* using (34).

The forward elimination may be continued as far as desired. The back substitution is equivalent to backward recurrence and continues until the initial point i is reached.

This computational scheme constitutes *Lozier's algorithm*.
Lozier gives the following stability theorem.

Theorem 8.6   *Let D have totally ranked basis $\{y^{(h)}\}$. Let D be (i, j)-factorizable $i \geq 0$, $0 \leq j \leq \sigma - 1$, and let the leading principal minor of $D(i)$ be nonzero. If $D_j^i = A_j^i B^{i+j}$, then B^{i+j} is totally separable. Furthermore if w has type j or less with respect to D, then z^{i+j} has type zero with respect to B^{i+j}.* ∎

The implication of this theorem is that when $j \geq$ type (w), the back substitution computation of the algorithm is stable, at least for r in a sufficiently large range.

We now determine the proper choice of the truncation parameter r. Let $\hat{B} \in \mathcal{D}_k$ denote the operator adjoint to B, i.e., $\hat{b}_s(n) = b_{\sigma-s}(n+s)$ (see A-VI). It is easy to show that

$$\hat{B}^p = (B_k^p)^{\mathrm{T}}, \qquad \hat{B}_0^{p,q} = (B_k^{p,q})^{\mathrm{T}}, \qquad \hat{B}_k^p = (B^{p+k})^{\mathrm{T}},$$

$$\hat{B}_k^{p,q} = (B_0^{p+k,q+k})^{\mathrm{T}}, \qquad q \geq p, \quad p \geq 0.$$

We have

Theorem 8.7   *Let $D_j^i = A_j^i B^{i+j}$ be the (i, j)-factorization of $D \in \mathcal{D}_\sigma$, $j < \sigma$, and let D be totally separable. Let w be a solution of $Dy = f$ with type $(w) \leq j$. Let $y_r^{i+j,r-1}$, z^{i+j} be the corresponding solutions of (32) and (35). Also let s satisfy $i + j + k - 1 \leq s < r$ and $u^{s+1} = (u(s+1), u(s+2), \ldots)$ be the solution of*

$$\hat{B}_k^{s-k+1} y^{s+1} = -\left[\begin{matrix} \hat{B}_0^{s-k+1,s} y^{s-k+1,s} \\ \cdots\cdots\cdots\cdots\cdots \\ 0 \end{matrix} \right], \qquad k = \sigma - j, \tag{8.36}$$

with $u^{s-k+1,s} := (0, 0, \ldots, 0, 1)^{\mathrm{T}}$.
  Then

$$y_r(s) = \sum_{m=s}^{r-1} u(m) z(m)/b_0(s), \tag{8.37}$$

where $b_0(s)$ is the $(m - i - j + 1)$st leading coefficient of B^{i+j}.

Proof   This is, essentially, the result of (A.16). ∎

Now assume the algorithm converges (conditions guaranteeing this will be discussed later),

$$\lim_{r \to \infty} y_r(n) = w(n), \qquad n \geq i + j.$$

Allowing $r \to \infty$ in (37) gives

$$w(s) = \sum_{m=s}^{\infty} u_s(m) z(m)/b_0(s), \qquad s \geq i+j+k-1.$$

If we denote the truncation error of the algorithm by

$$E_r(n) := w(n) - y_r(n),$$

then we have

$$E_r(n) = \sum_{s=n}^{\infty} u_n(s) z(s)/b_0(n), \qquad n \geq i+j+k-1.$$

It can be shown that when the second-order equation is appropriately scaled, this agrees with the truncation error estimate (6.18) for the Olver algorithm.

Cases in which the Lozier algorithm is applied to equations where the homogeneous equation has a fundamental set with σ distinct growth rates, as in the case of a totally separable operator, are the easiest to analyze from the point of view of convergence and stability. However, there are many operators of practical interest in which the solutions have similar asymptotic behavior, for instance, the class of problems in Section 7.4, to which we applied the Clenshaw averaging process. Lozier defines a more general class of operators, called *separable* operators, in which the solutions exhibit less than σ distinct rates of growth. These operators possess a basis which includes $\rho \leq \sigma$ disjoint subsets such that (i) neither of two distinct solutions in the same subset dominates the other; (ii) any two solutions taken from different subsets are such that one dominates the other; (iii) no nonzero linear combination of solutions from one subset is dominated by any solution in the subset. Such a basis is called *optimally ranked*. The reader will perceive that any Birkhoff equation (see Section B.2) is associated with such an operator, and the canonical sets can be grouped to form an optimally ranked basis. Thus, nearly all difference operators encountered in practical applications are separable.

The disjoint subsets of an optimally ranked basis can be arranged in order of increasing rate of growth. The solutions of the nonhomogeneous equation can be classified by comparing them with the solutions in an optimally ranked basis for the homogeneous equation. A particular solution $y(n)$ is said to be of type s if (i) $y(n)$ is not dominated by any solution in the first s subsets of the basis; (ii) $y(n)$ is dominated by every solution in the remaining $\rho - s$ subsets. As with totally separable operators the cases $s=0$, $s=\rho$ correspond to particular solutions for which forward recurrence and backward recurrence of the given equation are stable, respectively. For those intermediate cases $0 < s < \rho$, let j be the number of solutions in the first s subsets. It may be shown that if the operator is both separable and (i, j)-factorizable, then the associated forward and backward recurrence problems are stable in the sense that rounding errors are not propagated more rapidly than the solutions themselves, at least in the region appropriate to the given problem. Thus the solution of the boundary-value problem may be obtained in a stable manner by forward elimination without pivoting followed by back substitution. Further, under appropriate conditions, it can be shown that $y(n)$ is uniquely determined by j initial values and that the algorithm converges as $r \to \infty$. These conditions, which I will not exhibit here, have to do with the growth properties of certain principal minors of the Casorati determinant of the optimally ranked basis and reflect the condition that when one or more of the solutions in the Casoratian is replaced by a solution dominated by all the original solutions, then the resulting Casoratian is dominated by the original. Obviously when the required solution is 'sufficiently separated' asymptotically from the solutions in the basis, and those solutions from each other, this will be true.

Lozier discusses many more properties of this algorithm and presents an extensive section on numerical investigations.

9 The computation of $_3F_2(1)$

9.1 The recursion

Let a_i, $i = 1, 2, 3$, b_j, $j = 1, 2$, be complex parameters

$$a := a_1 + a_2 + a_3, \qquad b := b_1 + b_2.$$

For $\mathrm{Re}\,(a - b) < 0$, $b_j \neq 0, -1, -2, \ldots$, $_3F_2(1)$ is defined by the series

$$F \equiv {}_3F_2(1) \equiv {}_3F_2\begin{pmatrix} a_1 & a_2 & a_3 \\ & b_1 & b_2 \end{pmatrix} := \sum_{k=0}^{\infty} \frac{(a_1)_k (a_2)_k (a_3)_k}{(b_1)_k (b_2)_k k!}, \tag{9.1}$$

and for other values of a_i, b_j is defined to be the analytic continuation of this series. The definition gives no clear idea of what this region $\Omega \subset \mathscr{C}^5$ of analyticity is. In fact determining Ω is one of the problems I will address in this section.

$_3F_2(1)$ is one of the fundamental special functions of applied mathematics. Unlike Gauss' function of unit argument, $_2F_1(1)$, it cannot be expressed as a simple ratio of Gamma functions, although such formulas are possible for special values of the parameters; see Erdélyi *et al.* (1953, 4.4).

The function sometimes arises in the evaluation of a beta integral of Gauss's function, e.g.,

$$\int_{-1}^{1} (1 - t)^a (1 + t)^b P_\nu^{(\alpha,\beta)}(t)\,dt$$

$$= \frac{2^{a+b+1}(\alpha + 1)_\nu \Gamma(b + 1)}{\nu!\,\Gamma(a + b + 2)} \Gamma(a + 1)\,{}_3F_2\begin{pmatrix} -\nu, & \nu + \alpha + \beta + 1, & a + 1 \\ & \alpha + 1, & a + b + 2 \end{pmatrix},$$

where $P_\nu^{(\alpha,\beta)}$ is the Jacobi function.

If any a_i is zero or a negative integer, then the series terminates and the computation of F is trivial. Throughout I will assume this does not happen:

$$a_i \neq 0, -1, -2, \ldots. \tag{9.2}$$

For large k the general term in the series (1) behaves like $k^{a-b-1}(1 + O(k^{-1}))$. On the other hand the series with b_1 replaced by $b_1 + N$, N an integer, has a general term which behaves like $k^{a-b-N-1}(1+O(k^{-1}))$. In this series N can always be chosen large enough so that any desired accuracy may be obtained by taking a sufficient number of terms. (This statement is not so easy to justify as it may seem; see the discussion to follow.) Thus a formula relating F with a denominator parameter b_1+N for one or more values of N to F itself could serve as the basis of a computational algorithm.

The recursion is based on a special limiting case of the recursion relation for a generalized hypergeometric function (C.12). Replace z by z/λ and let $\lambda \to \infty$. After making an obvious identification of parameters ($q=3$, $p=2$, etc.) one finds that the recursion relation, originally one of fourth order, dramatically reduces to one of only second order:

$$A_0(n)y(n)+A_1(n)y(n+1)+A_2(n)y(n+2)=0, \qquad (9.3)$$

$$\left.\begin{array}{l}
A_0(n):=(n+b_1)(n+b_1+1)(n+b-a), \\
A_1(n):=(n+b_1+1)[-2(n+b_1)^2+(n+b_1)(2a-2-b_2)+a-1-a^*], \\
A_2(n):=(n+b_1+1-a_1)(n+b_1+1-a_2)(n+b_1+1-a_3),
\end{array}\right\} \tag{9.4}$$

where

$$a^*:=a_1a_2+a_1a_3+a_2a_3.$$

A solution of this equation is

$$w(n)=y^{(1)}(n):={}_3F_2\binom{a_1,\quad a_2,\quad a_3}{b_1+n,\quad b_2}.$$

Another (linearly independent) solution is

$$y^{(2)}(n):=\frac{\Gamma(n+b_1)}{\Gamma(n+b_1-b_2+1)}{}_3F_2\binom{1+a_1-b_2,\quad 1+a_2-b_2,\quad 1+a_3-b_2}{n+1+b_1-b_2,\quad 2-b_2}. \tag{9.5}$$

Straightforward series manipulations show

$$y^{(1)}(n) \sim 1+\frac{c_1}{n}+\frac{c_2}{n^2}+\cdots, \tag{9.6}$$

$$y^{(2)}(n) \sim n^{b_2-1}\left(1+\frac{d_1}{n}+\frac{d_2}{n^2}+\cdots\right). \tag{9.7}$$

Even when $y^{(2)}(n)$, as given by (5), is not defined the series on the right of (7) will be the Birkhoff series for a legitimate solution linearly independent of $y^{(1)}(n)$.

Because of the fact that $y^{(1)}$, $y^{(2)}$ are both of algebraic growth, computing F by the Miller algorithm—even providing a normalization were available—is not feasible. What we will do is to employ the recursion in the *backward direction* but with accurate initial values, $w(N+1), w(N)$.

Our first result concerns the domain of analyticity of F.

Theorem 9.1 *Let* $\mathbf{p} := (a_1, a_2, a_3, b_1, b_2)$ *be a vector in* $\mathscr{C}^5$ *and let* Ω *be the subset of* $\mathscr{C}^5$ *satisfying*

$$b_j \neq -m, \quad b-a \neq -m, \qquad m = 0, 1, 2, \ldots. \tag{9.8}$$

Then F is analytic for $\mathbf{p} \in \Omega$.

Proof The proof is quite simple. N may be chosen so the series definitions of $w(N+1)$, $w(N+2)$ converge uniformly, and hence are analytic, on any compact subset K of Ω. Since $A_0(n)$ cannot vanish, $w(N), w(N-1), \ldots, w(0) = F$ can be computed from (3) and all of these are analytic in K. Thus the equation serves to analytically continue F to any compact subset of $\mathscr{C}^5$. ∎

9.2 The algorithm; truncation error

The algorithm proceeds in four steps. I assume a_i, b_j are such that F exists and is nontrivial, i.e., (2) and (8) are satisfied. I also assume a_i, b_j are real. This assumption isn't necessary for the convergence of the algorithm but it simplifies the error analysis somewhat.

(1) Pick R_1, R_2 to be the smallest positive (≥ 0) integers such that

$$-R_1 < a_i, b_j \leq R_2,$$

and set

$$P := 2R_1 + 3R_2.$$

(2) Pick N (>1) an integer such that

$$\frac{KN^{1/2}}{(N-1)} (N)_{b_1+P} (\tfrac{4}{27})^N < \varepsilon, \tag{9.9}$$

where ε is the desired accuracy and

$$K := \frac{2\sqrt{(3\pi)} |\Gamma(b_2)|}{|\Gamma(a_1)\Gamma(a_2)\Gamma(a_3)|}.$$

(Note that a ballpark estimate for N can be obtained by approximating the left-hand side of (9) by $KN^{b_1+P-1/2}(\tfrac{4}{27})^N$.)

(3) Compute, as approximations to $w(N+P+j)$, $j = 0, 1$,

$$w^*(N+P+j) := \sum_{k=0}^{2N+R_1+2j} \frac{(a_1)_k (a_2)_k (a_3)_k}{(b_1+N+P+j)_k (b_2)_k k!}.$$

(4) Compute $w^*(n)$, $n = N+P-1, N+P-2, \ldots, 0$, from

$$w^*(n) = \frac{-A_1(n)w^*(n+1) - A_2(n)w^*(n+2)}{A_0(n)}.$$

Then

$$w^*(0) \approx F.$$

I will show that when N is determined via (9) then $|w^*(N+P+j) - w(N+P+j)| < \varepsilon$. Indeed, it can be shown that *if all computations are subsequently performed in exact arithmetic,*

$$w^*(0) = w(0) + \varepsilon O(N^\lambda), \qquad N \to \infty,$$

$$\lambda := \max(1, 2 - b_2).$$

(9.10)

I will have need of the following result which is an extension of an inequality due to Gautschi given in Luke (1975, p. 17).

Lemma *Let $b > a > 0$. Then*

$$g(n) := n^{b-a} \frac{\Gamma(n+a)}{\Gamma(n+b)} < 1, \qquad n = 1, 2, \ldots,$$

iff $\tau := (a-b)(a+b-1) \leq 0$.

Proof $\Rightarrow$: The formula in Erdélyi *et al.* (1953, vol. 1, p. 47 (3)) gives

$$g(n) = 1 + \frac{\tau}{2n} + O(n^{-2}), \qquad n \to \infty,$$

so given $\varepsilon > 0$ there is an n_0 such that

$$-\frac{\varepsilon}{n} < g(n) - 1 - \frac{\tau}{2n} < \frac{\varepsilon}{n}, \qquad n > n_0,$$

or

$$g(n) > 1 + \frac{(\tau/2) - \varepsilon}{n}, \qquad n > n_0.$$

If $\tau > 0$ I can choose $\varepsilon = \tau/2$ to get $g(n) > 1$, $n > n_0$, a contradiction.
$\Leftarrow$:

$$g(n+1)/g(n) = (n+1)^{b-a}(n+a)/n^{b-a}(n+b).$$

Let

$$\mu(x) = (x+1)^{b-a}(x+a)/x^{b-a}(x+b), \qquad x > 0.$$

Calculus shows that $\mu'(x)$ vanishes only for $\tau \neq 0$, $x = -ab(a-b)/\tau$, and this value of x is negative. Thus $\mu(x)$ is monotone—in fact, monotone decreasing to 1.

Thus

$$\mu(n) = \frac{g(n+1)}{g(n)} > 1,$$

or

$$g(n+1) > g(n), \qquad n \geqslant 1,$$

or

$$g(n+R) > g(n), \qquad R = 1, 2, 3, \ldots.$$

But by (24) $g(n+R) \to 1$ as $R \to \infty$, and this completes the proof. ∎

Now let

$$E(n) = \sum_{k=n+R_1+1}^{\infty} \frac{(a_1)_k (a_2)_k (a_3)_k}{(b_1+N+P)_k (b_2)_k k!}.$$

Then

$$|E(n)| < \frac{K}{2\sqrt{(3\pi)}} \Gamma(b_1+N+P)$$
$$\times \sum_{k=n}^{\infty} \frac{[(R_2+R_1+k)(R_2+R_1+k-1)\cdots(k+1)k!]^3}{k!^2 (N+P+R_1+k)\cdots(N+k+1)\Gamma(N+k+1)}.$$

Since $P = 3R_1 + 2R_2$ each linear factor in the numerator is dominated by a linear factor in the denominator. Thus

$$|E(n)| < \frac{K}{2\sqrt{(3\pi)}} \Gamma(b_1+N+P) \sum_{k=n}^{\infty} \frac{k!}{\Gamma(N+k+1)}.$$

This infinite series can be summed—in fact it is a Gaussian hypergeometric function of unit argument. The result is

$$|E(n)| < \frac{K\Gamma(b_1+N+P)n!}{2\sqrt{(3\pi)}(N-1)\Gamma(N+n)}.$$

Now let $n = 2N$, use the duplication and triplication formulas for the Gamma function and employ the lemma. The result for $j = 0$ (and hence $j = 1$) follows immediately.

Another way to proceed is to compute $w^*(0)$ for increasing the values of N as I have done in Table 9.1. Here I have chosen values of a_i, b_j for which F can be evaluated in closed form (see Erdélyi *et al.* (1953, vol. 1, p. 189 (5)). The results are self-explanatory, but a few remarks should be made. The computations were done in APL which has 16-figure precision. Increasing N increases the accuracy of $w^*(0)$ up to a point. After that the algebraic accumulation of round-off error becomes significant

Table 9.1   Computation of $_3F_2(1)$.

as	0.85, 0.36, 0.2	0.71, −0.36, 2.315	0.35, 0.73, 1.56
bs	1.49, 1.65	2.07, −0.605	0.62, −0.21
$a-b-1$	−2.73	0.2	1.23

N			
5	1.036 185 582 642 963	0.120 129 919 878 524	−0.547 178 366 121 112
7	1.036 185 591 161 948	0.119 911 412 572 423	−0.547 743 333 122 015
10	1.036 185 591 262 563	0.119 906 234 679 308	−0.547 763 490 410 554
12	1.036 185 591 262 818	0.119 906 212 239 439	−0.547 763 607 642 034
15	1.036 185 591 262 873	0.119 905 211 605 519	−0.547 763 611 425 356
20	1.036 185 591 262 936	0.119 906 211 596 063	−0.547 763 611 444 459
25	1.036 185 591 263 036	0.119 906 211 584 331	−0.547 763 611 444 145
TRUE	1.036 185 591 262 798	0.119 906 211 609 263	−0.547 763 611 446 791

and to further increase N produces a deterioration in accuracy in accordance with formula (10). The larger $2-b_2$ the worse the error accumulation, so the F value of column 1 can be computed more accurately than can the F values in columns 2 and 3. (The values of $\max(1, 2-b_2)$ are respectively 1, 2.605, 2.21.)

9.3   Computing the Beta function

The same considerations lead to an interesting algorithm for computing $B(x, y)$ for

$$1-x, 1-y, x+y-1 \neq 0, -1, -2, \ldots.$$

The computation is based on the evaluation of $_2F_1$ with unit argument and denominator parameter $c = n+1$. The appropriate recursion formula can be found from Erdélyi *et al.* (1953, vol. 1, p. 104).

(1)   Pick R_1, R_2 to be the smallest positive ($\geqslant 0$) integers such that

$$-R_1 < 1-x, 1-y \leqslant R_2,$$

and set

$$P := R_1 + 2R_2.$$

(2)   Pick N (>1) an integer such that

$$\frac{KN^{1/2}}{(N-1)} (N)_{1+P} (\tfrac{4}{27})^N < \varepsilon,$$

where ε is the desired accuracy and

$$K := \frac{2\sqrt{(3\pi)}}{|\Gamma(1-x)\Gamma(1-y)|}.$$

(3) Compute, as an approximation to $w(N+P+1)$,

$$w^*(N+P+1) := \sum_{k=1}^{2N+R_1} \frac{(1-x)_k(1-y)_k}{(N+P+1)_k k!}.$$

(4) Compute $w^*(n)$, $n = N+P, N+P-1, \ldots, 0$,

$$w^*(n) = \frac{(n+x)(n+y)w^*(n+1)}{(n+1)(n+x+y-1)}.$$

Then

$$B(x, y) \approx [(x+y-1)w^*(0)]^{-1}.$$

It can be shown that error accumulates like

$$w^*(0) = w(0) + \delta[1 + O(N^{-1})] + O(N\varepsilon(N)),$$

where $w^*(N+1) = w(N+1) + \delta$, δ being the initial error.

Since $\Gamma(1-x)$, $\Gamma(1-y)$ are not, strictly speaking, known, it is necessary to obtain lower bounds for them to use the previous result. This is easily done. For instance, for $a > 0$,

$$\frac{1}{\Gamma(a)} \leq 1.12917 \ldots.$$

9.4 Another $_3F_2(1)$

A recurrence for another important $_3F_2(1)$,

$$C(n) := \frac{(a)_n(b)_n}{(c)_n(n+\gamma)_n} {}_3F_2\left(\begin{matrix} n+\beta+1, & n+a, & n+b \\ 2n+\gamma+1, & n+c \end{matrix} \middle| 1\right),$$

$$\gamma = \alpha + \beta + 1,$$

which occurs in the expansion,

$$f(x) = \sum_{n=0}^{\infty} C(n)P_n^{(\alpha,\beta)}(x),$$

$$f(x) := {}_2F_1\left(\begin{matrix} a, & b \\ & c \end{matrix} \middle| \frac{1+x}{2}\right),$$

(9.11)

can be found as follows. The formulas Erdélyi et al. (1953, vol. 1, p. 56 (1)) show f satisfies

$$(1-x^2)y'' + [(2c-a-b-1) - (a+b+1)x]y' - aby = 0.$$

Putting the expansion (11) in the previous equation and using the differential equation (10.23) satisfied by $P_n^{(\alpha,\beta)}$ shows, after a little algebra, we must have

$$\sum_{n=0}^{\infty} C(n)\{[(2c-a-b-1+\alpha-\beta)+(\gamma-a-b)x]p_n'$$
$$-[n(n+\gamma)+ab]p_n\}=0. \quad \textbf{(9.12)}$$

Now

$$xp_n' = (xp_n)' - p_n,$$

so (12) can be written

$$\left.\begin{aligned}
&\sum_{n=0}^{\infty} C(n)[d_1 p_n' + d_2(xp_n)' + d_3 p_n] = 0,\\
&d_1 := 2c-a-b-1+\alpha-\beta,\\
&d_2 := \gamma-a-b,\\
&d_3 := a+b-\gamma-ab-n(n+\gamma).
\end{aligned}\right\} \quad \textbf{(9.13)}$$

Differentiating the relationship for xp_n obtainable from (10.21), substituting this and (10.30) into the expansion (13) and setting the coefficient of p_n' to 0 (which must be permissible since $\{p_n'\}$ itself is an orthogonal system) gives a recurrence of *second order* for $C(n)$:

$$A_0(n)y(n)+A_1(n)y(n+1)+A_2(n)y(n+2)=0, \quad \textbf{(9.14)}$$

where

$$\left.\begin{aligned}
&A_0(n):=-2(n+\gamma)(n+a)(n+b)/(2n+\gamma+1)(2n+\gamma),\\
&A_1(n):=e_1+e_2/(2n+\gamma+1)(2n+\gamma+3),\\
&A_2(n):=2(n+\alpha+2)(n+\beta+2)(n+\gamma+2-a)(n+\gamma+2-b)/\\
&\qquad\qquad\qquad (n+\gamma+1)(2n+\gamma+3)(2n+\gamma+4),\\
&e_1 := 2c-a-b-1-\frac{(\beta-\alpha)}{2},\\
&\qquad e_2 := \frac{(\beta-\alpha)}{2}(\gamma+1-2a)(\gamma+1-2b).
\end{aligned}\right\} \quad \textbf{(9.15)}$$

This is the same recurrence predicted by the Lewanowicz theory (see Section 11.2.1) and the recurrence is minimal. Lewanowicz has conjectured (private communication) that

$$_{p+2}F_{p+1}\left(\begin{matrix} n+a_{p+2} \\ 2n+c, \quad n+b_p \end{matrix} \,\middle|\, 1\right),$$

satisfies a recurrence of minimal order $p+1$. (It is known, in fact it follows from the formula in Appendix C, that this function satisfies a recurrence of order $p+2$.)

The recurrence (14) can be used to show that $C(0)$ cannot in general be expressed as a ratio of gamma functions.

$C(0)$, $C(1)$ may be computed by the method of this section and since $C(n)$ has asymptotically algebraic behavior in n (see Wimp (1982)) $C(n)$ may be computed in a stable fashion by using (14) in the forward direction.

10 Computations based on orthogonal polynomials

10.1 Preliminaries; properties of some orthogonal polynomials

10.1.1 Chebyshev polynomials

In this and subsequent chapters I will need to refer to properties of some important classical orthogonal polynomials, primarily the Chebyshev polynomials.

Let

$$T_n(x) := \cos n\theta, \qquad x = \cos \theta, \quad n \geq 0,$$

so

$$T_0(x) = 1, \qquad T_1(x) = x, \qquad T_2(x) = 2x^2 - 1, \ldots, \qquad \text{etc.}$$

$T_n(x)$ is called the *Chebyshev polynomial of the first kind* and satisfies

$$T_{n+1}(x) - 2xT_n(x) + T_{n-1}(x) = 0, \qquad n \geq 1. \tag{10.1}$$

Also the Rodrigues formula holds,

$$2^n (\tfrac{1}{2})_n T_n(x) = (-1)^n \sqrt{(1-x^2)} D^n [(1-x^2)^{n-1/2}], \tag{10.2}$$

and the explicit expressions

$$T_n(x) = \frac{[x + \sqrt{(x^2-1)}]^n + [x - \sqrt{(x^2-1)}]^n}{2}$$

$$= F\!\left(\begin{matrix} -n, & n \\ & \tfrac{1}{2} \end{matrix} \,\middle|\, \frac{1-x}{2}\right). \tag{10.3}$$

Let $f(x)$ be defined on $[-1, 1]$ with $f(\cos \theta) \in L[0, \pi]$. Then the coefficients

$$C(n) = \frac{2}{\pi} \int_{-1}^{1} \frac{f(x) T_n(x)}{\sqrt{(1-x^2)}} \, dx, \tag{10.4}$$

exist and the series

$$f(x) \sim \sum_{n=0}^{\infty}{}' C(n) T_n(x), \tag{10.5}$$

is called the *Chebyshev series for f(x)*. The series converges to $\frac{1}{2}(f(x^-) + f(x^+))$ if f is of bounded variation in a neighbourhood of x. If $f \in C^1[-1, 1]$, on the other hand, the series converges uniformly to $f(x)$ on $[-1, 1]$. Of course these are simply statements about Fourier series (see Rogosinski (1950)).

The polynomial

$$T_n^*(x) := T_n(2x - 1), \qquad 0 \leq x \leq 1, \tag{10.6}$$

is called the *shifted* Chebyshev polynomial.

The following properties of these polynomials are easily established.

$$T_n(-x) = (-1)^n T_n(x); \tag{10.7}$$

$$T_n(1) = 1; \qquad T_n(-1) = (-1)^n; \qquad T_{2n}(0) = (-1)^n;$$

$$T_{2n+1}(0) = 0; \qquad T_n^*(1) = 1; \qquad T_n^*(0) = (-1)^n; \tag{10.8}$$

$$T_{2n}(x) = T_n^*(x^2); \tag{10.9}$$

$$\int_{-1}^{1} (1 - x^2)^{-1/2} T_n(x) T_m(x)\, dx = \begin{cases} 0, & m \neq n, \\ \dfrac{\pi}{2}, & m = n \neq 0, \\ \pi, & m = n = 0; \end{cases} \tag{10.10}$$

$$(1 - x^2) y'' - x y' + n^2 y = 0, \qquad y = T_n(x); \tag{10.11}$$

$$T_n(x) = \frac{n}{2} \sum_{m=0}^{\leq n/2} \frac{(-1)^m (n - m - 1)!}{m!\,(n - 2m)!} (2x)^{n - 2m}. \tag{10.12}$$

The following properties will also be useful:

$$T_n(x) = \frac{n!}{(\frac{1}{2})_n} P_n^{(-1/2, -1/2)}(x), \tag{10.13}$$

in terms of the Jacobi polynomial (Section 10.1.2).

$$x^m T_n(x) = \frac{1}{2^m} \sum_{k=0}^{m} \binom{m}{k} T_{|n+m-2k|}(x); \tag{10.14}$$

$$\left.\begin{aligned} \int_0^x T_{2n}(t)\, dt &= \frac{1}{2} \left[\frac{T_{2n+1}(x)}{2n+1} - \frac{T_{|2n-1|}(x)}{2n-1} \right]; \\ \int_0^x T_{2n+1}(t)\, dt &= \frac{1}{2} \left[\frac{T_{2n+2}(x)}{2n+2} - \frac{T_{2n}(x)}{2n} \right] + \frac{(-1)^n}{4} \left(\frac{1}{n} + \frac{1}{n+1} \right), \\ \int_0^x T_1(t)\, dt &= \tfrac{1}{4}[T_2(x) + T_0(x)]; \end{aligned}\right\} \quad n > 0; \tag{10.15}$$

$$T_n(x) = \frac{\varepsilon_n}{4} \left[\frac{T'_{n+1}(x)}{n+1} - \frac{T'_{|n-1|}(x)}{n-1} \right], \qquad n \geq 0; \tag{10.16}$$

$$x^m T_n^*(x) = 2^{-2m} \sum_{k=0}^{2m} \binom{2m}{k} T_{|n+m-k|}^*(x); \tag{10.17}$$

$$\left.\begin{aligned}
\int_0^x T_n^*(t)\,dt &= \frac{1}{4}\left[\frac{T_{n+1}^*(x)}{n+1} - \frac{T_{|n-1|}^*(x)}{n-1}\right] - \frac{(-1)^n}{2(n^2-1)}, \\
&\qquad\qquad\qquad\qquad\qquad n = 0, 2, 3, \ldots, \\
\int_0^x T_1^*(t)\,dt &= \tfrac{1}{8}[T_2^*(x) - T_0^*(x)].
\end{aligned}\right\} \quad \textbf{(10.18)}$$

10.1.2   Jacobi polynomials

Let α, β be real, $\alpha > -1$, $\beta > -1$. The nth Jacobi polynomial is defined by

$$p_n(x) := P_n^{(\alpha,\beta)}(x) := \frac{(\alpha+1)_n}{n!} F\left(\begin{matrix} -n, & n+\gamma \\ & \alpha+1 \end{matrix}\middle|\,\frac{1-x}{2}\right)$$

$$= (-1)^n \frac{(\beta+1)_n}{n!} F\left(\begin{matrix} -n, & n+\gamma \\ & \beta+1 \end{matrix}\middle|\,\frac{1+x}{2}\right), \qquad \textbf{(10.19)}$$

where

$$\gamma := \alpha + \beta + 1. \qquad\qquad\qquad\qquad \textbf{(10.20)}$$

p_n satisfies the recursion

$$\left.\begin{aligned}
&2(n+1)(n+\gamma)(2n+\gamma-1)p_{n+1} \\
&\quad = (2n+\gamma)[(2n+\gamma-1)(2n+\gamma+1)x + (\alpha^2-\beta^2)]p_n \\
&\qquad -2(n+\alpha)(n+\beta)(2n+\gamma+1)p_{n-1}, \\
&\qquad\qquad\qquad\qquad\qquad\qquad\qquad n \geq 0, \\
&p_{-1}(x) = 0; \quad p_0(x) = 1; \quad p_1(x) = \frac{\alpha-\beta+(\gamma+1)x}{2}.
\end{aligned}\right\} \quad \textbf{(10.21)}$$

$$2^n n!\, p_n = (-1)^n (1-x)^{-\alpha}(1+x)^{-\beta} D^n[(1-x)^{\alpha+n}(1+x)^{\beta+n}], \qquad n \geq 0, \\ \textbf{(10.22)}$$

is called the Rodrigues' formula for p_n. The differential equation for p_n is

$$(1-x^2)y'' + [\beta - \alpha - (\gamma+1)x]y' + n(n+\gamma)y = 0. \qquad \textbf{(10.23)}$$

p_n satisfies the further differentiation formula

$$2^m D^m P_n^{(\alpha,\beta)}(x) = (n+\gamma)_m P_{n-m}^{(\alpha+m,\beta+m)}(x), \qquad 0 \leq m \leq n+1. \quad \textbf{(10.24)}$$

$P_n^{(\alpha,\beta)}(x)$ is a system of polynomials orthogonal on $[-1, 1]$ with respect to the weight function $(1-x)^\alpha(1+x)^\beta$ and

$$\int_{-1}^1 (1-x)^\alpha(1+x)^\beta p_n(x)p_m(x)\,dx = \begin{cases} 0, & m \neq n; \\ h(n), & m = n; \end{cases} \qquad \textbf{(10.25)}$$

where

$$h(n) = \frac{2^\gamma \Gamma(n+\alpha+1)\Gamma(n+\beta+1)}{n!\,(2n+\gamma)\Gamma(n+\gamma)}. \qquad\qquad \textbf{(10.26)}$$

A function $f(x)$ defined on $[-1, 1]$ and satisfying certain conditions (for these see Freud (1966)) possesses the expansion

$$f(x) = \sum_{n=0}^{\infty} C(n) p_n(x), \tag{10.27}$$

where

$$C(n) := \frac{1}{h(n)} \int_{-1}^{1} (1-x)^{\alpha}(1+x)^{\beta} f(x) p_n(x) \, dx. \tag{10.28}$$

If f has a continuous nth derivative the above integral may be integrated n times by parts using (22) and gives

$$C(n) = \frac{1}{2^n n! \, h(n)} \int_{-1}^{1} f^{(n)}(x)(1-x)^{\alpha+n}(1+x)^{\beta+n} \, dx. \tag{10.29}$$

Other useful formulas are

$$p_n = \nu_1 p'_{n+1} + \nu_2 p'_n + \nu_3 p'_{n-1}, \tag{10.30}$$

$$\left.\begin{aligned}
\nu_1 &:= 2(n+\gamma)/(2n+\gamma)(2n+\gamma+1), \\
\nu_2 &:= 2(\alpha-\beta)/(2n+\gamma-1)(2n+\gamma+1), \\
\nu_3 &:= -2(n+\alpha)(n+\beta)/(n+\gamma-1)(2n+\gamma)(2n+\gamma-1),
\end{aligned}\right\} \tag{10.31}$$

(see Askey and Gasper (1971)),

$$\int_{-1}^{1} (1-x)^{\rho}(1+x)^{\sigma} p_n(x) \, dx$$

$$= \frac{2^{\rho+\sigma+1}\Gamma(\rho+1)\Gamma(\sigma+1)\Gamma(n+\alpha+1)}{\Gamma(\rho+\sigma+2)n! \, \Gamma(\alpha+1)}$$

$$\times {}_3F_2\left(\begin{matrix} -n, & n+\gamma, & \rho+1 \\ \alpha+1, & \rho+\sigma+2 \end{matrix} \middle| 1\right). \tag{10.32}$$

(The formula given in the reference Erdélyi *et al.* (1954, vol. 2, p. 284 (3)) for this integral is incorrect.) In certain cases ($\rho = \alpha$, $\sigma = \beta$, etc.) the ${}_3F_2$ reduces and can be expressed in terms of Gamma functions.

The *Gegenbauer polynomials* $C_n^{\lambda}(x)$ are a specialization of the Jacobi polynomials,

$$C_n^{\lambda}(x) := \frac{(2\lambda)_n}{(\lambda+\frac{1}{2})_n} P_n^{(\lambda-1/2,\lambda-1/2)}(x), \qquad \lambda \neq 0. \tag{10.33}$$

Important formulas are

$$C_n^{\lambda}(x) = \frac{(2\lambda)_n}{n!} F\left(\begin{matrix} -n, & n+2\lambda \\ \lambda+\frac{1}{2} \end{matrix} \middle| \frac{1-x}{2}\right); \tag{10.34}$$

$$(n+1)C_{n+1}^\lambda(x) = 2(n+\lambda)xC_n^\lambda(x) - (n+2\lambda-1)C_{n-1}^\lambda(x) \qquad \textbf{(10.35)}$$

$$= \frac{(-1)^n(2\lambda)_n}{2^n n! (\lambda+\frac{1}{2})_n} (1-x^2)^{(1/2)-n}D^n[(1-x^2)^{n+\lambda-1/2}];$$

$$\textbf{(10.36)}$$

$$2(n+\lambda)C_n^\lambda(x) = \frac{d}{dx}[C_{n+1}^\lambda(x) - C_{n-1}^\lambda(x)], \qquad n \geq 1;$$

$$\textbf{(10.37)}$$

$$2\lambda C_0^\lambda(x) = \frac{d}{dx}C_1^\lambda(x);$$

$$\int_{-1}^{1} (1-x^2)^{\lambda-1/2}C_n^\lambda(x)C_m^\lambda(x)\,dx = \begin{cases} 0, & m \neq n; \\ h(n), & m = n; \end{cases} \qquad \textbf{(10.38)}$$

$$h(n) = \frac{(\sqrt{\pi})(2\lambda)_n\Gamma(\lambda+\frac{1}{2})}{(n+\lambda)n!\,\Gamma(\lambda)}. \qquad \textbf{(10.39)}$$

For $f(x)$ satisfying appropriate conditions

$$f(x) = \sum_{n=0}^{\infty} B(n)C_n^\lambda(x); \qquad \textbf{(10.40)}$$

$$B(n) := \frac{1}{h(n)} \int_{-1}^{1} (1-x^2)^{\lambda-1/2}f(x)C_n^\lambda(x)\,dx. \qquad \textbf{(10.41)}$$

10.2   Evaluation of finite sums of functions which satisfy a linear homogeneous recurrence

10.2.1   The algorithm

Much of the work in this chapter will involve the computation, for a fixed value of x, of sums of the form

$$f_N(x) := \sum_{n=0}^{N} A(n)p_n(x), \qquad \textbf{(10.42)}$$

where p_n satisfies a three-term recurrence

$$p_{n+1}(x) + a(n)p_n(x) + b(n)p_{n-1}(x) = 0, \qquad n \geq 1. \qquad \textbf{(10.43)}$$

($a(n)$ and $b(n)$ may depend on x.) This will include the case in which the p_n form a system of orthogonal polynomials with respect to some distribution. There is an ingenious way of computing f_N which avoids the computation of any of the p_n except p_0, p_1. The method is a nesting procedure similar to the procedure used to evaluate polynomials; it was first given by Clenshaw (1955) for the case when the $\{p_n\}$ are Chebyshev polynomials and was subsequently generalized by Luke (1975, p. 476).

The algorithm proceeds as follows. Compute the sequence $\{B(n)\}$,

$0 \leqslant n \leqslant N+2$, from

$$B(n) = -a(n)B(n+1) - b(n+1)B(n+2) + A(n),$$
$$n = N, N-1, \ldots, 0. \tag{10.44}$$

$$B(N+2) = B(N+1) = 0.$$

Then

$$f_N(x) = B(0)p_0(x) + B(1)[p_1(x) + a(0)p_0(x)]. \tag{10.45}$$

This formula is easily demonstrated. Simply solve for $A(n)$ from (44), substitute in the sum f_N, (42), and then employ the recurrence (43).

Luke (1975) and Smith (1965) are responsible for a similar nesting procedure for computing $f'_N(x)$. As before, compute the sequence $\{B(n)\}$ and then $\{C(n)\}$, $0 \leqslant n \leqslant N+1$, from

$$C(n) = -a(n)C(n+1) - b(n+1)C(n+2) + \delta(n),$$
$$n = N-1, N-2, \ldots, 0,$$
$$C(N+1) = C(N) = 0,$$

where

$$\delta(n) := B(n+1)a'(n) + B(n+2)b'(n+1).$$

Then

$$f'_N(x) = (B(0) + B(1)a(0))p'_0(x) + B(1)p'_1(x) + B(1)a'(0)p_0(x)$$
$$- C(0)p_0(x) - C(1)[p_1(x)a(0)p_0(x)].$$

For the important case when $p_n(x) = T_n(x)$, the algorithm can be summarized as follows:

$$f_N(x) = \sum_{n=0}^{N} A(n)p_n(x), \qquad f'_N(x) = \sum_{n=1}^{N} A(n)p'_n(x),$$
$$B(n) = 2xB(n+1) - B(n+2) + A(n), \qquad n = N, N-1, \ldots, 0,$$
$$B(N+2) = B(N+1) = 0,$$
$$C(n) = 2xC(n+1) - C(n+2) - 2B(n+1),$$
$$n = N-1, N-2, \ldots, 0,$$
$$C(N+1) = C(N) = 0,$$

and

$$f_N(x) = B(0) - xB(1), \qquad f'_N(x) = -B(1) - C(0) + xC(1).$$

Other important cases are given in Luke (1975).

The procedure is easily generalized to higher-order difference equations. Let

$$p_{n+1}(x) + a_0(n)p_n(x) + a_1(n)p_{n-1}(x) + \cdots + a_r(n)p_{n-r}(x) = 0, \qquad n > r,$$

and define $\{B(n)\}$, $0 \leqslant n \leqslant N + r + 1$, by

$$B(n) = -a_0(n)B(n+1) - a_1(n+1)B(n+2)$$
$$- a_2(n+2)B(n+3) - a_r(n+r)B(n+r+1) + A(n),$$
$$n = N, N-1, \ldots, 0,$$

$$B(N+r+1) = B(N+r) = \cdots = B(N+1) = 0.$$

Then

$$f_N(x) = \sum_{k=0}^{r} B(k)[p_k + a_0(k-1)p_{k-1} + a_1(k-1)p_{k-2} + \cdots$$
$$+ a_{k-1}(k-1)p_0].$$

(Here $a_i(j) = 0$ if either $i < 0$ or $j < 0$.)

Example 10.1 The Clenshaw method is highly effective in computing series of the form (42) when $p_n(x) \equiv w(n)$ is itself to be computed by the Miller algorithm. Let's consider the case where the difference equation is of second order (the theory of Chapter 4). A value of N is chosen which will guarantee the required accuracy in $w(n)$. Next $B(n)$ is computed from (44). Then

$$f_N(x) \approx w(0)\left[B(0) + B(1)\left(\frac{1}{r(0)} + a(0)\right)\right],$$

where $r(n) := y_N(n)/y_N(n+1)$.

But $r(0)$ may be computed by using the equation

$$r(n) = -a(n) - \frac{b(n)}{r(n+1)}, \qquad n = N-2, N-3, \ldots, 0,$$

$$r(N-1) = -a(N-1),$$

which requires only $(N-1)$ multiplication–divisions. This scheme is easily changed to accommodate an arbitrary normalization relation for $w(n)$. ∎

10.2.2 Error analysis; three-term recurrence

The algorithm for computing $f_N(x)$ in the previous section is so attractive computationally that the numerical analyst may be inclined to overlook its shortcomings. In certain cases, even for small values of N, its use can lead to disaster. The following is due to Elliott (1968).

Example 10.2 Consider

$$f_{12} = \sum_{n=0}^{6} \varepsilon_n J_{2n}(1).$$

To 10D, f_{12} is 1. We have

$$B(n) = 2nB(n+1) - B(n+2) + A(n),$$

$$A(n) = \begin{cases} 2, & n \text{ even}, \quad n \neq 0, \\ 1, & n = 0 \\ 0, & n \text{ odd}. \end{cases}$$

Suppose we compute the $B(n)$ in floating-point arithmetic to 10 S.F. using only single-precision operations. $B(n)$ for $n = 4(1)12$, being integers with 10 or less digits, may be computed exactly, while rounded values must be used for $B(0), \ldots, B(3)$. One finds

$$B(0) = -0.737\ 724\ 590\ 6 \quad (11),$$
$$B(1) = 0.128\ 281\ 876\ 7 \quad (12).$$

If, using values of $J_0(1)$, $J_1(1)$ correctly rounded to 10D, we compute f_{12} from (45), we get -30 for f_{12} rather than 1. ∎

Before we proceed to a detailed error analysis, I will remark that the source of numerical instability of the algorithm lies in the behavior of $p_n(x)$ with respect to another linearly independent solution, call it $q_n(x)$, of the recursion. If p_n is strongly dominated by q_n the algorithm will be unstable; the stronger the domination, the greater the instability. However, if p_n and q_n are about the same size, or, better yet, p_n dominates q_n, the algorithm will be stable.

How should one sum the series in those cases when $p_n(x)$ is, say, the minimal solution of the recursion? One method is to apply the simplified Miller algorithm to the third-order recurrence satisfied by f_n,

$$-\frac{b(n+2)}{A(n+1)} y(n) + y(n+1) \left[\frac{b(n+2)}{A(n+1)} - \frac{a(n+2)}{A(n+2)} \right]$$

$$+ y(n+2) \left[\frac{a(n+2)}{A(n+2)} - \frac{1}{A(n+3)} \right] + \frac{y(n+3)}{A(n+3)} = 0.$$

This recurrence has as a basis of solutions 1, f_n, and

$$g_n := \sum_{k=0}^{n} A(k) q_k(x).$$

Thus if 1 and q_n dominate p_n sufficiently the modified Miller algorithm stands a chance of converging.

Another possibility is to compute $p_n(x)$ directly from its recurrence using the modified Miller algorithm, and then form the sum directly.

We now proceed to Elliott's error analysis. Let barred quantities denote exact quantities computed approximately, so, e.g.,

$$\bar{A}(n) := A(n) + \delta A(n), \tag{10.46}$$

$\delta A(n)$ being the error in computing $A(n)$.

From the recurrence for $B(n)$ we find

$$\bar{B}(n) = (\bar{A}(n) - \bar{a}(n)\bar{B}(n+1) - \bar{b}(n+1)\bar{B}(n+2)) + r(n), \tag{10.47}$$

where $r(n)$ is the round-off error arising from rounding the quantity in parentheses. We rewrite this as

$$\bar{B}(n) = \bar{A}(n) - a(n)\bar{B}(n+1) - b(n+1)\bar{B}(n+2) + \eta(n) + r(n), \tag{10.48}$$

$$\eta(n) := -\bar{B}(n+1)\delta a(n) - \bar{B}(n+2)\delta b(n+1)$$

$$\approx -B(n+1)\delta a(n) - B(n+2)\delta b(n+1), \tag{10.49}$$

on neglecting higher-order terms.

We find from (48) and (49) that $\delta B(n)$ satisfies

$$y(n) + a(n)y(n+1) + b(n+1)y(n+2) = \delta A(n) + \eta(n) + r(n), \tag{10.50}$$

which is the same equation, with a different right-hand side, as that satisfied by $B(n)$. Thus we have

$$p_0 \delta B(0) + (p_1 + a(0)p_0)\delta B(1) = \sum_{n=0}^{N} (\delta A(n) + \eta(n)r(n))p_n.$$

The quantity $\bar{f}_N$ actually computed is

$$\bar{f}_N := [\bar{p}_0 \bar{b}(0) + (\bar{p}_1 + \bar{a}(0)\bar{p}_0)\bar{b}(1)] + s,$$

where s denotes the round-off error arising from computing the expression in brackets.

Finally we find

$$\delta f_N = \sum_{n=0}^{N} (\delta A(n) + \eta(n) + r(n))p_n + \xi + s, \tag{10.51}$$

where

$$\xi := \bar{B}(0)\delta p_0 + \bar{B}(1)\delta p_1 + [\bar{a}(0)\delta p_0 + p_0 \delta a(0)]\bar{B}(1).$$

In the previous example $\eta(n) = \delta A(n) = 0$ since the coefficients $a(n)$, $b(n)$, $A(n)$ are given exactly. Thus

$$\delta f_N = \sum_{n=0}^{N} r(n)p_n + \xi + s,$$

and this is *not* negligible because the errors $r(n)$ in rounding $B(n)$ for n small are enormous.

Since the sum of Chebyshev polynomials

$$f_N = \sum_{n=0}^{N} A(n)T_n(x), \qquad -1 \leqslant x \leqslant 1,$$

occurs so frequently let's apply the previous error analysis to it. It is easy to show that

$$B(n) = \sum_{k=n}^{N} A(k)U_{k-n}(x),$$

where

$$U_k(x) := \frac{\sin(k+1)\theta}{\sin\theta}, \qquad \theta := \arccos x,$$

is the Chebyshev polynomial of the second kind. In carrying out the analysis we assume the computation is done in fixed-point arithmetic and $|r(n)|$, $|s| \leqslant \frac{1}{2} \times 10^{-t}$. Let $\delta x = \gamma$. Then, since $a(n) = -2x$, $\delta a(n) = -2\gamma$. Since $b(n) = 1$, we assume $\delta b(n) = 0$. Thus

$$\eta(n) = 2\gamma B(n+1),$$

approximately. Since $p_0 = 1$, $p_1 = x$, we have $\delta p_0 = 0$, $\delta p_1 = \gamma$. Thus

$$\xi = -\gamma B(1),$$

approximately. We have from (51)

$$\delta f_N = \sum_{n=0}^{N} (\delta A(n) + r(n))T_n(x) + 2\gamma \sum_{n=0}^{N} B(n+1)T_n(x) - \gamma B(1) + s.$$

$$\textbf{(10.52)}$$

Now the second sum on the right-hand side may be summed in closed form,

$$\sum_{n=0}^{N} B(n+1)T_n(x) = \sum_{n=1}^{N} B(n)T_{n-1}(x)$$

$$= \sum_{n=1}^{N} \left(\sum_{k=n}^{N} A(k)U_{k-n}(x) \right) T_{n-1}(x)$$

$$= \sum_{k=1}^{N} A(k) \left(\sum_{n=1}^{k} U_{k-n}(x)T_{n-1}(x) \right)$$

$$= \sum_{k=1}^{N} \frac{A(k)}{2} (k+1)U_{k-1}(x),$$

the last step resulting from a known trigonometric series.

Thus (52) may be rewritten as

$$\delta f_N \approx \sum_{n=0}^{N} (\delta A(n) + r(n)) T_n(x) + \gamma \sum_{n=1}^{N} n A(n) U_{n-1}(x) + s.$$

Making use of the facts that $|T_n(x)| \le 1$, $|U_{n-1}(x)| \le n$, $-1 \le x \le 1$, and assuming that $|\delta A(n)|$ and $|\gamma|$ are less than $\frac{1}{2} \times 10^{-t}$ gives

$$|\delta f_N| \le \frac{1}{2} \times 10^{-t} \left\{ (2N+3) + \sum_{n=1}^{N} n^2 |A(n)| \right\}. \tag{10.53}$$

If the sum above is ignored this result agrees with Clenshaw's earlier (1955) partial error analysis. Clearly, the sum makes a significant contribution only when one is dealing with a very slowly convergent Chebyshev series in the first place, an unlikely situation considering what the algorithm is designed to do.

Oliver (1977, 1979) has made an in-depth study of the Clenshaw nesting procedure for Chebyshev polynomials. In the latter reference he compares the method to other polynomial evaluation schemes and derives error bounds which can be expressed in the following form: let

$$f_N(x) := \sum_{n=0}^{N} A(n) T_n(x) = \sum_{n=0}^{N} C(n) x^n, \tag{10.54}$$

and suppose errors are introduced at each stage of the recursion process so that the quantity actually computed is $\bar{f}_N(x)$. Then

$$|\bar{f}_N(x) - f_N(x)| \le \varepsilon \sum_{n=0}^{N} \rho_n(x) |C(n)|,$$

or

$$|\bar{f}_N(x) - f_N(x)| \le \varepsilon \sum_{n=0}^{N} \sigma_n(x) |A(n)|,$$

depending on which form (54) of the polynomial is chosen. The ρ_n, σ_n are error amplification factors and ε the accuracy parameter. Oliver develops expressions for these factors and discusses how they vary with x. Two of his many conclusions are that (i) the accuracy of the methods of Clenshaw and Horner are sensitive to values of x and the errors tend to reach their extreme at the endpoints of the interval, and (ii) when a polynomial has coefficients of constant sign or strictly alternating sign, converting into the Chebyshev form does not bring any improvement in accuracy of evaluation. Anyone involved with polynomial approximations for digital computers should consult these references.

Other work on polynomial evaluation schemes can be found in Gentleman (1969), Mesztenyi and Witzgall (1967), and Newbery (1974, 1975).

In analogy to matrix theory, one can define a *condition number* for the space $\mathcal{P}_{n-1}$ of polynomials of degree $\le n-1$ on $[a, b]$. This number reflects the difficulties to be expected in evaluating members of $\mathscr{P}_{n-1}$. Let $B \equiv B_n : \mathscr{R}^n \to \mathscr{P}_{n-1}$ be defined by

$$B(a_0, a_1, \ldots, a_{n-1}) = a_0 + a_1 x + a_2 x^2 + \cdots + a_{n-1} x^{n-1}.$$

The *condition* of the map B is

$$\sigma(n) = \text{cond}_\infty B = \|B\|_\infty \|B^{-1}\|_\infty,$$

where all norms are sup norms. $\sigma(n)$ depends on $[a, b]$ and n. Gautschi (1979d) has found that on $[-1, 1]$, $\sigma(n)$ grows like $(1 + \sqrt{2})^n$ and on $[0, 1]$ like $(1 + \sqrt{2})^{2n}$.

One can also define a condition number for the map $B' : \mathscr{R}^n \to \mathscr{P}_{n-1}$ where

$$B'(a_0, a_1, \ldots, a_{n-1}) = a_0 p_0(x) + a_1 p_1(x) + a_2 p_2(x) + \cdots + a_{n-1} p_{n-1}(x),$$

where $\{p_n(x)\}$ is, say, a class of orthogonal polynomials. Gautschi (1972b) has found that for the Legendre polynomials and Chebyshev polynomials on $[-1, 1]$ that number is $\leq n(2n-1)^{1/2}$ and $\leq (\sqrt{2})n$ respectively. It is interesting to contrast these numbers with $\sigma(n)$. Although there is still not general agreement about the relative computational merits of truncated expansions in orthogonal polynomials vs. those truncated expansions rearranged in powers of x, these results seem to favor using the former.

10.2.3 Converting one expansion into another

It is often desirable to convert from one series

$$f_N(x) := \sum_{n=0}^{N} A(n) p_n(x), \tag{10.55}$$

into another

$$f_N(x) = \sum_{n=0}^{N} A^*(n) q_n(x), \tag{10.56}$$

where $\{p_n\}$, $\{q_n\}$ are sequences of polynomials satisfying the three-term recurrences

$$p_{n+1}(x) + (a(n) + b(n)x)p_n(x) + c(n)p_{n-1}(x) = 0, \tag{10.57}$$
$$p_{-1}(x) = 0, \qquad p_0(x) = 1, \qquad 0 \leq n \leq N-1,$$

$$q_{n+1}(x) + (a^*(n) + b^*(n)x)q_n(x) + c^*(n)q_{n-1}(x) = 0, \tag{10.58}$$
$$q_{-1}(x) = 0, \qquad q_0(x) = 1, \qquad 0 \leq n \leq N-1.$$

The orthogonality of neither $\{p_n\}$ nor $\{q_n\}$ is necessary for the existence of these recurrences since we do not assume $c(n), c^*(n) \neq 0$ and it turns out that the algorithm I shall develop in this section, due to Salzer (1973), has many useful applications when neither $\{p_n\}$ nor $\{q_n\}$ is orthogonal. The result is a sort of lozenge algorithm involving a five-term recurrence relation which generalizes a previous result of Hamming (1962) for which $p_n(x) = x^n$, $q_n(x) = T_n^*(x)$.

First let us recall the Clenshaw algorithm for evaluating the sum (55). We establish the recurrence

$$B(n) = -(a(n) + b(n)x)B(n+1) - c(n+1)B(n+2) + A(n),$$
$$n = N, N-1, \ldots, 0, \quad \textbf{(10.59)}$$

$$B(N+2) = B(N+1) = 0.$$

Then

$$f_N(x) = B(0)p_0(x) + B(1)[p_1(x) + (a(0) + b(0)x)p_0(x)]$$
$$= B(0). \qquad \textbf{(10.60)}$$

Define

$$B(N-k) := \sum_{n=0}^{k} A^{(k)}(n)q_n, \qquad 0 \leqslant k \leqslant N. \qquad \textbf{(10.61)}$$

We shall devise a method to update $A^{(k)}(n)$ to $A^{(k+1)}(n)$, thus going from $B(N-k)$ to $B(N-k-1)$. This way $B(0)$ may ultimately be computed which, we see from (61), will yield $A^{(N)}(n) = A^*(n)$, thus giving the desired coefficients.

We start with

$$B(N) = A(N),$$
$$B(N-1) = A^{(1)}(0)q_0 + A^{(1)}(1)q_1, \qquad \textbf{(10.62)}$$

where

$$A^{(1)}(0) = -a(N-1)A(N) + A(N-1) + \frac{b(N-1)A(N)a^*(0)}{b^*(0)}, \qquad \textbf{(10.63)}$$

$$A^{(1)}(1) = b(N-1)A(N)/b^*(0).$$

Given (62) and

$$B(N-k+1) = \sum_{n=0}^{k-1} A^{(k-1)}(n)q_n,$$

we have

$$B(N-k-1) = -a(N-k-1)\sum_{n=0}^{k} A^{(k)}(n)q_n - b(N-k-1)x$$
$$\times \sum_{n=0}^{k} A^{(k)}(n)q_n$$
$$-c(N-k)\sum_{n=0}^{k-1} A^{(k-1)}(n)q_n + A(N-k-1). \qquad \textbf{(10.64)}$$

Writing (58) as

$$xq_n = -\frac{c^*(n)}{b^*(n)}q_{n-1} - \frac{a^*(n)}{b^*(n)}q_n - \frac{1}{b^*(n)}q_{n+1},$$

we find that (64) implies the $A^{(k)}(n)$ satisfy a recurrence which can be written as

$$A^{(k+1)}(n) = -A^{(k-1)}(n)c(N-k) + \frac{A^{(k)}(n-1)}{b^*(n-1)}b(N-k-1)$$

$$+A^{(k)}(n)\left[-a(N-k-1) + b(N-k-1)\frac{a^*(n)}{b^*(n)}\right]$$

$$+A^{(k)}(n+1)b(N-k-1)\frac{c^*(n+1)}{b^*(n+1)} + \delta_{n0}A(N-k-1),$$

$$0 \leqslant n \leqslant k+1, \qquad -1 \leqslant k \leqslant N-1, \qquad \textbf{(10.65)}$$

with $A^{(j)}(i) = 0$, $i < 0$ or $i > j$. (This convention makes the recursion yield the correct initial value $A^{(0)}(0) = A(N)$.)

Example 10.3 It is often required to convert the polynomial

$$f_N(x) = \sum_{n=0}^{N} A(n)x^n,$$

to an expansion in Chebyshev polynomials,

$$f_N(x) = \sum_{n=0}^{N} A^*(n)T_n(x).$$

We have $a(n) \equiv c(n) \equiv 0$, $b(n) \equiv -1$, $a^*(n) \equiv 0$, $b^*(n) \equiv -\varepsilon_n$, $c^*(n) \equiv 1$, and the very simple recursion

$$A^{(k+1)}(n) = \frac{A^{(k)}(n-1)}{\varepsilon_{n-1}} + \frac{A^{(k)}(n+1)}{\varepsilon_{n+1}} + \delta_{n0}A(N-k-1),$$

$$0 \leqslant n \leqslant k+1, \qquad -1 \leqslant k \leqslant N-1,$$

$$\varepsilon_n = \begin{cases} 1, & n=0; \\ 2, & n>0. \end{cases}$$

By performing computations for a sufficiently large value of N it is possible to compute to any desired accuracy the Chebyshev coefficients for a function analytic in $\bar{N}_1$ given its Taylor coefficients. A remarkable feature of the algorithm is that it essentially requires no multiplications, which is a drawback of the method based on basic series given by Fields and Wimp (1963). ∎

Example 10.4 This example, due to Salzer (1976), shows that a polynomial which interpolates to a given function $f(x)$ at points

$x_0, x_1, \ldots, x_N$ can be conveniently rearranged as a series of (in this case) Chebyshev polynomials.

Write the Newton divided-difference formula as

$$f_N(x) = f(x_0) + \sum_{n=1}^{N} \prod_{i=0}^{n} (x - x_i)[x_0, x_1, \ldots, x_n],$$

in the notation of Milne-Thomson (1960, p. 3). We have $A(n) = [x_0, x_1, \ldots, x_n]$, $q_n(x) = \prod_{i=0}^{n}(x - x_i)$, $1 \leq n \leq N$, $a(n) = x_n$, $b(n) \equiv -1$, $c(n) = 0$. Again, the algorithm is one of great simplicity:

$$A^{(k+1)}(n) = \frac{A^{(k)}(n-1)}{\varepsilon_{n-1}} - x_{N-k-1}A^{(k)}(n) + \frac{A^{(k)}(n+1)}{\varepsilon_{n+1}}$$

$$+ \delta_{n0}A(N-k-1),$$

$$0 \leq n \leq k+1, \qquad -1 \leq k \leq N-1.$$

Converting an interpolation series into a power series is another case of great interest.

Salzer (1976) contains applications to other sorts of interpolation series. ∎

Example 10.5 In certain physical problems (e.g., the calculation of feeding coefficients to produce optimum beam patterns) it is necessary to evaluate the coefficients $A(n)$ in the equation

$$T_N(ax + b) = \sum_{n=0}^{N} A(n)T_n(x). \tag{10.66}$$

Often it is required to do this for fairly large values of N. In such cases a direct approach to the problem, i.e., the evaluation of explicit polynomials with terms $a^i b^j$ and with integral coefficients, may not be practical. When $n = 24$, for instance, the polynomial has 91 terms and some of the coefficients are 14 digit integers. Salzer (1975) has shown that recurrence methods are a far superior way to approach the problem.

There are two approaches, forward and backward recurrence.

1. *Forward recurrence* Define

$$T_k(ax + b) = \sum_{n=0}^{k} A^{(k)}(n)T_n(x). \tag{10.67}$$

Since $T_0(ax + b) = 1$, $T_1(ax + b) = ax + b$, using the recurrence for T_n shows that

$$A^{(0)}(0) = 1, \qquad A^{(1)}(0) = b, \qquad A^{(1)}(1) = a,$$

$$A^{(k+1)}(n) = 2bA^{(k)}(n) + a[A^{(k)}(n+1) + (1 + \delta_{n1})A^{(k)}(n-1)]$$

$$- A^{(k-1)}(n), \qquad 0 \leq m \leq k-1, \quad 1 \leq k \leq N-1,$$

$$A^{(j)}(i) = 0, \qquad i < 0, \quad i > j.$$

Thus the $A(n)$ in (66) for all values of $N_0 \leq N$ are obtained as by-products of the computation of the required coefficients.

2. *Backward recurrence* The following five-term recurrence follows from the work in this section:

$$A^{(k+1)}(n) = bA^{(k)}(n)[2 - \delta_{N-1,k}]$$
$$+ [1 - \tfrac{1}{2}\delta_{N-1,k}]a(A^{(k)}(n+1)$$
$$+ [1 + \delta_{n,1}]A^{(k)}(n-1)) - A^{(k-1)}(n),$$
$$0 \leq n \leq k+1, \qquad 1 \leq k \leq N-1,$$

with

$$A^{(0)}(0) = 1, \qquad A^{(1)}(0) = 2b, \qquad A^{(1)}(1) = 2a.$$

Then $A(n) = A^{(N)}(n)$. Notice these As are not those of the previous method (1) and so the backward recurrence will not yield the Chebyshev coefficients for $T_k(ax+b)$ for $0 \leq k \leq N$ as does the forward recurrence method.

Two recent approximate methods have been developed for dealing with this problem. Drane (1964) gives a closed form approximation which helps avoid the loss of significant figures due to summing large terms which alternate in sign (which is appreciable when $N \sim 10$). He expresses

$$K(n) := \frac{2}{\pi} \int_{-1}^{1} (T_n(ax+b)T_n(x)/\sqrt{(1-x^2)}) \, dx,$$

in terms of the modified Bessel function I_1 and gives numerics for $N = 10$ to indicate a satisfactory agreement with exact results. Van der Maas (1954) treats the special case $b = 0$. ∎

11 Series solutions to ordinary differential equations

When one attempts to satisfy a linear differential equation with a series of the form

$$\sum C(n)p_n(z),$$

and the functions $\{p_n\}$ satisfy linear difference equations it is often found that the coefficients $C(n)$ satisfy linear *difference* equations. If the base functions $\{p_n\}$ are chosen appropriately so that the previous expansion converges rapidly then it is reasonable to expect that $C(n)$ will be a minimal solution of the difference equation it satisfies and therefore that it can be computed efficiently from the equation by the Miller or Olver algorithms or one of their variants. Such indeed is usually the case and research into this and related ideas has provided a highly effective body of techniques for approximating the solutions of differential equations, even nonlinear equations.

There are actually two topics of interest. The first is the derivation of the required difference equation for the coefficients $C(n)$. Work in this area has been done by many mathematicians, but the largest and most comprehensive theory is due to the Polish mathematician S. Lewanowicz. I discuss some of his findings in Section 11.2.3.

The second topic is demonstrating the convergence of the Miller (or related) algorithm. It seems nearly impossible to make general assertions about convergence. However, when $p_n(z) := (z-a)^n$ (the Taylor series case), an examination of the solutions of the adjoint *differential* equation can often be used to decide convergence of the computational scheme for $C(n)$ because this equation can be related in a simple way to the adjoint difference equation which, as we have seen, monitors the convergence of the Miller algorithm. This work is due primarily to Thacher, although I have included my own elaborations and modifications. The more difficult case, when the $\{p_n\}$ satisfy nontrivial difference equations, has been discussed primarily by Clenshaw, Elliott and Norton. Although explicit error estimates are difficult to come by, the computational techniques seem to be remarkably effective, particularly when the $\{p_n\}$ are the Chebyshev polynomials. Indeed exhaustive tables for the Chebyshev coefficients for the expansions of many important transcendental functions have been compiled using the techniques described in this chapter.

Since the computation of the coefficient $C(n)$ by means of its integral representation requires a detailed knowledge of the value of the defining function and in any event is numerically a highly unstable process, it is unlikely that the coefficients could have been obtained any other way.

The discussion of the second topic occupies the remainder of the chapter.

11.1 Taylor series solutions

The linear differential equation to be solved will be written in the form

$$\mathcal{L}[w(t)] := \sum_{\nu=0}^{\sigma} p_\nu(t)D^\nu w(t) = h(t), \tag{11.1}$$

where p_ν, h are polynomials,

$$p_\nu(t) := \sum_{j=0}^{m_\nu} p_{\nu j}t^j, \qquad p_{\sigma,m_\sigma} \neq 0,$$

$$h(t) := \sum_{j=0}^{\mu} h_j t^j.$$

Suppose we seek a solution $w(t)$ to (1) analytic at 0 satisfying

$$w^{(j)}(0) = a_j, \qquad 0 \leq j \leq \sigma - 1. \tag{11.2}$$

Let

$$w(t) := \sum_{j=0}^{\infty} w_j t^j. \tag{11.3}$$

Define

$$m := \sup_{\nu} (m_\nu - \nu).$$

Note

$$m \geq 0, \qquad m_\nu \leq m + \nu, \qquad \text{for every } \nu.$$

Now I assume

$$\mu < m.$$

In what follows it will be notationally convenient to make the interpretations

$$p_{\nu j} := 0, \quad \nu > \sigma, j < 0; \qquad j > m_\nu; \qquad w_k := 0, \quad k < 0. \tag{11.4}$$

Substituting the series (3) into (1) and selecting the coefficient of t^k gives

$$\sum_{r=0}^{\infty} w_r \sum_{\nu=0}^{r} \frac{p_{\nu,k+\nu-r}}{(r+1)_{-\nu}} = h_k, \qquad k \geq 0,$$

or, by (4),

$$\sum_{r=-m}^{\sigma} w_{r+k} \sum_{\nu=0}^{r+k} \frac{p_{\nu,\nu-r}}{(r+k+1)_{-\nu}} = h_k, \qquad k \geqslant 0. \tag{11.5}$$

(5) can be decomposed into two equations, the equation which, for $k = 0, 1, 2, \ldots, m-1$, comprises m initial conditions for the determination of $w_\sigma, w_{\sigma+1}, \ldots, w_{\sigma+m-1}$, and the recurrence relation of order $\sigma + m$,

$$\mathcal{U}[w_k] = \sum_{j=0}^{\sigma+m} w_{j+k} A_j(k) = 0, \qquad k \geqslant 0,$$

$$A_j(k) = \sum_{\nu=0}^{j+k} \frac{p_{\nu,\nu+m-j}}{(j+k+1)_{-\nu}}. \tag{11.6}$$

I will assume, in what follows, that $A_0(k) A_{\sigma+m}(k) \neq 0$, $k \geqslant 0$.

We now ask whether the Taylor series coefficients of $w(t)$ can be computed by applying the Miller algorithm of Chapter 4 to equation (6). As determined in that chapter the success of the Miller algorithm depends on the behavior of the solutions of the adjoint equation

$$\hat{\mathcal{U}}(w_k) := \sum_{j=0}^{\sigma+m} w_{j+k} \hat{A}_j(k) = 0, \qquad k \geqslant 0,$$

$$\hat{A}_j(k) = \sum_{\nu=0}^{\sigma} \frac{p_{\nu,j+\nu-\sigma}}{(\sigma+m+k+1)_{-\nu}}. \tag{11.7}$$

A little algebra shows that the differential operator $\hat{\mathcal{L}}$ which generates the adjoint equation in the sense that $\mathcal{L}$ generated $\mathcal{U}$ is

$$\hat{\mathcal{L}}[w(t)] = \sum_{\nu=0}^{\sigma} t^\nu D^\nu \left[t^{m+\nu} p_\nu\left(\frac{1}{t}\right) w(t) \right]. \tag{11.8}$$

When the series

$$w(t) = \sum_{k=0}^{\infty} w_k t^k,$$

is substituted into $\hat{\mathcal{L}}$ and the coefficient of t^k is selected we find

$$\partial_k \hat{\mathcal{L}}[w(t)] = \sum_{r=k-\sigma-m}^{k} w_r \sum_{\nu=0}^{\sigma} \frac{p_{\nu,m+r+\nu-k}}{(k+1)_{-\nu}},$$

(with the understanding that $w_r = 0$, $r < 0$). Thus letting $R(t)$ range over all possible polynomials of degree $\leqslant \sigma + m - 1$ in

$$\hat{\mathcal{L}}[w(t)] = R(t),$$

will generate all possible solutions of the adjoint recurrence $\hat{\mathcal{U}}(w_k) = 0$ provided

$$\sum_{\nu=0}^{\sigma} \frac{p_{\nu,m+\nu}}{(k+1)_{-\nu}} \neq 0, \qquad 0 \leqslant k \leqslant m + \sigma - 1,$$

which I will henceforth assume.

To complete the analysis of this situation we need

Definition 11.1 *The formal series $\sum_{k=0}^{\infty} f_k t^k$ (see Appendix C) is said to be* root-proper *if the (possibly infinite) limit* $\lim_{k \to \infty} |f_{k+1}/f_k|$ *exists.* ∎

I can now formulate a sufficient condition for the convergence of the Miller algorithm as applied to equation (6).

Theorem 11.1 *As $R(t)$ ranges over all nontrivial polynomials of degree $\leqslant \sigma + m - 1$ let $\mathcal{L}[w(t)] = R(t)$ have a root-proper formal series solution $w^{(1)}(t)$ such that there are $\sigma - 1$ other linearly independent root-proper solutions $\{w^{(i)}(t)\}$ with radius of convergence larger than $w^{(1)}(t)$.*

Then the Miller algorithm based on (6) and properly normalized converges and the quantities w_k so computed are the Taylor series coefficients for a solution of (1) holomorphic at 0 provided the relations (5) are satisfied for $0 \leqslant k \leqslant m - 1$.

Proof In the notation of A-VI I can write

$$\frac{y_N(k)}{y_N(r)} = \sum_{j=1}^{\sigma+m} \hat{y}^{(j)}(N) y^{(j)}(k) \bigg/ \sum_{j=1}^{\sigma+m} \hat{y}^{(j)}(N) y^{(j)}(r), \tag{11.9}$$

for any basis $\{\hat{y}^{(j)}\}$ of the adjoint equation. Now let $\hat{y}^{(1)}(k)$ correspond to the w_k generated by $w^{(1)}(t)$ and $\hat{y}^{(h)}(k)$ to those generated by $w^{(h)}(t)$, $2 \leqslant h \leqslant \sigma + m$, and let r (fixed) be such that $y^{(1)}(r) \neq 0$. Then $\{\hat{y}^{(h)}(k)\}$ is a basis for the adjoint equation. Since $w^{(1)}(t)$ is root-proper, $w_k \neq 0$ for k sufficiently large. One can easily confirm that $\hat{y}^{(1)}(k)$ dominates $\hat{y}^{(h)}(k)$, $2 \leqslant h \leqslant \sigma + m$. Thus the algorithm converges.

To show that the coefficients so generated form a Taylor series with a nonzero radius of convergence is left as an exercise. ∎

Example 11.1 (The incomplete Gamma function) In this example convergence of the Miller algorithm can be decided directly, without an appeal to Theorem 1. But the example provides an interesting and rare case of a three-term recurrence where initial values can actually be determined so as to generate a solution of prescribed asymptotic growth.

Consider the equation

$$xh'(x) + (a - x)h(x) = 1, \tag{11.10}$$

with the general solution

$$h(x) = cx^{-a}e^x + e^x\Gamma(a)\gamma^*(a, x),$$

$$\gamma^*(a, x) := \frac{x^{-a}}{\Gamma(a)}\int_0^x e^{-u}u^{a-1}\, du.$$ **(11.11)**

The substitution

$$t := 1 - \frac{x}{\eta}, \qquad w(t) := h[\eta(1-t)],$$

makes the origin an ordinary point and the differential equation becomes

$$\mathscr{L}[w(t)] = (t-1)w'(t) + [\eta t + (a-\eta)]w(t) = 1.$$ **(11.12)**

We have

$$p_0(t) = \eta t + (a-\eta), \quad p_1(t) = t-1; \qquad m = \sigma = 1.$$

The recursion relation (6) is

$$\mathscr{U}(w_k) = \eta w_k + (k+1+a-\eta)w_{k+1} - (k+2)w_{k+2} = \delta_{0,k+1},$$

$$k \geqslant -1, \qquad w_{-1} = 0.$$ **(11.13)**

Also

$$\hat{\mathscr{L}}[w(t)] = t^2(1-t)w'(t) + [\eta + (a+1-\eta)t - 2t^2]w(t),$$ **(11.14)**

and

$$\hat{\mathscr{U}}(w_k) = (k+2)w_k - (a-\eta+k+2)w_{k+1} - \eta w_{k+2}.$$ **(11.15)**

The theory of Section B.2 shows that the recursion

$$\hat{\mathscr{U}}(w_k) = 0$$

has a fundamental set $\{w_k^{(1)}, w_k^{(2)}\}$ with the asymptotic behavior

$$w_k^{(1)} = k^{k+\theta_1}(-\eta/e)^k[1 + O(k^{-1})],$$ **(11.16)**

$$w_k^{(2)} = k^{\theta_2}[1 + O(k^{-1})].$$ **(11.17)**

Thus the Miller algorithm, properly normalized, will converge. (Note in this case it is not necessary to solve the adjoint differential equation.)

Using the recursion relation for the Charlier polynomials $c_n(x; a)$ (see Erdélyi *et al.* (1953) vol. 2, p. 227 (8)) enables one to solve the recurrence (15) explicitly. One solution is found to be

$$u_k^{(1)} = (a+1)_k(-\eta)^{-k}\Phi(1-a; -a-k; \eta).$$ **(11.18)**

(For the above definition to make sense it is necessary to require that $-a-k \neq 0, -1, -2, \ldots$, but the definition is easily modified if this condition is violated.)

Another solution may be determined by the techniques of Wimp

(1967) to be

$$u_k^{(2)} = \frac{(k+1)!}{(a+2)_k} \Phi(a; a+k+2; -\eta).$$
(**11.19**)

Obviously $w_k^{(j)}$ can be identified with a constant multiple of $u_k^{(j)}$, $j = 1, 2$. The series $\sum u_k^{(1)} t^k$ is a formal series solution (not convergent) to the equation

$$\hat{\mathscr{L}}[w(t)] = c_0 + c_1 t,$$
(**11.20**)

and the series $\sum u_k^{(2)} t^k$ is a series convergent for $|t| < 1$ (for suitable c_0, c_1).

It is interesting to compute those values of c_0, c_1 which will produce the solution of the recurrence of algebraic growth, (17). It is readily found that

$$c_0 = \eta \Phi(a; a+2; -\eta), \qquad c_1 = (a+1)\Phi(a; a+1; -\eta),$$

produce $u_k^{(2)}$ and thus generate the solution of the adjoint equation (20) analytic at 0.

Obviously Theorem 1 applies and I can infer that the Miller algorithm based on (13) converges. (The solutions of (20) can be displayed explicitly as Horn functions, Φ_1.)

The result of applying the Miller algorithm for $\eta = 5$, $a = \frac{11}{2}$, $Cw_{15} = 1$, $Cw_{16} = 0$ is shown in Table 11.1, whose data are due to Thacher (1972).

Table 11.1   Computation of the incomplete Gamma function $w_0 = e^5 \Gamma(5.5) \gamma^*(5.5, 5)$.

k	Approx. Cw_k	Approx. w_k
15	1.000 00	−0.000 007 88
14	−3.100 00	0.000 024 44
13	11.990 00	−0.000 094 53
12	−41.053 00	0.000 323 66
11	133.806 50	−0.001 054 93
10	−406.282 15	0.003 203 12
9	1 147.566 82	−0.009 047 40
8	−2 922.941 25	0.023 596 30
7	7 153.620 39	−0.056 399 03
6	−15 519.136 60	0.122 352 62
5	30 189.946 10	−0.238 017 04
4	−51 831.904 60	0.408 641 89
3	76 838.660 30	−0.605 794 75
2	−95 252.585 90	0.750 969 84
1	93 729.489 10	−0.738 961 77
0	−66 219.881 10	0.522 076 47
C	−126 839.430 00	

(True $w_0 = 0.522\ 076\ 473\ 93$)

We need no initial condition to determine C because the first equation in (13) requires

$$(a - \eta)w_0 - w_1 = 1.$$

This equation implies that $c = 0$ in (11) and thus w_k provides the Taylor coefficients for the function

$$w(t) = e^{\eta(1-t)}\Gamma(a)\gamma^*(a; \eta(1-t)).$$

Since

$$w_0 = e^\eta \Gamma(a)\gamma^*(a; \eta),$$

this algorithm provides a very efficient way of computing the incomplete Gamma function $\gamma^*(a, \eta)\Gamma(a)$ for all complex values of a, η.

Note that

$$w(1) = \frac{1}{a} = 0.181\,818\,18 = \sum_{k=0}^{\infty} w_k.$$

The value obtained by summing the entries in the last column in the table is $0.181\,811\,01$. ∎

Example 11.2   The equation

$$xh'(x) + (a - x - 1)h(x) = -x,$$

has the general solution

$$h(x) = cx^{1-a}e^x + e^x x^{1-a}\,\Gamma(a, x),$$

$$\Gamma(a, x) = \int_x^\infty e^{-u}u^{a-1}\,du.$$

Since $e^x x^{1-a}\Gamma(a, x) \sim 1$, $x \to \infty$, it is appropriate to make the substitution

$$t := 1 - \frac{\eta}{x}, \qquad w(t) := h[\eta/(1-t)].$$

The resulting differential equation for $w(t)$ is

$$(1-t)^2 w'(t) - [(\eta + 1 - a) - (1-a)t]w(t) = -\eta,$$

which leads to the recurrence

$$\mathcal{U}(w_k) = (k+1-a)w_k - (2k+3+\eta - a)w_{k+1} + (k+2)w_{k+2} = 0,$$

$$k \geq 0,$$

$$w_1 + (a - 1 - \eta)w_0 = -\eta.$$

The boundary condition at ∞ yields the additional condition

$$\sum_{k=0}^{\infty} w_k = 1.$$

The adjoint difference equation is

$$(k+2)w_k - (2k+5+\eta-a)w_{k+1} + (k+3-a)w_{k+2} = 0,$$

and from the work in Section 5.2 we see this has solutions

$$w_k^{(1)} = (k+1)! \, \Psi(k+2-a, 1-a; \eta),$$
$$w_k^{(2)} = \Phi(k+2-a, 1-a; \eta),$$

the latter by Kummer's transformation being essentially a Laguerre polynomial. The leading terms for the Birkhoff series for these functions are

$$k^\theta e^{\pm 2\sqrt{(k\eta)}}.$$

Thus the corresponding differential equation $\hat{\mathcal{L}}[w(t)] = R(t)$ has all non-trivial solutions holomorphic in $|t| < 1$ and Theorem 1 fails to demonstrate convergence. However the algorithm *does* converge, as a direct analysis based on Theorem 4.1(3) will show.

The results of a computation for $\eta = 5$, $a = \frac{11}{2}$, again due to Thacher, are shown in Table 11.2. The value obtained for $w_0 = e^\eta \eta^{1-a} \Gamma(a, \eta)$ is accurate to 1×10^{-8}. ∎

Other examples given in Thacher's paper are the computation of the Struve function $\mathbf{H}_\nu(x)$ from

$$x^2 y''(x) + x y'(x) + (x^2 - \nu^2) y(x) = (2\nu+1)(x/2)^{\nu+1}/\Gamma(\tfrac{3}{2})\Gamma(\nu+\tfrac{3}{2}),$$

the exponential integral $E_n(x)$ from

$$x(n+x)y'(x) + [n - (n+x)^2]y(x) = -(n+x)^2,$$

the modulus of the Bessel functions from the third-order equation,

$$x^3 y'''(x) + x(4x^2 - m)y'(x) + my(x) = 0,$$

and the complete elliptic integral $K(x)$ from

$$x(1-x)y''(x) + (1-2x)y'(x) - \frac{y(x)}{4} = 0.$$

I will present one more example of this technique.

Example 11.3 (Thacher (1979)) The differential equation

$$(1-t)^2 w''(t) - 2(1+\sigma-t)w'(t) - \frac{4\nu^2-1}{4} w(t) = 0, \qquad \textbf{(11.21)}$$

Table 11.2   Computation of the incomplete Gamma function $w_0 = e^5 5^{-4.5} \Gamma(5.5, 5)$.

k	Approx. Cw_k	Approx. w_k
10	1.000 000 00	−0.000 000 26
9	4.555 555 56	−0.000 001 17
8	21.222 222 2	−0.000 005 43
7	123.666 667	−0.000 031 66
6	1 082.259 26	−0.000 277 11
5	25 325.148 1	−0.006 484 53
4	−518 841.000	0.132 849 78
3	3 024 516.16	−0.774 430 55
2	−8 693 887.62	2.226 079 08
1	13 770 297.9	−3.525 899 29
0	−11 514 115.6	2.948 201 42
C	−3 905 471.14	

(True $w_0 = 2.948\,201\,420$)

is satisfied by

$$w(t) := \sqrt{\left(\frac{2\sigma}{\pi(1-t)}\right)} \, e^{\sigma/(1-t)} K_\nu\left[\frac{\sigma}{1-t}\right]. \tag{11.22}$$

Assuming

$$w(t) := \sum_{k=0}^{\infty} w_k t^k, \tag{11.23}$$

leads to the three-term recurrence for w_k,

$$(k + \nu + \tfrac{1}{2})(k - \nu + \tfrac{1}{2})y(k) - 2(k+1)(k+1+\sigma)y(k+1)$$
$$- (k+1)(k+2)y(k+2) = 0.$$

This is just the recurrence (5.19) which is satisfied by a multiple of $\Psi(k + \nu + \tfrac{1}{2}, 2\nu + 1; 2\sigma)$. In fact the series (23) is a consequence of the generating function (see Erdélyi *et al.* (1954, vol. 1, p. 283 (6))),

$$\Psi\left(a, c; \frac{t}{1-x}\right) = (1-x)^a \sum_{k=0}^{\infty} \frac{(a)_k(a+1-c)_k}{k!} \Psi(k+a, c; t)x^k.$$

Since the desired solution (22) of (21) $\to 1$ as $t \to 1$, the necessary normalization is $1 = \sum_{k=0}^{\infty} w_k$ and $w_0 = \sqrt{(2\sigma/\pi)}e^\sigma K_\nu(\sigma)$. The algorithm for the computation of w_0 is thus the same as the method for the computation of K_ν presented in Section 5.2. ■

11.2 The construction of general recurrence relations for the coefficients of Gegenbauer series

11.2.1 Introduction and basic formulas

Let $f(x)$ be a function defined in $[-1, 1]$ which has the uniformly convergent expansion

$$f(x) = \sum_{n=0}^{\infty} B(n)C_n^{\lambda}(x), \tag{11.24}$$

$B(n) \equiv B(n, f)$, and, in addition, let f satisfy the linear differential equation of order σ,

$$\sum_{\nu=0}^{\sigma} p_{\nu}(x)D^{\nu}f(x) = p(x), \qquad p_{\sigma} \not\equiv 0, \tag{11.25}$$

where $p_0, p_1, \ldots, p_{\sigma}$ are polynomials and where the Gegenbauer coefficients of the function p are known. I will also assume that $f^{(\sigma)}(x)$ has a uniformly convergent Gegenbauer series. In these sections I will give a method for constructing a linear recurrence for $B(n)$ with coefficients which are rational functions of n. Once such a recurrence is found it is often a trivial matter to compute $B(n)$ by one of the several techniques discussed in the previous chapters.

Most of the results given here on such recurrences were obtained by Lewanowicz (1976) who generalized earlier work of Paszkowski (1975). The latter treated the case $\lambda = 0$ (the Chebyshev polynomials). (Much earlier, Clenshaw had derived certain mixed recurrences. This work is discussed in Section 11.2.2.)

Only in certain cases will the algorithm we present yield a recurrence of optimum order. Lewanowicz has given optimum recurrence relations for both Gegenbauer and Jacobi polynomial expansions (1976, 1980). The algorithms for constructing such recurrences are very complicated, and to keep matters simple I prefer to present a method which sometimes leads to less than optimal recurrences and does not treat the most general case, i.e., Jacobi polynomials, but which is easy to apply and covers the most frequently sought polynomial expansions, i.e., Chebyshev and Legendre series.

Recall

$$B(n, f) = \frac{1}{h(n)} \int_{-1}^{1} (1-x^2)^{\lambda-1/2} C_n^{\lambda}(x)f(x) \, dx,$$

$$h(n) = (\sqrt{\pi})(2\lambda)_n \Gamma(\lambda + \tfrac{1}{2})/(n+\lambda)n! \, \Gamma(\lambda).$$

The case $\lambda = 0$, because of the necessity of distinguishing between $n = 0$ and $n > 0$ in (24), (25) causes notational problems. Let the Chebyshev series for f be denoted

$$f(x) = \sum_{n=0}^{\infty}{}' C(n, f) T_n(x).$$

It is somewhat easier to work with the related coefficients defined by

$$E(n, f) := \begin{cases} (n+\lambda)^{-1} B(n, f), & \lambda \neq 0, \\ C(n, f), & \lambda = 0. \end{cases} \tag{11.26}$$

We will also have a use for coefficients with negative order, $n < 0$. If $2\lambda = m$, m a nonnegative integer, then we will take

$$E(-n, f) := \begin{cases} 0, & 1 \leqslant n \leqslant m-1, \\ E(n-m, f), & n \geqslant m, \end{cases} \tag{11.27}$$

while if 2λ is not an integer, we take

$$E(-n, f) := 0, \qquad n \geqslant 1. \tag{11.28}$$

Now assume that f' can be expanded in a uniformly convergent series of Gegenbauer polynomials. In view of the formulas (24) and (10.37) we find that

$$2(n+\lambda) E(n, f) = E(n-1, f') - E(n+1, f'), \qquad n \geqslant 1. \tag{11.29}$$

Similarly, from the recurrence relation (10.35) we see that

$$E(n, xf) = \tfrac{1}{2}[\alpha(n) E(n-1, f) + \beta(n) E(n+1, f)], \qquad n \geqslant 1, \tag{11.30}$$

where

$$\alpha(n) := \begin{cases} n/(n+\lambda), & \lambda \neq 0, \\ 1, & \lambda = 0, \end{cases} \tag{11.31}$$

$$\beta(n) := 2 - \alpha(n).$$

In what follows, we will need a generalization of (30). Define for an arbitrary function μ of the variable n

$$\mu^+(n) := \mu(n+1), \qquad \mu^-(n) := \mu(n-1).$$

By induction, one finds

$$E(n, x^l f) = 2^{-l} \sum_{j=0}^{l} \alpha_{lj}(n) E(n-l+2j, f), \qquad n, l \geqslant 0. \tag{11.32}$$

where

$$\alpha_{lj}(n) := \binom{l}{j}, \qquad \lambda = 0,$$

$$\alpha_{00}(n) := 1,$$

$$\alpha_{lj}(n) := \begin{cases} \alpha(n)\alpha_{l-1,0}^-(n), & j = 0, \\ \alpha(n)\alpha_{l-1,j}^-(n) + \beta(n)\alpha_{l-1,j-1}^+(n), & 1 \le j \le l-1, \quad l \ge 1, \\ \beta(n)\alpha_{l-1,l-1}^+(n), & \lambda \ne 0. \end{cases}$$

$$\left. \vphantom{\begin{cases}\\\\\\\\\end{cases}} \right\}$$

(11.33)

The desired recurrence relation is of the form

$$\sum_{k=0}^{\rho} A_k(n) E(n+k, f) = \mu(n), \qquad n \ge 0, \tag{11.34}$$

where $\mu(n)$, $A_k(n)$ are known functions of n, $A_k(n)$ being rational. In addition there are one or more normalizing conditions of the form

$$\sum_{n=0}^{\infty} u(n) B(n) = v. \tag{11.35}$$

Such conditions, along with the recurrence (34), can often serve as the basis for the computation of $E(n)$ via the techniques of the previous chapters.

Before I begin the discussion of the derivation of (34) I will first devote a section to a simple method due to Clenshaw for obtaining *mixed*, rather than *pure*, recurrence relations. These recurrences involve $B(n)$ and the coefficients

$$E^{(i)}(n) := E(n, f^{(i)}), \qquad 0 \le i \le \sigma. \tag{11.36}$$

(I will also use the notation $E(n)$, $E'(n)$, $E''(n)$ instead of $E^{(0)}(n)$, $E^{(1)}(n)$, $E^{(2)}(n)$, etc.) Though these recurrences are not of the form (34) they also can often be used in a straightforward manner to compute the $E(n)$. Because such recurrences are not simple linear difference equations, however, it is much more difficult to make definitive statements about the convergence of algorithms based on their use. I shall confine myself primarily to the derivation of the algorithms and presenting exploratory numerical computations.

11.2.2 The algorithms of Clenshaw and Elliott

These recurrences were first explored by Clenshaw (1957) in a famous paper (for the case of Chebyshev polynomials) and subsequently generalized by Elliott (1960) to Gegenbauer polynomials. Clenshaw (1962) used these recurrences to tabulate the Chebyshev coefficients for a number of important mathematical functions satisfying linear differential equations. Oliver (1969) has studied the error involved for the Chebyshev case.

Let f satisfy the differential equation (25). Since the Gegenbauer coefficients on both sides are equal, we have

$$\sum_{k=0}^{\sigma} E(n, p_k f^{(k)}) = E(n, p), \qquad n \geqslant 0. \tag{11.37}$$

This same equation holds for negative n. But (27), (28) imply that then the above relation is either (i) for $2\lambda = m$, m a nonnegative integer, equivalent to the relation obtained from (37) by substituting $-(n+m) \geqslant 0$ for n, or (ii) trivial.

Because of the expression (32) for $E(n, x^l f)$ we may write the above as

$$E(n, p_i f^{(i)}) = \mathcal{L}_0^{(i)}(E^{(i)}(n)), \qquad 0 \leqslant i \leqslant \sigma, \tag{11.38}$$

where the operator $\mathcal{L}_0^{(i)}$ is defined by

$$\mathcal{L}_0^{(i)}(E^{(i)}(n)) := \sum_{j=-d_i}^{d_i} \lambda_j^{(i)}(n) E^{(i)}(n+j), \tag{11.39}$$
$$d_i := \partial p_i,$$

and the $\lambda_j^{(i)}$ are rational functions of n. If $\lambda \neq 0$ then by virtue of (31)–(33) we have, in particular,

$$\lambda_{-d_i}^{(i)}(n) = p_{i0}(n - d_i + 1)_{d_i} / (n + \lambda - d_i + 1)_{d_i},$$
$$\lambda_{d_i}^{(i)}(n) = p_{i0}(n + 2\lambda)_{d_i} / (n + \lambda)_{d_i},$$

where p_{i0} is defined by

$$p_i(x) = p_{i0} x^{d_i} + O(x^{d_i - 1}).$$

If $\lambda = 0$ (the Chebyshev case) then the $\lambda_j^{(i)}$ are constant and have the property

$$\lambda_j^{(i)} = \lambda_{-j}^{(i)}, \qquad 1 \leqslant j \leqslant d_i,$$
$$\lambda_{-d_i}^{(i)} = \lambda_{d_i}^{(i)} = p_{i0}.$$

Equations (38) and (39) imply

$$\sum_{i=0}^{\sigma} \mathcal{L}_0^{(i)} E^{(i)}(n) = E(n, p), \qquad n \geqslant 0. \tag{11.40}$$

This equation and the relations

$$2(n + \lambda) E^{(i-1)}(n) = E^{(i)}(n-1) - E^{(i)}(n+1), \qquad 1 \leqslant i \leqslant \sigma, \tag{11.41}$$

constitute a system of $\sigma + 1$ linear difference equations involving the unknown sequences $\{E^{(i)}(n)\}$, $0 \leqslant i \leqslant \sigma$, and serve as the backbone of the Clenshaw and Elliott algorithms.

In general boundary conditions on the solution $f(x)$ will be given at $x = 0$ or $x = \pm 1$. For these points we have

$$C_n^\lambda(1) = \Gamma(n + 2\lambda)/\Gamma(2\lambda)n!,$$
$$C_n^\lambda(-1) = (-1)^n C_n^\lambda(1),$$
$$C_{2n+1}^\lambda(0) = 0, \quad C_{2n}^\lambda(0) = (-1)^n \Gamma(n + \lambda)/n! \Gamma(\lambda), \quad \lambda \neq 0.$$

Thus from the differential equation and the associated boundary conditions we obtain a set of linear equations in the unknowns $E^{(i)}(n)$, $0 \leq i \leq \sigma$. Clenshaw (1957) has developed two methods for solving these equations.

Method I This is a recurrence method. One assumes $E^{(i)}(n) = 0$, $n \geq N$, for some suitable N. One then assigns arbitrary values to $E^{(i)}(N)$ and the equations (40), (41) are used to generate $E^{(i)}(n)$, $n = N-1, N-2, \ldots, 0$. This is done in the following systematic way. Pick $n = N$ in (41) and compute $E^{(i)}(N-1)$, $1 \leq i \leq \sigma$, and then take n suitably in (39), (40) to compute $E(N-1)$. The process is then repeated with $n = N-2$ in (41), etc. Of course the $E(n)$ ultimately computed will not satisfy the boundary conditions. If $p(x) \equiv 0$, one may then attempt to take linear combinations of solutions for different values of N or different arbitrary starting values to satisfy the initial conditions. If $p(x) \neq 0$, then one may generate different solutions of the homogeneous analogue of (40) (taking $E(n, p) = 0$) as well as (40) itself and attempt to satisfy the boundary conditions by taking suitable linear combinations of the homogeneous solutions added to the nonhomogeneous solution. Unfortunately, every case presents its own peculiarities, so it is hardly possible to give a general treatment of the algorithm.

Method II This is an iterative method. It starts with an initial guess for $\{E(n)\}$ which satisfies the boundary conditions. From these values (41) is used to compute $\{E^{(i)}(n)\}$, $1 \leq i \leq \sigma$, and then (40) to obtain corrected values of $\{E(n)\}$. As the reader is probably aware, such schemes often do not converge, or converge slowly. Since Method I is often quite rapidly convergent, the iterative method is perhaps most useful in correcting small errors due to rounding which arise in the application of Method I.

I will give a number of examples.

Example 11.4 (The exponential integral) Let

$$-Ei(-w) := \int_w^\infty \frac{e^{-t}}{t} \, dt, \quad w > 0.$$

This function has the asymptotic expansion

$$-Ei(-w) \sim \frac{e^{-w}}{w}\left[1 - \frac{1}{w} + \frac{2!}{w^2} - \frac{3!}{w^3} + \cdots\right], \qquad w \to \infty,$$

and so it occurs to us to define

$$-Ei(-w) := \frac{e^{-w}}{w}\, y(x), \qquad x := \frac{1}{w}.$$

Thus $y(x)$ satisfies

$$x^2 y' + (1 + x)y = 1. \qquad (11.42)$$

I will solve the above equation in the range $0 \leqslant x \leqslant 1$ which will correspond to a w-range $1 \leqslant w \leqslant \infty$. The natural polynomials for effecting the solution are the shifted Chebyshev polynomials $T_n^*(x)$. (Alternatively one could simply make the substitution $x = (t+1)/2$ in (42).)

Using the properties of the shifted polynomials given in Section 10.1.1 we find that (40) takes the form

$$\tfrac{1}{16}\{E'(|n-2|) + 4E'(|n-1|) + 6E'(n) + 4E'(n+1) + E'(n+2)\}$$
$$+ \tfrac{3}{2}E(n) + \tfrac{1}{4}E(|n-1|) + \tfrac{1}{4}E(n+1) = E(n, 1), \qquad (11.43)$$

since

$$\frac{d}{dx} f(x) = \frac{1}{2} \frac{d}{dt} f\!\left(\frac{t+1}{2}\right).$$

Also (41) implies

$$4nE(n) = E'(n-1) - E'(n+1), \qquad (11.44)$$

and, of course,

$$y(x) = \sum_{n=0}^{\infty}{}' E(n)T_n^*(x).$$

Equation (44) may be used to eliminate $E'(n+2)$, $E'(|n-2|)$ from (43) and we find the equations may be written

$$E(n-1) = E(n+1) - \frac{1}{n}[E'(n+1) + 2E'(n) + E'(n-1) + 6E(n)],$$

$$E'(n-1) = E'(n+1) + 4nE(n), \qquad n \geqslant 1.$$

For $n = 0$ the left-hand side of (43) may be written

$$K = 2E'(0) + 2E'(1) + 6E(0).$$

All the equations used to compute $E(n)$, $E'(n)$ are homogeneous. Thus if we multiply the $E(n)$ so generated by

$$C = \frac{4E(n, 1)}{K} = \frac{8}{K},$$

Table 11.3

		Trial		Final
n	$E(n)$		$E'(n)$	$E(n)$
0	+707 936		−417 812	+1.515 744
1	− 89 622		+162 224	−0.191 887
2	+ 17 516		− 59 324	+0.037 503
3	− 4 238		+ 22 096	−0.009 074
4	+ 1 173		− 8 468	+0.002 511
5	− 357		+ 3 328	−0.000 764
6	+ 117		− 1 328	+0.000 251
7	− 41		+ 520	−0.000 088
8	+ 15		− 180	+0.000 032
9	− 5		+ 40	−0.000 011
10	+ 1		0	+0.000 002

the resulting coefficients will satisfy all the equations, at least to the prescribed accuracy.

The numerical example shown in Table 11.3 (for $N = 11$) is due to Clenshaw. We take

$$E(10) = 1, \quad E'(10) = 0, \quad E(11) = E'(11) = E(12) = E'(12) = \cdots = 0.$$

C is found to be 2.141 705 (−6). The correct value of $y(0)$ is 1, while the computed value of $\frac{1}{2}E(0) - E(1) + E(2) - \cdots$ is 0.999 995. The series yields

$$\tfrac{1}{2}E(0) + E(1) + E(2) + \cdots = 0.596 347$$

and gives a value of $-Ei(-1)$ of 0.219 384. The true value is 0.219 383 9. ■

Example 11.5 Here we wish to find Chebyshev coefficients for the solution of

$$y'' - 2(1 + 2x^2)y = 0, \quad y(0) = 1, \quad y'(0) = 0,$$

in the range $[0, 1]$. Obviously y is even, so we can use the polynomials T_n in the range $[-1, 1]$.

The necessary recursion relationships are

$$E(|n - 2|) = E''(n) - 4E(n) - E(n + 2), \quad n \geq 1, \tag{11.45}$$

$$2nE'(n) = E''(n - 1) - E''(n + 1),$$

$$2nE(n) = E'(n - 1) - E'(n + 1), \quad n \geq 1. \tag{11.46}$$

The two conditions are that y be even, and that

$$E(0) = 2(1 + E(2) - E(4) + E(6) - \cdots), \tag{11.47}$$

a consequence of the equation $y(0) = 1$.

The system is overdetermined in the sense that the attempt to satisfy both the first equation of (45),

$$2E(2) = E''(0) - 4E(0),$$

and (46) will fail. It is here that taking a linear combination of solutions is useful. For any sequence $\{E(n)\}$ generated from (45), (46) let the residual R be defined by

$$R(E(n)) := 4E(0) + 2E(2) - E''(0).$$

Let $\alpha(n)$ be one trial solution and $\beta(n)$ another. We take as the true solution the linear combination

$$E(n) := A\alpha(n) + B\beta(n),$$

where A, B are determined from

$$AR(\alpha(n)) + BR(\beta(n)) = 0,$$

$$A(\tfrac{1}{2}\alpha(0) - \alpha(2) + \alpha(4) - \cdots) + B(\tfrac{1}{2}\beta(0) - \beta(2) + \beta(4) - \cdots) = 1.$$

(Note this averaging process is precisely the technique which was applied to ordinary homogeneous difference equations in Section 7.4.) The following computations, due to Clenshaw, give results corresponding to $\alpha(12) = 1$, $\alpha''(12) = \alpha''(14) = \cdots = 0$, $\beta(12) = 1$, $\beta''(12) = 10$, $\beta''(14) = \beta''(16) = \cdots = 0$. The results are given in Table 11.4.

Table 11.4

n	$\alpha(n)$	$\alpha'(n)$	$\alpha''(n)$	$\beta(n)$	$\beta'(n)$	$\beta''(n)$	$E(n)$
0	+8 400 031		+3 009 728	+6 396 121		6 839 338	3.506 766
1		−1 500 176			423 744		
2	− 630 092		+6 010 080	− 129 882		5 991 850	0.850 389
3		+1 020 192			943 272		
4	+ 130 417		− 111 072	+ 115 257		332 218	0.105 208
5		− 23 144			21 216		
6	− 2 648		+ 120 368	+ 1 072		120 058	0.008 722
7		+ 8 632			8 352		
8	+ 543		− 480	+ 513		3 130	0.000 543
9		− 56			144		
10	− 4		+ 528	+ 6		538	0.000 027
11		+ 24			24		
12	+ 1		0	+ 1		10	0.000 001

Table 11.5

n	(1) E(n)	(1) E'(n)	(1) E''(n)	(2) E(n)	(2) E'(n)	(2) E''(n)	(3) E(n)	(3) E'(n)	(3) E''(n)	(4) E(n)	(4) E'(n)	(4) E''(n)	True E(n)
0	2.0000	8.0000		3.2500	14.3334		3.4852	15.6074		3.5056	15.7214		3.5068
1	0	2.0000	3.0000	0.6667	5.9585	4.1874	0.8333	6.9153	4.3460	0.8495	7.0083	4.3566	0.8504
2	0	0	0.3333	0.0417	0.8335	0.8542	0.0969	1.2273	0.9480	0.1047	1.2768	0.9552	0.1052
3	0	0	0	0	0.0417	0.0792	0.0064	0.1227	0.1105	0.0085	0.1392	0.1138	0.0087
4	0	0	0	0	0	0.0030	0.0002	0.0072	0.0082	0.0005	0.0105	0.0092	0.0005
5	0	0	0	0	0	0	0	0.0002	0.0004	0	0.0005	0.0006	0.0000
6	0												0.0000
7	0												
8	0												
9	0												
10	0												

One problem in using this method is that the starting values of $\alpha(n), \beta(n)$ must be chosen so that the values of $\alpha(n), \beta(n)$ generated must not be too close together. Otherwise serious numerical instabilities may result. This indeed is what we discovered for the averaging algorithms described in Section 7.4.

This system may also be solved iteratively. Taking initial values $E(0) = 2$, $E(n) = 0$, $n \geqslant 1$, is sufficient to start the process. Then $E''(n)$ is found from (9) and $E'(n), E(n)$ from (10), except for $E(0)$, which is given by (11). This is repeated until agreement is reached to the desired number of figures.

The results, compiled by Clenshaw, are displayed in Table 11.5.

One advantage the iterative method has over computation by recurrence is that the number of terms in the expansion need not be specified in advance. As successive iterates are computed, the required coefficients necessary to define the series for $f(x)$ to the accuracy indicated by the computed value of $E(0)$ automatically appear in the tabulation. The disadvantage is the slowness or lack of convergence alluded to earlier. ■

Example 11.6 This example is due to Elliott (1960).

We seek the expansion of e^{x^2} on $[-1, 1]$ in terms of Legendre polynomials. The appropriate differential equation is

$$y' - 2xy = 0, \qquad y(0) = 1.$$

The equations (40), (41) yield

$$(n + \tfrac{1}{2})E'(n) - [nE(n-1) + (n+1)E(n+1)] = 0,$$

which we will use in the form

$$E(n-1) = \frac{1}{2n}[(2n+1)E'(n) - 2(n+1)E(n+1)], \qquad n \geqslant 1,$$

along with

$$E'(n-1) = E'(n+1) + (2n+1)E(n), \qquad n \geqslant 1.$$

Also, we have $E(2n+1) = E'(2n) = 0$ for all n.

Since the equations are homogeneous the values of $E(n)$ may be multiplied by a constant to satisfy the boundary condition which is

$$\sum_{n=0}^{N} (n + \tfrac{1}{2})E(n)P_n(0) = 1.$$

Elliott tabulates the coefficients for $N = 13$, and the results are accurate to 5D. ■

11.2.3 The Lewanowicz construction

Before I can proceed to the construction of an unmixed recurrence for $E(n)$ it is necessary to give technical definitions and properties of some important difference operators acting on doubly infinite sequences. In what follows script letters will denote difference operators, lower-case Roman letters z, w, sequences, and Greek letters scalars. $z(n)$ denotes the nth member of the sequence z, $n \in Z$.

Definition 11.2 *Let $\mathscr{C}_S$ denote the linear space of complex doubly infinite sequences. A difference operator $\mathscr{L}$ is a mapping of $\mathscr{C}_S$ into $\mathscr{C}_S$ such that the nth member of $\mathscr{L}[z]$, $\mathscr{L}(z)(n)$ is given by*

$$\mathscr{L}(z)(n) = \sum_{j=0}^{r} \lambda_j(n) z(n + \mu + j), \qquad z \in \mathscr{C}_S, \tag{11.48}$$

where $r \equiv r(\mathscr{L}) \geqslant 0$ and $\mu = \mu(\mathscr{L})$ are integers. r is called the order *of the operator $\mathscr{L}$.* ■

Clearly every difference operator is linear. Furthermore we will be concerned here exclusively with difference operators whose coefficients $\lambda_j(n)$ are rational functions of n. The set of these operators we denote by **L**.

The shift operator $\mathscr{E}$ and the identity operator $\mathscr{I}$ are defined by

$$\mathscr{E}(z)(n) = z(n+1), \qquad \mathscr{I}(z)(n) = z(n).$$

$\mathscr{E}$ can be iterated in the obvious way, $\mathscr{E}^m(z)(n) = z(n+m)$. ϑ will denote the zero operator, $\vartheta(z)(n) = 0$. $r(\vartheta)$, $\mu(\vartheta)$ have no definite meaning but it is convenient to take $r(\vartheta) = -1$. Equality, addition and scalar multiplication of operators, defined in an obvious way, make **L** into a linear space. Composition is defined by

Definition 11.3 *Let $\mathscr{L}, \mathscr{M} \in \mathbf{L}$. $\mathscr{P} = \mathscr{L} \cdot \mathscr{M}$ means*

$$\mathscr{P}(z) = \mathscr{L}(w) \qquad for \ w = \mathscr{M}(z), \quad z \in \mathscr{C}_S. \quad ■$$

This construction makes the space **L** a ring with unity $\mathscr{I}$. Composition is not in general commutative. If $\mathscr{L}\mathscr{M} = \mathscr{M}\mathscr{L}$ we say $\mathscr{L}$ and $\mathscr{M}$ commute. Any two difference operators with $\lambda_j(n) = $ constant commute.

If α is a rational function of n and $\mathscr{M} \in \mathbf{L}$, we write

$$\alpha(n).\mathscr{M} = \mathscr{L}\mathscr{M}, \qquad \mathscr{L}(z)(n) = \alpha(n)z(n).$$

It is easy to see that for any nonzero operators $\mathscr{L}$ and $\mathscr{M}$,

$$r(\mathscr{L} + \mathscr{M}) \leqslant \max\{r(\mathscr{L}) + \mu(\mathscr{L}), r(\mathscr{M}) + \mu(\mathscr{M})\}$$

$$- \min\{\mu(\mathscr{L}), \mu(\mathscr{M})\}, \qquad \mathscr{L} \neq -\mathscr{M}, \tag{11.49}$$

$$\mu(\mathscr{L}+\mathscr{M}) \geqslant \min(\mu(\mathscr{L}), \mu(\mathscr{M})), \tag{11.50}$$

$$r(\mathscr{L}\mathscr{M}) = r(\mathscr{L}) + r(\mathscr{M}), \tag{11.51}$$

$$\mu(\mathscr{L}\mathscr{M}) = \mu(\mathscr{L}) + \mu(\mathscr{M}). \tag{11.52}$$

Now let $l := r(\mathscr{L})$, $m := r(\mathscr{M})$ and let $\lambda_i(n)$, $\mu_i(n)$ be the coefficients of $\mathscr{L}$ and $\mathscr{M}$ respectively. We have a sharp inequality in (49) if and only if

$$\mu(\mathscr{L}) = \mu(\mathscr{M}), \qquad \lambda_0(n) + \mu_0(n) \equiv 0; \tag{11.53}$$

$$l + \mu(\mathscr{L}) = m + \mu(\mathscr{M}), \qquad \lambda_l(n) + \mu_m(n) \equiv 0, \tag{11.54}$$

whereas the sharp inequality in (50) occurs only where (53) is satisfied.

Definition 11.4   *Let $\mathscr{L} \in \mathbf{L}$, ρ a given function of n, and $z \in \mathscr{C}_S$. The relation*

$$\mathscr{L}(z)(n) = \rho(n), \qquad n \geqslant m, \quad m \text{ an integer,}$$

is called a linear recurrence relation (for z). The order of the recurrence is the order of $\mathscr{L}$. ∎

Now let

$$\mathscr{D} := \mathscr{E}^{-1} - \mathscr{E}.$$

Definition 11.5   *Let $\mathscr{L} \in \mathbf{L}$ be given and let there exist for a nonzero $\mathscr{A} \in \mathbf{L}$ a $\mathscr{Q} \in \mathbf{L}$ such that $\mathscr{A}\mathscr{L} = \mathscr{Q}\mathscr{D}$. The set of all such operators will be denoted by $\mathbf{A}(\mathscr{L})$. The operator $\mathscr{A}^* \in \mathbf{A}(\mathscr{L})$ will be called a minimum operator if for every $\mathscr{A} \in \mathbf{A}(\mathscr{L})$ we have $r(\mathscr{A}^*) \leqslant r(\mathscr{A})$.* ∎

I now define operators $\mathscr{B}_i$, $i \geqslant 0$, by

$$\mathscr{B}_i := \begin{cases} \mathscr{D}, & i = 0; \\ (n+\lambda)^{-1}[\alpha_i(n)\mathscr{E}^{-1} - \beta_i(n)\mathscr{E}], & i \geqslant 1, \end{cases} \tag{11.55}$$

where

$$\alpha_i(n) := (n+\lambda+i-1)_2, \qquad \beta_i(n) = (n+\lambda-i)_2, \tag{11.56}$$

and operators $\mathscr{S}_{ij}$ and $\mathscr{P}_i$ by

$$\mathscr{S}_{ij} := \begin{cases} \mathscr{I}, & i < j; \\ \mathscr{B}_i\mathscr{B}_{i-1}\cdots\mathscr{B}_j, & i \geqslant j \geqslant 0; \end{cases} \tag{11.57}$$

$$\mathscr{P}_i := \mathscr{S}_{i-1,0}, \qquad i \geqslant 0. \tag{11.58}$$

One can easily show that

$$\mathscr{S}_{ij} = \mathscr{B}_i\mathscr{S}_{i-1,j} = \mathscr{S}_{i,j+1}\mathscr{B}_j, \qquad i \geqslant j \geqslant 0, \tag{11.59}$$

$$\mathscr{P}_i = \begin{cases} \mathscr{S}_{i-1,j}\mathscr{P}_j, & i \geqslant j \geqslant 0; \\ \mathscr{B}_{i-1}\mathscr{P}_{i-1}, & i \geqslant 1. \end{cases} \tag{11.60}$$

Let $\gamma_0, \gamma_1, \ldots,$ be polynomials in n defined by

$$\gamma_0(n) = 1, \quad \gamma_i(n) = (n+\lambda-i+1)_{2i-1}, \quad i \geq 1. \tag{11.61}$$

Then (56) and (61) give

$$\gamma_{i+1} = \alpha_i \gamma_i^- = \beta_i \gamma_i^+, \quad i \geq 1. \tag{11.62}$$

Formula (29) can be rewritten

$$\mathscr{D}E^{(i)}(n) = 2(n+\lambda)E^{(i-1)}(n), \quad i > 0, \tag{11.63}$$

and this allows us to obtain a basic relationship between $E(n)$ and the coefficients $E^{(i)}(n)$, $i > 0$. We have

Lemma

$$\mathscr{P}_i E^{(i)}(n) = 2^i \gamma_i(n) E(n). \tag{11.64}$$

Proof The proof is by induction on i. By (58) and (61) we have $\mathscr{P}_0 = \mathscr{I}$, $\gamma_0 \equiv 1$ and this gives (64) for $i = 0$.

Since $\mathscr{P}_1 = \mathscr{B}_0 = \mathscr{D}$ and $\gamma_1 = n+\lambda$, for $i = 1$ (64) assumes the form

$$\mathscr{D}E'(n) = 2(n+\lambda)E(n). \tag{11.65}$$

Assume (64) holds for $i_0 \geq 1$. Equations (60), (55), (62) and (65) give

$$
\begin{aligned}
\mathscr{P}_{i+1}E^{(i+1)}(n) &= \mathscr{B}_i \mathscr{P}_i E^{(i+1)}(n) \\
&= 2^i \mathscr{B}_i \gamma_i(n)E'(n) \\
&= 2^i(n+\lambda)^{-1}[\alpha_i(n)\gamma_i^-(n)E'(n-1) - \beta_i(n)\gamma_i^+(n)E'(n+1)] \\
&= 2^i(n+\lambda)^{-1}\gamma_{i+1}(n)\mathscr{D}E'(n) \\
&= 2^{i+1}\gamma_{i+1}(n)E(n),
\end{aligned}
$$

and the proof is complete. ∎

I now proceed with the construction of the recurrence.

It is easy to show that the differential equation (25) is equivalent to the equation

$$\sum_{k=0}^{\sigma} (q_k f)^{(k)} = p, \tag{11.66}$$

$$q_k := \sum_{j=k}^{\sigma} (-1)^{j-k}\binom{j}{j-k}p_j^{(j-k)}, \quad 0 \leq k \leq n. \tag{11.67}$$

Thus,

$$\sum_{k=0}^{\sigma} E^{(k)}(n, q_k f) = E(n, p).$$

Let the operator $\mathscr{P}_\sigma$ defined by (58) act on both sides of the previous equation. Using (60) shows

$$\sum_{k=0}^{\sigma} \mathscr{S}_{\sigma-1,k} \mathscr{P}_k E^{(k)}(n, q_k f) = \mathscr{P}_\sigma E(n, p).$$

The lemma then gives

$$\sum_{k=0}^{\sigma} 2^k \mathscr{S}_{\sigma-1,k} [\gamma_k(n) E(n, q_k f)] = \mathscr{P}_\sigma E(n, p). \tag{11.68}$$

Since the q_i are polynomials I may define operators $\mathscr{L}_0, \mathscr{L}_1, \ldots, \mathscr{L}_\sigma$ such that

$$\mathscr{L}_i E(n, f) = \gamma_i(n) E(n, q_i f), \qquad 0 \le i \le \sigma, \tag{11.69}$$

and the result can be summed up in the following.

Theorem 11.2 *Let f satisfy the differential equation (25) and let its σth derivative possess a Gegenbauer series uniformly convergent on $[-1, 1]$.*
Then

$$\mathscr{L} E(n, f) = \pi_\sigma(n), \qquad n \ge 0, \tag{11.70}$$

where

$$\mathscr{L} := \sum_{k=0}^{\sigma} 2^k \mathscr{S}_{\sigma-1,k} \mathscr{L}_k, \qquad \pi_\sigma(n) := \mathscr{P}_\sigma E(n, p), \tag{11.71}$$

$E(n, f)$ being the Gegenbauer coefficients for f (as defined by (26)). ■

The actual construction of the recurrence proceeds as follows:

(i)   define operators $\mathscr{L}_0, \mathscr{L}_1, \ldots, \mathscr{L}_\sigma$ to satisfy (69);
(ii) form operators $\mathscr{S}_{\sigma-1,k}, 0 \le k \le n$, according to

$$\mathscr{S}_{\sigma-1,\sigma} = \mathscr{I}, \qquad \mathscr{S}_{\sigma-1,k} = \mathscr{S}_{\sigma-1,k+1} \mathscr{B}_k, \qquad k = \sigma-1, \sigma-2, \ldots, 0,$$

where $\mathscr{B}_k$ is defined by (55). Then determine $\mathscr{L}$ by (71);
(iii) define functions $\pi_0, \pi_1, \ldots, \pi_\sigma$ by

$$\pi_0(n) = E(n, p), \qquad \pi_k(n) = \mathscr{B}_{k-1} \pi_{k-1}(n), \qquad 1 \le k \le n.$$

Then $\pi_\sigma(n)$ forms the right-hand side of (70).

The example that follows is due to Lewanowicz (1976).

Example 11.7 Consider the Lommel function,

$$f(x) := (ax)^{1-\mu}s_{\mu\nu}(ax), \qquad -1 \leqslant x \leqslant 1, \quad a \neq 0,$$

$$\mu + \nu, \mu - \nu \neq -1, -3, -5, \ldots.$$

f satisfies

$$x^2 f'' + bxf' + (a^2 x^2 + c)f = a^2 x^2,$$

$$b := (2\mu - 1), \qquad c := (\mu - 1)^2 - \nu^2.$$

I will consider the Chebyshev case, $\lambda = 0$. We have $\sigma = 2$, $p_0 = a^2 x^2 + c$, $p_1 = bx$, $p_2 = x^2$, and

$$q_0 = a^2 x^2 + d, \qquad q_1 = (b - 4)x, \qquad q_2 = x^2,$$

$$d := 2 + c - b.$$

Then

$$\mathcal{L}_0 C(n, f) = \gamma_0(n)C(n, q_0 f)$$

$$= \tfrac{1}{4}[a^2 C(n-2) + 2(a^2 + 2d)C(n) + a^2 C(n+2)],$$

$$\mathcal{L}_1 C(n, f) = \gamma_1(n)C(n, q_1 f)$$

$$= \tfrac{1}{2}(b-4)n[C(n-1) + C(n+1)],$$

$$\mathcal{L}_2 C(n, f) = \gamma_2(n)C(n, q_2 f)$$

$$= \tfrac{1}{4}(n-1)_3[C(n-2) + 2C(n) + C(n+2)],$$

where $\gamma_i(n)$ are polynomials given by (62), $\lambda = 0$. According to (55) we have

$$\mathcal{B}_0 = \mathcal{D}, \qquad \mathcal{B}_1 = (n+1)\mathcal{E}^{-1} - (n-1)\mathcal{E},$$

and step (ii) yields

$$\mathcal{S}_{12} = \mathcal{I}, \qquad \mathcal{S}_{11} = \mathcal{B}_1, \qquad \mathcal{S}_{10} = (n+1)\mathcal{E}^{-2} - 2n\mathcal{I} + (n-1)\mathcal{E}^2,$$

$$\mathcal{L} = \tfrac{1}{4}a^2(n+1)\mathcal{E}^{-4} + ((n+1)[(n+b-4)(n-1)+d] + \tfrac{1}{2}a^2)\mathcal{E}^{-2}$$

$$+ 2n\left(n^2 - \frac{a^2}{4} - d - 1\right)\mathcal{I}$$

$$+ ((n-1)[(n-b+4)(n+1)+d] - \tfrac{1}{2}a^2)\mathcal{E}^2 + \tfrac{1}{4}a^2(n-1)\mathcal{E}^4.$$

Now the function $p = a^2 x^2$ on the right-hand side has the representation

$$p(x) = \tfrac{1}{2}a^2(T_0(x) + T_2(x)),$$

so

$$\pi_0(n) = C(n, p) = \begin{cases} a^2, & n = 0; \\ \dfrac{a^2}{2}, & |n| = 2; \\ 0, & |n| = 1, 2, 3, 4, \ldots. \end{cases}$$

We also have

$$\pi_1(n) = \begin{cases} \dfrac{a^2}{2}, & |n| = 1, 3; \\ 0, & |n| = 0, 2, 4, 5, \dots; \end{cases}$$

$$\pi_2(n) = \begin{cases} a^2, & |n| = 0, 2; \\ \dfrac{5a^2}{2}, & |n| = 4; \\ 0, & |n| = 1, 3, 5, 6. \end{cases}$$

Since f is even all the odd coefficients vanish and we find the coefficients $W(n) := C(2n)$ satisfy the fourth-order recurrence

$$a^2(2n+1)W(n-2) + 2(2(2n+1)[(2n+b-4)(2n-1)+d]+a^2)$$
$$\times W(n-1) + 4n[4(4n^2-d-1)-a^2]W(n)$$
$$+ 2[2(2n-1)[(2n-b+4)(2n+1)+d]-a^2]W(n+1)$$
$$+ a^2(2n-1)W(n+2)$$
$$= \begin{cases} 4a^2, & n = 0, 1; \\ 10a^2, & n = 2; \\ 0, & n = 3, 4, 5. \end{cases} \blacksquare$$

Clearly, the recurrence (70) may be rewritten in the form

$$\sum_{k=-d}^{d} \tau_k(n) E(n+k) = \mu(n),$$

and one can show that

$$d = \max_{\substack{0 \leqslant j \leqslant \sigma \\ p_{\sigma-j} \neq 0}} (d_{\sigma-j} + j).$$

Further the functions τ_k are polynomials in n of degree at most $2\sigma - 1$ and have the properties

$$\tau_{-k}(n) = -\tau_k(-n),$$
$$\tau_{-d} \neq 0, \qquad \tau_d \neq 0.$$

As previously remarked the order of this recurrence, $2d$, may not be the lowest possible. Lewanowicz has given a formula for the order $2d^*$ of the optimal recurrence which is very complicated. It may be shown that

$$\max_{\substack{0 \leqslant k \leqslant \sigma \\ p_k \neq 0}} (d_k - k) \leqslant d^* \leqslant d.$$

In the previous example the order of the recurrence obtained was $\max (d_i + i) = 8$ while the optimal recurrence must have order $\geqslant 4$.

For the Chebyshev case Lewanowicz establishes the following:

Theorem 11.3 *Let the conditions of Theorem 2 be satisfied and* $\lambda = 0$. *Then the previous algorithm leads to a recurrence relation of lowest order if and only if* $p_\sigma(1) \neq 0$, $p_\sigma(-1) \neq 0$. ∎

The recurrence relation of the example is optimal, since $p_2 = x^2$.

Example 11.8 Let

$$f(x) := \int_0^{\pi/2} (1 - x^2 \sin^2 t)^{1/2} \, dt, \qquad -1 \leqslant x \leqslant 1.$$

f satisfies

$$x(x^2 - 1)f'' + (x^2 - 1)f' - xf = 0.$$

Let $W(n) := E(2n)$. In the previous reference it is shown that $W(n)$ satisfies the recurrence

$$(2n + \lambda + 1)(2n - 3)(2n - 1)^2 W(n - 1)$$
$$-2(\lambda + 2)(2n + \lambda)(4n^2 + 4n\lambda - 1) W(n)$$
$$-(2n + \lambda - 1)(2n + 2\lambda + 1)^2(2n + 2\lambda + 3) W(n + 1) = 0.$$

In $E(n)$ this recurrence has order 4 and is optimal. The recurrence obtained from the previous algorithm is of order 6. Luke (1969, vol. 2, p. 29 (2)) expresses $W(n)$ in terms of a $_3F_2$ of the kind in Section 9.4. Theorem C.2 applied to this function provides the same recurrence for $E(n)$ as the algorithm in this section. Lewanowicz's recurrence of optimal order also follows from a recurrence given for the general $_3F_2(1)$ in Section 9.4. ∎

Lewanowicz has applied the same techniques to the computation of recurrence relations for the coefficients of expansions in Bessel functions (1979a) and to modified moments (1979b).

11.3 Chebyshev series solutions for nonlinear differential equations

While the techniques of the previous section may be extended to certain simple nonlinear equations, the resulting complexity of the procedure is a severe drawback since one has to form nonlinear functions of Gegenbauer series $\sum B(n) C_n^\lambda(x)$. A much better procedure, developed by Clenshaw and Norton (1963) for Chebyshev series and subsequently studied by Norton (1964), is based on the use of Picard iteration and instead of using infinite series of Chebyshev polynomials and employing orthogonality with respect to *integration* uses finite sums of Chebyshev polynomials and orthogonality with respect to *summation*.

Let N be an integer ≥ 1 and

$$x_{Ns} \equiv x_s := \cos \frac{\pi s}{N}, \qquad 0 \leq s \leq N. \tag{11.72}$$

It is well known that

$$\sum_{s=0}^{N}{}'' T_j(x_s) T_k(x_s) = \begin{cases} 0, & j, k < N, & j \neq k; \\ \dfrac{N}{2}, & j = k, & 0 < j < n; \\ N, & j = k = 0 \quad \text{or} \quad j = k = n; \end{cases} \tag{11.73}$$

see Luke (1969, vol. 1, p. 310). (The double prime notation means that both the first and the last term of the sum are to be halved.)

A consequence of (73) is that if a polynomial $p(x)$ of degree $\leq n$ is written in the form

$$p(x) := \sum_{n=0}^{N}{}'' C(n) T_n(x), \tag{11.74}$$

we have the following formula for the coefficients $C(n)$,

$$C(n) = \frac{2}{N} \sum_{k=0}^{N}{}'' p(x_k) x_{n \times k}, \qquad 0 \leq n \leq N. \tag{11.75}$$

The generalized Jacobi polynomials also satisfy an orthogonality property with respect to summation (Luke, 1969, vol. 1, p. 307). Thus the work in this section can be generalized to include expansions in these polynomials and their limiting cases such as Laguerre and Hermite polynomials. It is advisable however to present here only the simplest case.

If, instead of a polynomial, $p(x)$ is taken to be a function with suitable smoothness properties and the coefficients $C(n)$ are formed from the formula (75) these coefficients will approximate, with increasing accuracy as $N \to \infty$, the true Chebyshev coefficients of p. For a fixed N the truncated sum (74) with a single prime will then provide an approximation to $p(x)$ on $-1 \leq x \leq 1$.

For simplicity I will discuss the solution of the first-order equation

$$y' = f(x, y), \qquad y(0) = \eta. \tag{11.76}$$

(Higher-order equations can be represented by a vector system of first order, and the technique to follow may be applied in a straightforward manner to such a system.)

The technique is based on the Picard iteration sequence, that sequence of functions $y_k(x)$ generated from (76) with $y_0(x) := \eta$ and

$$y_k(x) := \eta + \int_0^x f(x, y_{k-1}(x)) \, dx, \qquad k \geq 1. \tag{11.77}$$

Now let N be fixed and assume we have a Chebyshev series representation of degree N to the iterate $y_{k-1}(x)$:

$$y_{k-1}(x) := \sum_{n=0}^{N}{}' A^{(k-1)}(n) T_n(x). \tag{11.78}$$

We now wish to calculate a similar Chebyshev series for $f(x, y_{k-1})$ given an algorithm for computing the value of $f(x, y)$ for any point (x, y). The summation formula (75) provides a method for obtaining the new coefficients, or, at least, approximations to them. Further, summation over functional values has the advantage that nonlinear operations which are difficult to apply to Chebyshev series, which the procedure of the previous section would require, can be entirely circumvented.

The values of $f(x_s, y_{k-1}(x_s))$ are computed from f and (78) for $0 \leq s \leq N$ and used in the formula for the coefficients (75) to obtain the expansion

$$f(x, y_{k-1}(x)) = \sum_{n=0}^{N}{}' A^{(k)}(n)' T_n(x). \tag{11.79}$$

(The computation of y_k is best accomplished by the algorithm of Section 10.2.) The integration required of the subsequent calculation of the Chebyshev series of $y_k(x)$ from (79) is then performed using the relationships

$$\left.\begin{array}{l} 2nA^{(k)}(n) = A^{(k)}(n-1)' - A^{(k)}(n+1)', \qquad 1 \leq n \leq N, \\[2mm] 2(N+1)A^{(k)}(N+1) = A^{(k)}(N)', \\[2mm] A^{(k)}(0) = 2\left[\eta - \sum_{n=1}^{N} \xi_n A^{(k)}(n)\right], \end{array}\right\}$$

where $\hspace{8cm}$ **(11.80)**

$$\xi_n = \begin{cases} (-1)^{n/2}, & n \text{ even}; \\ 0, & n \text{ odd}; \end{cases}$$

to obtain the series

$$y_k(x) = \sum_{n=0}^{N}{}' A^{(k)}(n) T_n(x). \tag{11.81}$$

The process is then repeated until the desired agreement is obtained among the coefficients $A_n^{(k)}$ for $k = K, K+1$; $0 \leq n \leq N$. The procedure is very straightforward.

Most of the arithmetic in the method occurs in the calculation of the Chebyshev series for $f(x, y_{k-1})$ from the series for y_{k-1}. About $2N^2$ multiplications are required in each cycle. The minimum value of N required to represent the solution to a given degree of accuracy is not known, *a priori*, but, as suggested by Clenshaw and Norton, it seems realistic to choose a moderate value of N, say $N = 4$, to start with and then introduce further coefficients, i.e., $A^{(k)}(N+1)$ from (80), into the series (81) as necessary to improve the solution. One way of judging whether new coefficients need to be incorporated is by the magnitude of the coefficient $A^{(k)}(N)$.

In general an upper bound for the error of $y_k(x)$ is

$$\sum_{n=0}^{N}{}'\,|A^{(k)}(n) - C(n)| + \sum_{k=N+1}^{\infty} |C(k)|,$$

where $C(n)$ is the true Chebyshev coefficient. Although $C(n)$ is not generally known, the second term in the above formula can be estimated by observing the rate of convergence of the $A^{(k)}(n)$ as $n \to \infty$ and the first term may be regarded as a sum of quantities containing one-signed rounding errors.

Example 11.9

$$y' + y = 0, \qquad y(0) = 1.$$

Take $N = 5$, $y_0(x) = 1$. Table 11.6, taken from Clenshaw and Norton, gives selected values of $A_n^{(k)}$ and, for comparison, the true Chebyshev coefficients for e^{-x}. At $k = 12$ the coefficients have settled down to 6D. (These are the entries in the next-to-last column of the table.)

Since $y_5(x)$ is a rearrangement of the truncated Taylor series for e^{-x} it

Table 11.6 Approximations to e^{-x}.

n	$A^{(0)}(n)$	$A^{(1)}(n)$	$A^{(2)}(n)$	$A^{(3)}(n)$	$A^{(4)}(n)$	$A^{(5)}(n)$	$A^{(12)}(n)$	$C(n)$
0	$+2$	$+2$	$+\frac{5}{2}$	$+\frac{5}{2}$	$+\frac{81}{32}$	$+\frac{81}{32} = +2.531\,250$	$+2.532\,020$	$+2.532\,132$
1	0	-1	-1	$-\frac{9}{8}$	$-\frac{9}{8}$	$-\frac{217}{192} = -1.130\,208$	$-1.130\,268$	$-1.130\,318$
2	0	0	$+\frac{1}{4}$	$+\frac{1}{4}$	$+\frac{13}{48}$	$+\frac{13}{48} = +0.270\,833$	$+0.271\,483$	$+0.271\,495$
3	0	0	0	$-\frac{1}{24}$	$-\frac{1}{24}$	$-\frac{17}{384} = -0.044\,271$	$-0.044\,335$	$-0.044\,337$
4	0	0	0	0	$+\frac{1}{192}$	$+\frac{1}{192} = +0.005\,208$	$+0.005\,473$	$+0.005\,474$
5	0	0	0	0	0	$-\frac{1}{1920} = -0.000\,521$	$-0.000\,547$	$-0.000\,543$
6								$+0.000\,045$
7								$-0.000\,003$

is clear the maximum error occurs at $x = -1$ and is given by

$$e - \sum_{n=0}^{5}{}' (-1)^n A^{(5)}(n) = e - 2.716\,67$$

$$= 0.001\,62.$$

In contrast, comparing the coefficients $A^{(12)}(n)$ with $C(n)$ shows the error of $y_{12}(x)$ cannot exceed

$$\sum_{n=0}^{5}{}' |A^{(12)}(n) - C(n)| + \sum_{n=6}^{\infty} |C(n)| = 0.000\,17.$$

Thus continuing the iteration on k from 5 to 12 gives a significant gain in accuracy even though the degree of approximation has not been increased. ■

Example 11.10

$$y' = y^2, \qquad y(-1) = \tfrac{2}{5}.$$

The fact that the boundary condition is given at -1 rather than 0 is easily accommodated by replacing the last equation of (80) by

$$A^{(k)}(0) = 2[\eta + A^{(k)}(1) - A^{(k)}(2) + \cdots + (-1)^N A^{(k)}(N+1)].$$

The true solution is $y(x) = (\tfrac{3}{2} - x)^{-1}$.

Here the computations (taken again from the above reference) were done with $N = 30$. The coefficients showed no change in the 11th decimal place after 22 iterations. The resulting coefficients, given as $C(n)$ in Table

Table 11.7 Solution of $y' = y^2$, $y(-1) = \tfrac{2}{5}$.

n	$C(n)$	n	$10^{11}C(n)$
0	1.788 854 382 00	14	251 722
1	0.683 281 573 00	15	96 149
2	0.260 990 337 00	16	36 726
3	0.099 689 438 00	17	14 028
4	0.038 077 977 00	18	5 358
5	0.014 544 492 99	19	2 047
6	0.005 555 501 97	20	782
7	0.002 122 012 93	21	299
8	0.000 810 536 81	22	114
9	0.000 309 597 51	23	44
10	0.000 118 255 73	24	17
11	0.000 045 169 67	25	6
12	0.000 017 253 28	26	2
13	0.000 006 590 17	27	1

11.7, agree with the known Chebyshev coefficients for $y(x)$,

$$C(n) = \frac{4}{\sqrt{5}} \left(\frac{3 - \sqrt{5}}{2} \right)^n,$$

to within 10^{-11}. ∎

Although differential equations of higher order may be expressed as first-order systems, the frequent occurrence of second-order equations in applied problems indicates the usefulness of a direct method, which can be formulated as follows.

Write the equation as

$$y'' = f(x, y, y').$$

From the Chebyshev series for y_{k-1} and y'_{k-1} evaluated at the points x_s we may compute the values of $f(x_s, y_{k-1}(x_s), y'_{k-1}(x_s))$ and thus, from the summation formula (75) compute the coefficients $A^{(k)}(n)''$ in

$$f(x, y_{k-1}, y'_{k-1}) = \sum_{n=0}^{N} {}' A^{(k)}(n)'' T_n(x).$$

Then, using the relationships

$$2n A^{(k)}(n)' = A^{(k)}(n-1)'' - A^{(k)}(n+1)'',$$

$$2n A^{(k)}(n) = A^{(k)}(n-1)' - A^{(k)}(n+1)',$$

we can calculate $A^{(k)}(n)'$, $A^{(k)}(n)$ in the expansions

$$y'_k = \sum_{n=0}^{N+1} {}' A^{(k)}(n)' T_n(x), \qquad y_k = \sum_{n=0}^{N+2} {}' A^{(k)}(n) T_n(x).$$

The coefficients $A(0)'$, $A(0)$ and $A(1)$ are determined from the boundary conditions.

Example 11.11 Consider van der Pol's equation

$$\frac{d^2 y}{dt^2} + (y^2 - 1) \frac{dy}{dt} + y = 0,$$

$$y(-\tfrac{1}{4}) = 0, \qquad y(\tfrac{1}{4}) = 2.$$

Writing $t = x/4$, $Y(x) = y(x/4)$, gives

$$Y'' = \tfrac{1}{4}(1 - Y^2) Y' - \tfrac{1}{16} Y,$$

$$Y(-1) = 0, \qquad Y(1) = 2.$$

The boundary conditions yield for $A^{(k)}(0)$ and $A^{(k)}(1)$ the expressions

$$A^{(k)}(0) = 2 - 2(A^{(k)}(2) + A^{(k)}(4) + A^{(k)}(6) + \cdots),$$

$$A^{(k)}(1) = 1 - (A^{(k)}(3) + A^{(k)}(5) + A^{(k)}(7) + \cdots).$$

Table 11.8 Solution of van der Pol's equation.

n	$C(n)$
0	+2.068 066 318 39
1	+1.023 980 677 83
2	−0.032 794 540 43
3	−0.024 855 749 86
4	−0.001 366 854 43
5	+0.000 901 078 63
6	+0.000 136 531 83
7	−0.000 026 407 95
8	−0.000 008 721 72
9	+ 378 65
10	+ 443 60
11	+ 25 56
12	− 18 63
13	− 3 07
14	+ 60
15	+ 21
16	− 1
17	− 1

Eleven iterations are sufficient to achieve 11D accuracy, and the results appear tabulated as $C(n)$ in Table 11.8. ∎

Clenshaw and Norton (1963) note that the algorithm of this section may not converge even when $y(x)$ is entire. For example, the procedure works for the problem

$$y'' + \lambda^2 y = 0, \qquad y(-1) = 0, \quad y(1) = 1,$$

only when $|\lambda| < \pi/2$.

Norton (1964) discusses a procedure based on Newton iteration which secures convergence for a larger class of problems and also increases the rate of convergence for problems which can be solved by the present method.

12 Multidimensional recursion algorithms; general theory

12.1 Background: convergence properties of complex sequences

For a study of multidimensional recursion algorithms we will need some material on the convergence properties of complex sequences.

I begin with a few definitions.

Definition 12.1 *A convergent not ultimately constant sequence* $\{s(n)\}$ *possesses* pth-*order convergence,* $p > 1$, *to its limit* s *if*

$$|s(n+1) - s| = O(|s(n) - s|^p). \quad \blacksquare \qquad (12.1)$$

It is easy to show that this is precisely equivalent to the statement that

$$s(n) - s = O(\Lambda^{p^n}), \qquad (12.2)$$

for some $0 < \Lambda < 1$.

The case $p = 2$ is called *quadratic convergence*.

The above definition is a bit lacking in precision since any pth-order convergent sequence is also qth-order convergent for $1 < q \leqslant p$. To remedy this the following definition is sometimes used.

Definition 12.2 *A convergent not ultimately constant sequence* $\{s(n)\}$ *possesses* exactly pth order $(p > 1)$ convergence, *to its limit* s *if*

$$|s(n+1) - s| = \xi(n) |s(n) - s|^p, \qquad n \geqslant n_0, \qquad (12.3)$$

where $\xi(n)$ *is a sequence bounded and bounded away from zero,* $n \geqslant n_0$. $\blacksquare$

This is equivalent to the statement that

$$|s(n) - s| = \eta(n)\Lambda^{p^n}, \qquad (12.4)$$

for some positive sequence $\{\eta(n)\}$ bounded and bounded away from zero. To see this take $\{s(n)\}$ to be a convergent not ultimately constant sequence satisfying (3) and write $r(n) := s(n) - s$. By induction one finds

$$|r(n)| = \eta(n)\Lambda^{p^n}, \qquad n \geq n_0,$$

$$\eta(n) = \exp\left\{-p^n \sum_{k=n}^{\infty} \frac{\ln|\xi(k)|}{p^{k+1}}\right\}, \qquad n \geq n_0,$$

$$\Lambda = |r(n_0)| \exp \sum_{k=n_0}^{\infty} \frac{\ln|\xi(k)|}{p^{k+1}}.$$

Clearly $\eta(n)$ is bounded and bounded away from zero. Also $\Lambda < 1$ since $\{s(n)\}$ converges and $\Lambda \neq 0$ or else $r(n_0) = 0$ which, by (4), yields $s(n) =$ constant, $n \geq n_0$. The converse is trivial.

The infimum of Λ for which (4) holds is called the *degree* of convergence.

We finally have

Definition 12.3 *A sequence $s(n)$ possesses* linear convergence *to its limit s if*

$$s(n) - s = O(\lambda^n),$$

for some $0 < \lambda < 1$. ∎

The reader should note that this definition of linear convergence admits a far wider class of sequences than that given in Wimp (1981, p. 6).

It is trivial to show that if the sequence $\{s(n+1) - s(n)\}$ possesses pth-order (respectively, linear) convergence so does $\{s(n)\}$.

As an example I derive a useful result about a Newton–Raphson iteration sequence.

Theorem 12.1 *Let $A > 0$, $x(0) \neq 0$, $x(0) \neq A$. Then the iteration sequence*

$$x(n+1) = \frac{x^2(n) + A^2}{2x(n)}, \qquad n \geq 0,$$

converges exactly quadratically with degree

$$K_1(x(0), A) := \begin{cases} |x(0) - A|/(A + x(0)), & x(0) > 0; \\ |x(0) + A|/(A - x(0)), & x(0) < 0. \end{cases} \qquad \textbf{(12.5)}$$

Proof Define

$$\varepsilon(n) := \frac{x(n) - A}{x(n) + A}.$$

$\varepsilon(n)$ satisfies

$$\varepsilon(n+1) = \varepsilon^2(n), \qquad n \geq 0,$$

or

$$\varepsilon(n) = \varepsilon(0)^{2^n}, \qquad n \geq 0.$$

Thus if $x(0) > 0$,

$$|x(n) - A| = \mu(n)\varepsilon(0)^{2^n},$$

where $\mu(n) := |x(n) + A|$ is bounded since $x(n)$ is convergent, and bounded away from 0 since $x(n) > 0$. The case $x(0) < 0$ is handled similarly. ∎

12.2   Introduction; invariants

Let $\mathbf{f} : \Omega \times J^0 \to \Omega \subset \mathscr{C}^p$, where f is continuous on Ω. In this and the next two chapters we shall study computational algorithms based on the following system,

$$\mathbf{x}(n+1) = \mathbf{f}[\mathbf{x}(n), n], \qquad n \geq 0. \tag{12.6}$$

p is called the *dimension* of the system.

Such systems (or recurrences), particularly in the case where $\mathbf{f}$ is independent of n, have been studied by many writers. For a fairly complete bibliography see Wimp (1981).

Of course, the general σth-order linear difference equation may be written in the form (6) but we shall here be concerned primarily with those situations in which f is nonlinear.

When $\mathbf{f}$ is independent of n the system (6) is called *autonomous*. In this section up to the point where I introduce the concept of an *invariant* (Definition 4) I will assume the system (6) is autonomous.

The sequence of points

$$\mathcal{O} := \{\mathbf{x}(n)\}_{n=0}^{\infty},$$

is called the *trajectory* of the system. If $\mathbf{x}(n) \to \mathbf{X}$ the system is said to be convergent. Then (6) defines an algorithm for the computation of $\mathbf{X}$.

When the autonomous system converges and Ω is closed, then its limit $\mathbf{X}$ will be a solution of the equation

$$\mathbf{f}(\mathbf{X}) = \mathbf{X},$$

i.e., a fixed point of the function $\mathbf{f}$.

With the exception of Section 12.3, I shall be concerned exclusively with convergent systems.

The problems involved with the computational format expressed by (6) are, roughly, twofold:

(i)   given $\mathbf{f}$, to 'identify' $\mathbf{X}$;

(ii) given an 'interesting' computational quantity $\mathbf{X}$ (usually defined by an integral or by a functional equation) to determine an algorithm of the type (6) for the computation of **X**.

The second item above is, admittedly, rather vague. Given any point in $\mathscr{C}^p$ trivial algorithms can be devised which converge to **X**. But these will involve **X** explicitly. Generally (6) will not involve **X** explicitly and the rapidity of convergence will be vastly greater than any facilely contrived algorithm. In fact the algorithms that we shall study are among the most rapidly convergent in all numerical analysis.

As is well known, an autonomous process will converge if **f** is continuous on a compact set $K \subset \mathscr{C}^p$, $\mathbf{x}(0) \in K$, and the mapping **f** is contracting,

$$\|\mathbf{f}(\mathbf{x}) - \mathbf{f}(\mathbf{y})\| \leqslant M \|\mathbf{x} - \mathbf{y}\|, \qquad M < 1, \quad \mathbf{x}, \mathbf{y} \in K.$$

(Obviously convergence will result if **f** is differentiable, $\|\mathbf{f}'(\mathbf{X})\| < 1$ and $\mathbf{x}(0)$ is sufficiently close to **X**.) Thus from the point of view of the convergence behavior of the algorithm there seems to be no reason to make a distinction between the one-dimensional case $p = 1$ and the higher-dimensional case $p > 1$.

However, it is important to observe that when $p = 1$ the limit X can never be a variable function of the initial value $x(0)$ provided $x(0)$ is sufficiently close to X. At most it is a piecewise constant function of $x(0)$. For $p > 1$ on the other hand the mapping $\mathbf{f}(\mathbf{x})$ may well have an infinity of fixed points and the limit **X** usually depends, often in a transcendental fashion even when **f** is algebraic, on $\mathbf{x}(0)$. The implications of this observation are profound, for such multidimensional algorithms, the earliest examples of which are very old, can serve as a way to compute important higher transcendental functions even for simple choices of **f**. For instance,

$$\mathbf{f}(x, y) = \begin{bmatrix} \dfrac{x + y}{2} \\ \sqrt{(xy)} \end{bmatrix}, \tag{12.7}$$

gives a famous algorithm associated with the Landen transformation of elliptic integrals variously attributed to Gauss or Lagrange. (For an excellent historical discussion of some of these algorithms see Carlson (1971).) Note that although the fixed points of the mapping are easily found to be points lying on $x = y$ this knowledge provides no way of determining the limit, if it exists, of the algorithm. Determining the limit **X** is one of the major problems encountered in the study of multidimensional algorithms.

The following concept is helpful in the study of (6).

Definition 12.4 *Let* $F: \Omega \times J^0 \to \mathscr{C}$, *F continuous on* Ω, *and have the* *property*

$$F[\mathbf{f}(\mathbf{x}), n+1] = \phi(n)F[\mathbf{x}, n],\tag{12.8}$$

$\mathbf{x} \in \Omega,$ *for some* $\phi: J^0 \to \mathscr{C},$ $\phi \neq 0.$ *Then F is called an* invariant *for the algorithm.* ∎

Notice that

$$F[\mathbf{x}(n+1), n+1] = \phi(n)F[\mathbf{x}(n), n],$$

or

$$F[\mathbf{x}(n), n] = \prod_{k=0}^{n-1} \phi(k)F[\mathbf{x}(0), 0],$$

so

$$F[\mathbf{x}(0), 0] = \lim_{n \to \infty} F[\mathbf{x}(n), n] \Big/ \prod_{k=0}^{n-1} \phi(k).\tag{12.9}$$

If $\mathbf{x}(n) \to \mathbf{X}$ converges and $F[\mathbf{X}, n]$ has a limit as $n \to \infty$, say $F[\mathbf{X}, \infty]$, then the product in (9) also converges and

$$F[\mathbf{X}, \infty] = \prod_{n=0}^{\infty} \phi(n)F[\mathbf{x}(0), 0].\tag{12.10}$$

This relation plus, perhaps, some additional information about $\mathbf{X}$ may enable one to determine $\mathbf{X}$. Even when $\mathbf{x}(n)$ does not converge the relation (9) may be very helpful.

Another possibility is that one is given an already existing F satisfying (8). One may then be able to construct an algorithm for the computation of quantities related to F.

I will give an example of each of these ideas in turn.

Example 12.1 Consider the algorithm of the arithmetic and harmonic means:

$$x(n+1) = \frac{x(n)+y(n)}{2}, \quad y(n+1) = \frac{2}{\dfrac{1}{x(n)} + \dfrac{1}{y(n)}},$$

$$n \geqslant 0, \quad x(0), y(0) > 0.$$

That the algorithm converges follows from a result (Theorem 13.1) to be established later. It is clear that $x(n)$, $y(n)$ have a common limit X. (This is the additional information required about $\mathbf{X}$ mentioned above.)

An invariant for the algorithm is $F(x, y) = xy$. Thus

$$X^2 = x(0)y(0), \quad X = \sqrt{[x(0)y(0)]}.$$

Table 12.1 Computation of $\sqrt{3}$.

n	$x(n)$	$y(n)$
0	1	3
1	2	1.5
2	1.75	1.714 285 714
3	1.732 142 857	1.731 958 763
4	1.732 050 810	1.732 050 805

A simple computation (Table 12.1) will indicate the astonishing kind of precision to be encountered in these algorithms. $x(4)$ is off by 2 and $y(4)$ by 3 units in the last place, respectively. ■

Example 12.2 (Carlson (1975)) Sometimes an integral serves as the invariant for the algorithm. I shall give many, many examples of this later. In the meantime, let $x, y > 0$; and

$$F(x, y) := \frac{2}{\pi} \int_0^{\pi/2} \ln\left(x \sin^2 \theta + y \cos^2 \theta\right) d\theta. \tag{12.11}$$

Put $\theta = \phi/2$ in (11) and break the integral up as follows:

$$F(x, y) = \frac{1}{\pi} \int_0^{\pi/2} \ln\left[x + y - \frac{x-y}{2} \cos \phi\right] d\phi$$
$$+ \frac{1}{\pi} \int_{\pi/2}^{\pi} \ln\left[x + y - \frac{x-y}{2} \cos \phi\right] d\phi.$$

Replacing ϕ by $\pi - \phi$ in the second integral and recombining gives

$$F(x, y) = \frac{1}{\pi} \int_0^{\pi/2} \ln\left[\left(\frac{x+y}{2}\right)^2 - \left(\frac{x-y}{2}\right)^2 \cos^2 \phi\right] d\phi$$
$$= \frac{1}{\pi} \int_0^{\pi/2} \ln\left[\left(\frac{x+y}{2}\right)^2 \sin^2 \phi + xy \cos^2 \phi\right] d\phi,$$

so

$$F\left[\left(\frac{x+y}{2}\right)^2, xy\right] = 2F(x, y), \tag{12.12}$$

and thus the integral is an invariant for the following algorithm:

$$x(n+1) = \left(\frac{x(n)+y(n)}{2}\right)^2; \qquad y(n+1) = x(n)y(n),$$
$$n \geq 0, \qquad x(0) := x > 0, \qquad y(0) := y > 0.$$

Also the algorithm can be used to compute F for, letting

$$\alpha(n):=\frac{\sqrt{x(n)}+\sqrt{y(n)}}{2}, \qquad \beta(n):=\frac{\sqrt{x(n)}-\sqrt{y(n)}}{2}, \qquad \delta(n):=\frac{\beta(n)}{\alpha(n)},$$

we find

$$\alpha(n+1)=\alpha^2(n), \qquad \beta(n+1)=\beta^2(n), \qquad \delta(n+1)=\delta^2(n), \qquad \textbf{(12.13)}$$

so $\delta(n)\rightarrow 0$ since $|\delta(0)|<1$. Also

$$\frac{y(n)}{x(n)}=\left(\frac{1-\delta(n)}{1+\delta(n)}\right)^2\rightarrow 1, \qquad \frac{\alpha^2(n)}{x(n)}\rightarrow 1. \qquad \textbf{(12.14)}$$

Using the formula (12) shows

$$F(x, y)=2^{-n}F(x(n), y(n))$$
$$=2^{-n}\ln(x(n))+2^{-n}F\left(1, \frac{y(n)}{x(n)}\right).$$

By continuity and the fact that $F(1, 1)=0$,

$$F(x, y)=\lim_{n\rightarrow\infty} 2^{-n}F(x(n), y(n))$$
$$=\lim_{n\rightarrow\infty} 2^{-n}\ln x(n),$$

which provides a way to compute F.

Actually F may be determined explicitly. Note

$$\lim_{n\rightarrow\infty} 2^{-n}\ln x(n)=2\lim_{n\rightarrow\infty} 2^{-n}\ln\alpha(n).$$

But by (13) $\alpha(n)=\alpha(0)^{2^n}$ so

$$F(x, y)=\lim_{n\rightarrow\infty} 2^{-n}[2^n\ln\alpha(0)]=\alpha(0)$$
$$=2\ln\left(\frac{\sqrt{x}+\sqrt{y}}{2}\right).$$

(14) reveals that convergence is quadratic. For more details on the computation of F, see Carlson (1975). ∎

Example 12.3 Consider the elliptic integral,

$$H(\lambda)\equiv H(a, b, c; \lambda):=\int_{\lambda}^{\infty}[(t+a)(t+b)(t+c)]^{-1/2}\,dt, \qquad \lambda, a, b, c \text{ real},$$

where $\lambda+a$, $\lambda+b$, $\lambda+c$ are nonnegative and at most one is 0.

To develop an algorithm for the computation of H we need a

Lemma (Carlson (1978)) *Let λ, a, b, c be as above and*

$$\Lambda:=\lambda+(\lambda+a)^{1/2}(\lambda+b)^{1/2}+(\lambda+a)^{1/2}(\lambda+c)^{1/2}+(\lambda+b)^{1/2}(\lambda+c)^{1/2}.$$

Then

$$H(\lambda) = 2H(\Lambda).$$

Proof See Carlson (1978). ∎

Now let $x(0)$ be such that $x(0) + a$, $x(0) + b$, $x(0) + c$ are nonnegative and at most one is 0. Define

$$x(n+1) = x(n) + (x(n) + a)^{1/2}(x(n) + b)^{1/2} + (x(n) + a)^{1/2}(x(n) + c)^{1/2}$$
$$+ (x(n) + b)^{1/2}(x(n) + c)^{1/2}, \qquad n \geq 0. \tag{12.15}$$

The function $F(x) = H(x)$ is then an invariant for the algorithm and $\phi(n) \equiv \frac{1}{2}$. Thus

$$2^n H(x(n)) = H(x(0)). \tag{12.16}$$

Now $x(n)$ is a divergent sequence (in fact, $x(n) = O(4^n)$). Let us consider the asymptotic expansion of $H(T)$, $T \to \infty$,

$$H(T) = \int_T^\infty t^{-3/2} \left[\left(1 + \frac{a}{t} \right) \left(1 + \frac{b}{t} \right) \left(1 + \frac{c}{t} \right) \right]^{-1/2} dt$$

$$= \int_T^\infty t^{-3/2} \left[1 + \frac{a+b+c}{t} + \frac{ab + ac + bc}{t^2} + \cdots \right]^{-1/2} dt$$

$$= 2T^{-1/2} + \frac{(a+b+c)}{3} T^{-3/2}$$

$$+ \tfrac{1}{20}(3(a^2 + b^2 + c^2) + 2(ab + ac + bc))T^{-5/2} + \cdots, \tag{12.17}$$

by a straightforward power series argument and term-by-term integration.

Thus, if we define

$$x(n) := 4^n w(n),$$

we get

$$2^n H(x(n)) \approx 2w(n)^{-1/2},$$

and

$$H(x(0)) \approx 2w(n)^{-1/2}, \qquad n \to \infty.$$

The process converges linearly but very rapidly, the error being $O(4^{-n})$. The inclusion of additional terms from the series (17) will make the process even more rapidly convergent, $O(16^{-n})$ for example if two terms are retained.

The algorithm is very well suited to floating-point computations.

By using the formulas in Erdélyi *et al.* (1953, vol. 1, p. 115 (5); vol. 2,

p. 318 (5)) we find the special value

$$H(0, 1, z; 0) = \frac{2}{\sqrt{z}} K\left(\sqrt{\left(1 - \frac{1}{z}\right)}\right), \qquad z > 1.$$

In the following table the computation for $H(0, 1, \frac{4}{3}; 0)$ is displayed. Tables (Abramowitz and Stegun (1964, p. 608)) give the value $H = 2.919\,805\,263\,412\,678$.

Table 12.2 Computation of $H(0, 1, \frac{4}{3}; 0) = \sqrt{3}\, K(0.25)$.

n	$2w(n)^{-1/2}$
5	2.922 170 871 689 267
10	2.919 807 571 369 595
15	2.919 805 265 666 542
20	2.919 805 263 414 881
25	2.919 805 263 412 683

It is easy to show this algorithm converges for all complex $x(0)$, a, b, c, provided each lies in the right-half plane. ∎

There is a geometrical concept closely related to invariance which is also useful.

Definition 12.5 *Let* $G : \mathscr{C}^p \to \mathscr{C}$ *be continuous on the closure* Ω *of some open set containing* $\{x(n)\}$. *Let* $S \subset \Omega$ *be the surface defined by*

$$G(\mathbf{x}) = 0, \qquad \mathbf{x} \in S, \tag{12.18}$$

with $\mathbf{x}(0) \in S$.
S is called an invariant surface *for the algorithm (6) if*

$$G(\mathbf{x}) = G(f(\mathbf{x})), \qquad \mathbf{x} \in S. \quad ∎ \tag{12.19}$$

If S is such a surface, then $\mathbf{x}(n) \in S$ and

$$G(\mathbf{x}(n)) = G(\mathbf{x}(0)) = 0,$$

so

$$G(\mathbf{X}) = G(\mathbf{x}(0)) = 0, \tag{12.20}$$

and $\mathbf{X}$ may often be found from the above equation.
For example

$$G(x, y) := y - \frac{x(0)y(0)}{x},$$

defines an invariant surface (curve) for the algorithm of Example 1. Thus from (20),

$$YX = x(0)y(0),$$

and since we know $Y = X$, this determines X.

Obviously, given an invariant F for the algorithm of the form $F(\mathbf{x})$ one can determine an invariant surface by taking $G(\mathbf{x}) = F(\mathbf{x}) - F(\mathbf{x}(0))$.

The concept of invariant surfaces is most useful in the analysis of two-dimensional algorithms (Chapter 13).

When the system (6) is autonomous

$$\mathbf{x}(n+1) = \mathbf{f}(\mathbf{x}(n)), \tag{12.21}$$

several theorems are available to assess the convergence of the algorithm.

Theorem 12.2 *Let* $\mathbf{f}$ *satisfy the Lipschitz condition*

$$\|\mathbf{f}(\mathbf{x}) - \mathbf{f}(\mathbf{y})\| \le \lambda \|\mathbf{x} - \mathbf{y}\|, \qquad 0 \le \lambda < 1,$$

for all vectors $\mathbf{x}, \mathbf{y} \in N_\rho(\mathbf{x}(0))$.
Let $\mathbf{x}(0)$ *satisfy*

$$\|\mathbf{f}(\mathbf{x}(0)) - \mathbf{x}(0)\| \le (1 - \lambda)\rho.$$

Then

$$\|\mathbf{x}(n) - \mathbf{x}(0)\| \le \rho, \qquad n \ge 0,$$

and $\mathbf{x}(n)$ *converges to a fixed point of* $\mathbf{f}$ *unique in* $N_\rho(\mathbf{x}(0))$.

Proof See Isaacson and Keller (1966, p. 110ff.). ∎

(The choice of norm $\|\cdot\|$ is irrelevant.)

Verifying Lipschitz continuity of a vector-valued function can be quite difficult. If it is known that the equation

$$\mathbf{f}(\mathbf{x}) = \mathbf{x} \tag{12.22}$$

has a root, and $\mathbf{f}$ has a derivative,

$$D\mathbf{f} := \left[\frac{\partial f_i}{\partial x_j}\right],$$

the conditions for convergence are simpler.

Theorem 12.3 *Let the equation* (22) *have a root* $\mathbf{X}$, *let* $\mathbf{f} \in C^1$ *in a ball centered at* $\mathbf{X}$ *and* $\|D\mathbf{f}\| < 1$ *there. Then if* $\mathbf{x}(0)$ *is sufficiently close to* $\mathbf{X}$ *the algorithm will converge* (*to* $\mathbf{X}$).

Proof This is a straightforward generalization of the standard result for $p = 1$; see Householder (1953, p. 118ff.). ∎

It must be admitted that these theorems are usually not adequate to decide the convergence of those algorithms encountered in practice; for instance, in Gauss' algorithm (7) it is easily found that

$$\|D\mathbf{f}\| = \frac{1}{2}\left[1 + \max\left(\sqrt{\left(\frac{y}{x}\right)}, \sqrt{\left(\frac{x}{y}\right)}\right)\right], \qquad x, y > 0,$$

in the $\|\cdot\|_1$ norm. Thus $\|D\mathbf{f}\| \geq 1$ everywhere. This example is rather typical; convergence must usually be decided by *ad hoc* methods.

An enormous amount of research has taken place in the past dozen years or so on the convergence properties of the more general systems $\mathbf{x}(n+1) = \mathbf{f}[\mathbf{x}(n), n]$; it is impossible in the space that I have to give references to this material. Most of the work has been done in Russia and other Eastern European countries. *Math Reviews* is a good source of information.

12.3 Divergence; strange attractors

When the algorithm $\mathbf{x}(n+1) = \mathbf{f}[\mathbf{x}(n), n]$ does not converge, the behavior of the trajectory of the algorithm, $\mathcal{O} := \{\mathbf{x}(n)\}_{n=0}^{\infty}$, can be bizarre, to put it mildly. Such sequences arise, for example, in discrete time autonomous dynamical systems of the form

$$\mathbf{x}(t+1) = \mathbf{f}[\mathbf{x}(t)],$$

and often model systems with an irregular nonperiodic 'chaotic' time evolution. Examples of such systems, which appear frequently in physics, chemistry and biology, are the properties of smoke rising from a smoke-stack or convection currents in the atmosphere. The trajectory may be dense in certain regions in p space, or periodic ($\mathbf{x}(n+r) = \mathbf{x}(n)$, $n > n_0$), or may have a finite number of limit points. Even for the simplest systems, characterizing these limit points or the closure of $\bigcup \mathbf{x}(n)$ may be extraordinarily difficult. Nothing is known in generality about such systems; even for simple cases, most available information is conjectural and based on numerical computations, albeit sometimes extensive.

Of particular interest are sets called *attractors*.

Definition 12.6 *A set S is an attractor for the trajectory $\{\mathbf{x}(n)\}$ if*

$$\lim_{n \to \infty} d(\mathbf{x}(n), S) = 0.$$

(Here d is the usual distance function between sets.) ∎

The curious phenomenon exhibited in the following example is to be expected in divergent algorithms.

Example 12.4 (The Hénon attractor) Consider

$$x(n+1) = y(n) + 1 - Ax^2(n),$$
$$y(n+1) = Bx(n),$$

(12.23)

where $A > 0$, $0 < B < 1$; see Hénon (1976), Feit (1978), Curry (1979).

The mapping $\mathbf{f} = (y + 1 - Ax^2, Bx)$ has two fixed points, (x_1, y_1), (x_2, y_2),

$$x_1 := \frac{B - 1 + \sqrt{[(1-B)^2 + 4A]}}{2A}, \qquad y_1 = Bx_1,$$

$$x_2 := \frac{B - 1 - \sqrt{[(1-B)^2 + 4A]}}{2A}, \qquad y_2 = Bx_2.$$

The transformation maps vertical lines into horizontal lines and horizontal lines into parabolas opening to the left.

Let's call an initial point $(x(0), y(0))$ *divergent* if $(x(n), y(n)) \to (-\infty, -\infty)$. Define

$$Q := \{(x, y) \mid x < -1 - \sqrt{(1 + 4A)}, \ y < 0\}.$$

The following statements are easy to establish:

(i) $f(Q) \subset Q$;
(ii) the set of all divergent points is

$$D := \bigcup_{n=0}^{\infty} \mathbf{f}^{-n}(Q);$$

(iii) there is a compact set K bounded by a polygon (depending on A and B) such that if $(x(0), y(0))$ is not divergent, then there is an n_0 such that $(x(n), y(n)) \in K$ for all $n > n_0$.

The set K above acts as an attractor. Depending on the values of A, B and $(x(0), y(0))$ trajectories may or may not converge toward the attractor. This is because the quadratic term in (23) may eventually dominate the computations.

Let us consider what happens when the parameter A varies. For $A = 0$ all points in the plane converge to $(1/(1-B), B/(1-B))$, the unique fixed point of $\mathbf{f}$. When $A < (\frac{3}{4})(1-B)^2$ the rightmost fixed point of f lies in the region $|x| < (1-B)/2A$, where both eigenvalues of $D\mathbf{f}$ are less than 1 in absolute value and this point is an attractor. When $A = (\frac{3}{4})(1-B)^2$ the larger eigenvalue has absolute value 1 and a bifurcation takes place. The trajectory now has two limit points, the points in the trajectory alternating between arbitrarily small neighborhoods of these points. Such an orbit is called a *period two attractor*. This continues until $A = (1-B)^2 + (1+B)^2/4$, when an attracting orbit of period four appears. As A continues to increase further bifurcations take place and attracting orbits of periods $8, 16, \ldots$, appear. The A intervals corresponding to these orbits appear to decrease in size geometrically.

There are also encountered ranges of values of A in which the orbit is not a periodic attractor, but rather, the trajectory appears to be dense on a lamina of curves which are, apparently, parabolas. As A moves to the right this set coalesces into a periodic attractor. For instance this happens between 1.225 and 1.24, the resulting trajectory having a period seven attractor. As A continues to grow, the periodic attractor swells into a nonperiodic attractor of the aforementioned type. Figure 12.1 shows four situations corresponding to A values 1.309, 1.3087, 1.3085, 1.3, $B = 3$, and initial conditions $x(0) = y(0) = 0$. ∎

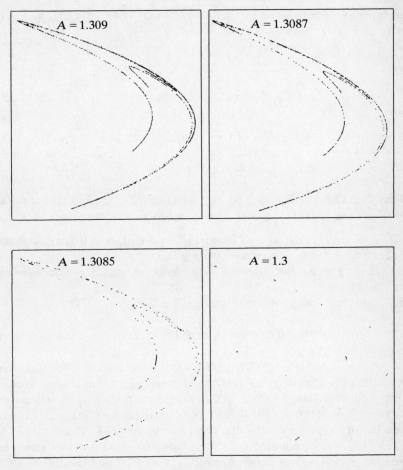

Fig. 12.1

It has been found that nonperiodic attractors have peculiar properties. Trajectories may wander about the attractor in an apparently erratic manner and the attractors are often highly sensitive to initial conditions. Such an attractor has come to be called a 'strange attractor', in the terminology of Ruelle and Takens (1971).

Based on the previous example I will formulate some definitions which are useful in describing such systems.

Denote by $\mathbf{f}^n$ the nth iterate of $\mathbf{f}$, $\mathbf{f}^n = \mathbf{f}(\mathbf{f}(\cdots \mathbf{f}))$. Let $D\mathbf{f}^n$ be the $p \times p$ matrix $\{\partial f_i^n / \partial x_j\}$.

Definition 12.7   *The algorithm*

$$\mathbf{x}(n+1) = \mathbf{f}(\mathbf{x}(n)), \tag{12.24}$$

has sensitive dependence on initial condition *for a set U if, for each* $\mathbf{x}(0) \in U$,

(i) $\mathbf{x}(n)$ *is bounded;*
(ii)

$$\lim_{n \to \infty} \frac{1}{n} \ln \|D\mathbf{f}^n\|_{\mathbf{x} = \mathbf{x}(0)} := C_{\mathbf{x}(0)} > 0. \quad \blacksquare \tag{12.25}$$

This means, roughly, that $x_1(n)$, $x_2(n)$ will not remain close together even when $\mathbf{x}_1$, $\mathbf{x}_2$ are iterations based on different initial values $\mathbf{x}_1(0)$, $\mathbf{x}_2(0) \in U$, which are close together.

Definition 12.8   *A bounded set $A \subset \mathscr{C}^p$ is called a* strange attractor *for the algorithm (24) if A is contained in a ball U with the properties:*

(i)   *if $\mathbf{x}(0) \in U$, then $\mathbf{x}(n) \in U$ and $\lim_{n \to \infty} d(\mathbf{x}(n), A) = 0$ (i.e., A is an attractor for each trajectory starting in U);*
(ii)   *the algorithm has sensitive dependence on initial condition when $\mathbf{x}(0) \in U$;*
(iii)   *there is an $\mathbf{x}(0) \in A$ such that $\{\mathbf{x}(n)\}$ is dense in A.* $\blacksquare$

This last condition, the indecomposability criterion, guarantees that A cannot be made smaller.

Given the fact that a set A is an indecomposable attractor, the question of whether the algorithm has sensitive dependence on initial condition in some ball containing the set will determine whether A is a strange attractor. This in turn can be determined by computing $C \equiv C_{\mathbf{x}(0)}$. If $C < 0$ then the orbit has only attracting fixed points; otherwise it is strange.

Feit (1978) has computed C for the algorithm of the previous example for a range of A values between, roughly, 0 and 2.5 and certain selected values of B. For $B = 0.3$, the first positive value of A occurs at $A = 1.058\,04\ldots$. To the right of this point there is a scattering of subintervals on which $C<0$, each, of course, corresponding to the presence of a periodic attractor. Furthermore, for almost any A interval chosen at random in $(1.058\ldots, 4.2)$ a computer search found a subinterval in which $C<0$ and where, consequently, a periodic attractor existed.

Such situations also arise in autonomous dynamical systems represented by differential equations

$$\mathbf{x}'(t) = \mathbf{F}[\mathbf{x}(t)].$$

It was, in fact, by considering a system of three coupled first-order differential equations due to Lorenz (1963), that Hénon formulated the algorithm of this example.

This topic is one on which fervent research is being conducted and published proceedings and colloquia on strange attractors appear regularly. Unfortunately the subject is tangential to our main object of study so I can't pursue it in detail. The interested reader should consult the very readable article by Ruelle (1980). A more technical discussion by the same author has appeared in Ruelle (1981) and contains numerous references to current work. Another readable survey is Feigenbaum (1980). Applications to population models are given in Guckenheimer *et al.* (1977). Another current survey, this with many beautiful pictures, is in Marzec and Spiegel (1980).

12.4 Mean values

In most of the known two-dimensional iterative algorithms the construction of the double sequence $\mathbf{x}(n) = \{x(n), y(n)\}$ is based on generalized mean value functions which are generalizations of the well-known arithmetic or geometric means.

Definition 12.9 *A continuous function* $M: R^+ \times R^+ \to R^+$ *is called a mean if, for every* $x, y \in R^+$,

$$x \wedge y \leqslant M(x, y) \leqslant x \vee y.$$

If, in addition, $M(\lambda x, \lambda y) = \lambda M(x, y)$, $\lambda > 0$, *then* M *is said to be homogeneous. If* $M(x, y) = M(y, x)$, M *is said to be symmetric.* ∎

Let $f: R^+ \to R^+$ be continuous and strictly monotonic. Then, obviously,

$$M(x, y) = f^{-1}\left(\frac{f(x) + f(y)}{2}\right), \tag{12.26}$$

is a mean. M is then called a *quasiarithmetic mean* and f its *generating function*. (See Hardy *et al.* (1952).) Some well-known examples of quasiarithmetic means are given in Table 12.3.

Table 12.3   Mean value functions.

M	f	Name
$\dfrac{x+y}{2}$	x	Arithmetic
$\sqrt{(xy)}$	$\ln x$	Geometric
$2xy/(x+y)$	$1/x$	Harmonic
$\left(\dfrac{x^c+y^c}{2}\right)^{1/c}$	x^c	Root-power
$\dfrac{1}{c}\ln\left(\dfrac{e^{cx}+e^{cy}}{2}\right)$	$e^{cx},\ c>0$	Exponential

That certain of these means satisfy inequalities is a consequence of a generalized inequality involving mean values, see Hardy *et al.* (1952, p. 26), which is as follows. Let

$$\mathbf{a}:=(a_1, a_2, \ldots, a_n), \qquad a_j>0,$$

and

$$M_p(\mathbf{a}):=\begin{cases} \left[\dfrac{1}{n}\sum_{j=1}^{n}a_j^p\right]^{1/p}, & p\neq 0, \pm\infty; \\[3mm] \left(\displaystyle\prod_{j=1}^{n}a_j\right)^{1/n}, & p=0; \\[3mm] \min_j a_j, & p=-\infty; \\[2mm] \max_j a_j, & p=\infty. \end{cases} \tag{12.27}$$

Theorem 12.4   *For* $-\infty\leqslant p\leqslant q\leqslant\infty,$

$$M_p(\mathbf{a})\leqslant M_q(\mathbf{a}). \quad\blacksquare \tag{12.28}$$

When $p=0$, $q=1$, this provides the well-known inequality of the arithmetic and geometric mean.

As I shall show, not all means, in fact, not even all homogeneous means, can be formed with a generating function as in (26).

I shall require the following

Lemma 1   (Jensen's equation)   *Let f be locally integrable and*

$$f\left(\frac{x+y}{2}\right)=\frac{f(x)+f(y)}{2}. \tag{12.29}$$

Then

$$f(x) = ax + b. \tag{12.30}$$

Proof Letting $x \to 2x$, $y \to 2y$ gives

$$f(x + y) = \frac{f(2x)}{2} + \frac{f(2y)}{2}.$$

But $f(2x)/2 = f(x) - f(0)/2$. Letting $g(x) := f(x) - f(0)$ gives the equation

$$g(x + y) = g(x) + g(y) \tag{12.31}$$

(Cauchy's equation). These operations are reversible, so (31) is, in fact, equivalent to (29).

The integrability of g implies

$$yg(x) = \int_0^{x+y} g(t)\, dt - \int_0^x g(t)\, dt - \int_0^y g(t)\, dt.$$

The right-hand side is symmetric in x and y so interchanging them shows

$$yg(x) = xg(y),$$

or

$$\frac{g(x)}{x} = \frac{g(y)}{y} = \text{constant},$$

which is, essentially, (30).

This sly proof is due to Shapiro (1973). ■

Lemma 2 *Let f, g be generating functions and*

$$f^{-1}\left(\frac{f(x) + f(y)}{2}\right) = g^{-1}\left(\frac{g(x) + g(y)}{2}\right), \qquad x, y \in R^0. \tag{12.32}$$

Then for some $a \neq 0$ and b,

$$g(x) = af(x) + b.$$

Proof Let $f(x) = u$, $f(y) = v$, $x = f^{-1}(u)$, $y = f^{-1}(v)$. Then (32) can be written

$$F\left(\frac{u + v}{2}\right) = \frac{F(u) + F(v)}{2}, \qquad F := gf^{-1},$$

which is Jensen's equation. An application of Lemma 1 finishes the proof. ■

Theorem 12.5 *If M is a homogeneous quasiarithmetic mean, then M is a root-power or geometric mean.*

Proof Write $f(\lambda x) := g_\lambda(x)$, $g_\lambda^{-1}(x) := \lambda^{-1}f^{-1}(x)$. Then

$$f^{-1}\left(\frac{f(\lambda x)+f(\lambda y)}{2}\right) = \lambda f^{-1}\left(\frac{f(x)+f(y)}{2}\right),$$

becomes

$$g_\lambda^{-1}\left[\frac{g_\lambda(x)+g_\lambda(y)}{2}\right] = f^{-1}\left[\frac{f(x)+f(y)}{2}\right],$$

and Lemma 2 shows f satisfies the equation

$$f(\lambda x) = af(x) + b.$$

Let $\lambda = e^\alpha$, $x = e^{\alpha u}$, $F(y) = f(e^{\alpha y})$. Then F satisfies

$$F(u+1) = aF(u) + b.$$

We have

$$F(u) = B\omega(u)a^u + \frac{b}{1-a},$$

$a \neq 1$, $\omega(u)$ an arbitrary function of period 1. (I will consider only the case $a \neq 1$. The case $a = 1$ is even simpler.) Thus

$$f(x) = B\omega\left(\frac{\ln x}{\ln \lambda}\right)x^{\ln a/\ln \lambda} + \frac{b}{1-a}.$$

Since f must be independent of λ, we take $\omega \equiv 1$, $a = \lambda^r$, $b = (1-\lambda^r)c$, to get

$$f(x) = Bx^r + c.$$

But this generates the root-power means. (The case $a = 1$ gives the geometric means.) ∎

We now give an example of a homogeneous mean which is not a quasiarithmetic mean.

Theorem 12.6 *The function*

$$\begin{cases} L(x, y) := \dfrac{x-y}{\ln x - \ln y}, & x \neq y, \\ \\ \quad\quad := x, & x = y, \end{cases} \qquad (12.33)$$

is a nonquasiarithmetic mean (called the logarithmic mean).

Proof Note that L is homogeneous. To show L is a mean, we will establish the inequality of Carlson (1972):

$$\sqrt{(xy)} < (xy)^{1/4} \frac{\sqrt{x}+\sqrt{y}}{2} < L(x, y) < \left(\frac{\sqrt{x}+\sqrt{y}}{2}\right)^2 < \frac{x+y}{2}, \qquad x \neq y.$$

$$(12.34)$$

If $\lambda > 0$ the inequality of the arithmetic and geometric means gives

$$\lambda^2 + \lambda(x+y) + \left(\frac{x+y}{2}\right)^2 > \lambda^2 + \lambda(x+y) + xy > \lambda^2 + 2\lambda\sqrt{(xy)} + xy,$$

so

$$\int_0^\infty \left(\lambda + \frac{x+y}{2}\right)^{-2} d\lambda < \int_0^\infty [(\lambda + x)(\lambda + y)]^{-1} d\lambda,$$

$$< \int_0^\infty (\lambda + \sqrt{(xy)})^{-2} d\lambda.$$

Evaluating the integrals and inverting gives

$$\sqrt{(xy)} < L(x, y) < \frac{x+y}{2}.$$

(Actually, this result is sufficient to prove the theorem.) Now replace x by $\sqrt{x}$, y by $\sqrt{y}$ to get

$$(xy)^{1/4} < \frac{2(\sqrt{x}-\sqrt{y})}{\ln x - \ln y} < \frac{\sqrt{x}+\sqrt{y}}{2}, \qquad (12.35)$$

and multiply by $(\sqrt{x}+\sqrt{y})/2$. This shows the inner inequality in (34) and the outer inequalities follow by once again using the inequality of the arithmetic and geometric means. ∎

Repeating in an obvious way the process which led to (34) gives

$$(xy)^{2^{-n-1}} \prod_{k=1}^n \alpha_k(x, y) < L(x, y) < \alpha_n(x, y) \prod_{k=1}^n \alpha_k(x, y), \qquad x \neq y,$$

$$\alpha_k(x, y) := \frac{x^{2^{-k}} + y^{2^{-k}}}{2},$$

empty products being interpreted as 1. Note the left-hand side above is increasing in n, the right-hand side decreasing. Letting $n \to \infty$ gives

Carlson's result,

$$L(x, y) = \prod_{k=1}^{\infty} \alpha_k(x, y).$$

Putting $y = 1$ gives the interesting product

$$\ln x = (x-1) \prod_{k=1}^{\infty} \frac{2}{1+x^{2^{-k}}}, \qquad x > 0.$$

13 Two-dimensional algorithms

13.1 General remarks; invariant curves

In these sections I will study iterative algorithms of the form

$$x(n+1) = f[x(n), y(n)], \quad y(n+1) = g[x(n), y(n)], \quad n \geq 0. \quad (13.1)$$

Such algorithms have an obvious geometric interpretation. Consider the transformation T of the plane into itself,

$$x' = f(x, y), \quad y' = g(x, y). \quad (13.2)$$

Let $y = \phi(x)$ be the equation of a curve γ. If γ is mapped into itself by the above transformation, i.e., $T(\phi) = \phi$, then γ is called an *invariant curve* or *self-conjugate curve* for the transformation; see Kuczma (1968, Ch. 14). Analytically this means that $y = \phi(x)$ implies $y' = \phi(x')$, i.e.,

$$\phi[f(x, \phi(x))] = g(x, \phi(x)).$$

This equation is called the *equation of invariant curves*.

Now let $(x(0), y(0)) \in \gamma$. The algorithm (1) defines the transition from the nth point $P(n) := (x(n), y(n))$ to the $(n+1)$st point $P(n+1) = (x(n+1), y(n+1))$. (Note that self-conjugate curves always exist, for let $\gamma(0)$ be an arbitrary arc connecting $P(0)$ and $P(1)$ and let $\gamma(1)$ be the image under the above mapping of $\gamma(0)$. Continuing in this way we get a system of arcs $\gamma(0), \gamma(1), \gamma(2), \ldots$, whose union is (obviously) a self-conjugate curve. However if, as is usually the case, we wish to determine whether or not $P(n)$ converges and if so to what, the above construction is of no help since the construction requires a knowledge of $P(n)$.)

It may be that a self-conjugate curve γ is known *a priori*. In such a (rare) case the limit of $P(n)$, if it exists, is the intersection (or one of the intersections) of γ with one of the fixed points of the mapping (2).

Example 13.1 (Tricomi (1966)) Consider the algorithm formed by the weighted arithmetic means

$$x(n+1) = (1-p)x(n) + py(n),$$
$$y(n+1) = qx(n) + (1-q)y(n), \quad n \geq 0.$$

A straight line γ of slope μ in the (x', y') plane

$$y' = \mu x' + c,$$

is carried into a straight line in the (x, y) plane whose equation is

$$(1 - q - p\mu)y = [(1-p)\mu - q]x + c.$$

Consequently γ will be self-conjugate iff $\mu = -q/p$. The fixed points of the mapping consist of points $y = x$. Thus starting at the point $P(0) = (x(0), y(0))$ the algorithm will converge to the common limit of $x(n)$ and $y(n)$,

$$X = (qx(0) + py(0))/(p+q). \quad \blacksquare$$

Since f, g in many algorithms are defined by means the following result is useful.

Theorem 13.1 *Let the functions f, g in the algorithm (1) be symmetric means with $g \leqslant f$ (respectively, $f \leqslant g$). Let $x(0), y(0) > 0$. Then $\lim_{n \to \infty} x(n) = X$, $\lim_{n \to \infty} y(n) = Y$ exist and*

$$x(0) \wedge y(0) \leqslant Y \leqslant X \leqslant x(0) \vee y(0)$$

(respectively,

$$x(0) \wedge y(0) \leqslant X \leqslant Y \leqslant x(0) \vee y(0)).$$

Proof I will establish only the first case. The second is similar. Since f, g are symmetric I may assume $y(0) \leqslant x(0)$. Then $y(n) \leqslant x(n)$, $n \geqslant 0$. But because f, g are means $y(n+1) \geqslant y(n)$, $x(n+1) \leqslant x(n)$, so

$$y(0) \leqslant y(n) \leqslant y(n+1) \leqslant x(n+1) \leqslant x(n) \leqslant x(0).$$

$y(n)$, $x(n)$ are bounded, monotone and thus convergent. $\blacksquare$

In Example 12.8, f, g are the arithmetic and harmonic means respectively and we know (Theorem 12.4) that $g \leqslant f$. Thus the algorithm converges.

Sometimes two-dimensional algorithms result from one-dimensional iterative sequences in the following manner. Suppose $x(n)$ is defined by

$$x(n+1) = \phi(x(n)) + \psi(x(n)). \tag{13.3}$$

Let

$$y(n) := \psi(x(n)).$$

Then the algorithm (3) may be written

$$x(n+1) = \phi(x(n)) + y(n),$$

$$y(n+1) = y(n)\psi[\phi(x(n)) + y(n)]/\psi(x(n)).$$

Example 13.2 Let α be an integer >1 and

$$x(n+1) = \left(1-\frac{1}{\alpha}\right)x(n) + \frac{1}{\alpha}y(n),$$

$$y(n+1) = y(n)\left[\left(1-\frac{1}{\alpha}\right) + \frac{y(n)}{\alpha x(n)}\right]^{1-\alpha}.$$

(13.4)

It is easily seen that this algorithm can be derived from the Newton iterative sequence

$$x(n+1) = \left(1-\frac{1}{\alpha}\right)x(n) + \frac{c}{\alpha}x(n)^{1-\alpha},$$

which converges to $X = c^{1/\alpha}$, $c>0$. Thus (4) converges to $X = Y = y(0)^{1/\alpha}x(0)^{1-(1/\alpha)}$ provided $y(0)>0$, $x(0)>0$. ∎

13.2 Evaluation of certain infinite products

Two-dimensional iteration schemes can be adapted to the computation of a class of infinite products that arise in the theory of theta functions (Gatteschi (1969/70), Slater (1966, Ch. 7), Allasia and Bonardo (1980)).
Consider a two-dimensional algorithm

$$x(n+1) = x(n)f\left[\frac{y(n)}{x(n)}\right], \qquad y(n+1) = x(n+1)g\left[\frac{y(n)}{x(n)}\right].$$

(13.5)

Putting

$$z(n) := y(n)/x(n), \qquad n \geq 0,$$

(13.6)

and assuming that $\lim_{n\to\infty} x(n)$ exists, which we denote as usual by X, gives

$$X = x(0)\prod_{n=0}^{\infty} f(z(n)),$$

(13.7)

where

$$z(n+1) = g(z(n)).$$

(13.8)

Thus the algorithm provides a means of computing the infinite product (7). However it is not clear in general how to construct such an algorithm for evaluating a given infinite product, nor whether the algorithm offers computational advantages over the straightforward evaluation of X by means of partial products. I shall investigate these ideas, at least for some simple examples, in this section.
The fixed points of the transformation

$$x' = xf\left(\frac{y}{x}\right), \qquad y' = x'g\left(\frac{y}{x}\right),$$

if such exist, are the locus of points $y = \alpha x$ or the origin, $x = y = 0$. Thus if the algorithm converges and, say, f, g are continuous in a suitable region, either $Y = 0$ or $Y = \alpha X$. In most cases it turns out the latter prevails and $\alpha = 1$. This fact can often be put to computational advantage.

In this section we shall investigate an interesting generalization of the algorithm based on the arithmetic–harmonic mean,

$$x(n+1) = \frac{x(n) + y(n)}{2}, \quad y(n+1) = \frac{2}{\dfrac{1}{x(n)} + \dfrac{1}{y(n)}}, \quad n \geq 0.$$

Theorem 1 shows that this algorithm converges. Since $z(n) = y(n)/x(n)$ satisfies

$$z(n+1) = 4z(n)/(1 + z(n))^2,$$

we see $z(n) \to 1$ and $X = Y$. The equation (7) and the above give

$$X = x(0) \lim_{N \to \infty} \left\{ \prod_{n=0}^{N} \frac{z(n+1)}{z(n)} \right\}^{-1/2}$$

$$= x(0)z(0)^{-1/2} \lim_{N \to \infty} z(N+1)^{-1/2} = \sqrt{[x(0)y(0)]}.$$

A modified algorithm in which $y(n)$ is computed with the latest available information about $x(n)$ is

$$x(n+1) = \frac{x(n) + y(n)}{2}, \quad y(n+1) = \frac{2x(n+1)y(n)}{x(n+1) + y(n)}, \quad n \geq 0.$$

In fact, I shall follow Gatteschi's discussion of the slightly more general algorithm

$$x(n+1) = \xi x(n) + (1-\xi)y(n), \quad y(n+1) = \frac{x(n+1)y(n)}{\eta x(n+1) + (1-\eta)y(n)},$$

$$n \geq 0, \quad \textbf{(13.9)}$$

where ξ, η are arbitrary real numbers, $|\xi\eta| < 1$, η being such that the algorithm is defined, i.e., $\eta x(n+1) + (1-\eta)y(n) \neq 0$. Substituting the expression for $x(n+1)$ into that for $y(n+1)$ and referring to (8) shows

$$f(z(n)) = \xi + (1-\xi)z(n), \quad g(z(n)) = \frac{z(n)}{(1-\xi\eta)z(n) + \xi\eta}.$$

The relationship $z(n+1) = g(z(n))$ becomes

$$z(n+1) = \frac{z(n)}{(1-\xi\eta)z(n) + \xi\eta},$$

or

$$\frac{1}{z(n+1)} - 1 = \xi\eta\left(\frac{1}{z(n)} - 1\right),$$

and thus

$$\frac{1}{z(n)}-1=\xi^n\eta^n\left(\frac{1}{z(0)}-1\right),$$

$$\frac{1}{z(n)}=1-\left(1-\frac{1}{z(0)}\right)\xi^n\eta^n.$$

We have

$$f(z(n))=\frac{1-(1-1/z(0))\xi^{n+1}\eta^n}{1-(1-1/z(0))\xi^n\eta^n},$$

and

$$X=x(0)\prod_{n=0}^{\infty}\frac{1-(1-x(0)/y(0))\xi^{n+1}\eta^n}{1-(1-x(0)/y(0))\xi^n\eta^n}, \qquad (13.10)$$

provided the infinite product converges. Since $y=x$ is an invariant curve for the transformation $x'=\xi x+(1-\xi)y$, $y'=x'y/(\eta x'+(1-\eta)y)$, $y(n)$ will also converge when $x(n)$ does and we then have $Y=X$.

Define a, b, q by

$$\frac{x(0)}{y(0)}:=1-b, \qquad \xi:=\frac{a}{b}, \qquad \eta:=\frac{q}{\xi}.$$

If

$$1-aq^n\neq 0, \qquad 1-bq^n\neq 0, \qquad (13.11)$$

the infinite product (10) will converge. But

$$\frac{x(n)}{y(n)}=1-bq^n, \qquad \frac{x(n+1)}{y(n)}=1-aq^n, \qquad (13.12)$$

and

$$\eta x(n+1)+(1-\eta)y(n)=\eta x(n)\frac{1-aq^n}{1-bq^n}+\frac{(1-\eta)x(n)}{1-bq^n}$$

$$=x(n)\frac{1-bq^{n+1}}{1-bq^n}. \qquad (13.13)$$

Thus if (11) holds equations (12) and (13) show the denominator in (9) cannot be zero so the algorithm is defined.

I now wish to establish that

$$\lim_{n\to\infty}\frac{X-x(n)}{y(n)-Y}=\frac{1-\xi}{(1-\eta)\xi}, \qquad \eta\xi>0. \qquad (13.14)$$

A little algebra shows

$$\frac{x(n+1)-x(n)}{y(n)-y(n+1)}=\frac{(1-\xi)x(n+1)}{(1-\eta)\xi y(n+1)},$$

so

$$\lim_{n\to\infty} \frac{x(n+1)-x(n)}{y(n)-y(n+1)} = \frac{1-\xi}{(1-\eta)\xi}.$$

I now require a

Lemma *Let* $u(n)$, $v(n)$ *be real sequences. Then*

(i) *if* $v(n+1)/v(n)$ *is bounded away from* 1 *for* n *sufficiently large, then*

$$\lim_{n\to\infty} \frac{u(n)}{v(n)} = c \Rightarrow \lim_{n\to\infty} \frac{\Delta u(n)}{\Delta v(n)} = c;$$

(ii) *if* $u(n)$, $v(n)$ *are null sequences and* $v(n)$ *is strictly monotone for* n *sufficiently large, then*

$$\lim_{n\to\infty} \frac{\Delta u(n)}{\Delta v(n)} = c \Rightarrow \lim_{n\to\infty} \frac{u(n)}{v(n)} = c.$$

Proof These results are consequences of the Toeplitz limit theorem; see Brezinski (1977, p. 12ff.). ∎

To apply the lemma let $u(n) = x(n) - X$, $v(n) = Y - y(n)$; then $v(n)$ is strictly monotone if $y(n)$ is. But

$$\frac{y(n)}{y(n-1)} = \frac{1-aq^{n-1}}{1-bq^n} = 1 + b\xi(\eta-1)q^{n-1} + \cdots.$$

Since $q > 0$ this means that $y(n)$ is ultimately either increasing or decreasing.

(14) shows that when $q > 0$ an approximation to X, better than either $x(n)$ or $y(n)$, can be obtained by taking $t(n)$ as defined by

$$\frac{t(n)-x(n)}{y(n)-t(n)} := \frac{1-\xi}{(1-\eta)\xi},$$

or

$$t(n) := px(n) + (1-p)y(n),$$

$$p := \frac{(1-\eta)\xi}{1-\xi\eta}.$$

In Table 13.1, taken from Gatteschi's paper, the algorithm (9) is used to compute the product

$$\prod_{n=0}^{\infty} \frac{1-0.2(0.5)^n}{1-0.5(0.5)^n} = 2.252\ 052\ 4.$$

Table 13.1   Computation of a product.

n	$x(n)$	$y(n)$	$t(n)$
0	1.000 000 000 0	2.000 000 000 0	2.200 000 000 0
1	1.600 000 000 0	2.133 333 333 3	2.239 999 999 9
2	1.919 999 999 9	2.194 285 714 3	2.249 142 857 2
3	2.084 571 428 4	2.223 542 857 0	2.251 337 142 7
4	2.167 954 285 5	2.237 888 294 8	2.251 875 096 7
5	2.209 914 691 0	2.244 992 702 1	2.252 008 304 3
6	2.230 961 497 6	2.248 528 123 6	2.252 041 448 8
7	2.241 501 473 1	2.250 291 675 1	2.252 049 715 5
8	2.246 775 594 2	2.251 172 415 4	2.252 051 779 6
9	2.249 413 686 8	2.251 612 527 2	2.252 052 295 3

A slight modification of the above algorithm allows for the computation of a famous infinite product which occurs in many different contexts in applied mathematics. As before let $g(z) = z/[(1-q)z+q]$. But now let $f(z) = \xi + (1-\xi)/z$. Then

$$f[z(n)] = 1 - (1-\xi)\left(1 - \frac{1}{z(0)}\right)q^n,$$

and the result is the algorithm

$$x(n+1) = x(n)\frac{\xi y(n) + (1-\xi)x(n)}{y(n)},$$

$$y(n+1) = x(n+1)\frac{y(n)}{qx(n) + (1-q)y(n)}, \qquad n \geqslant 0,$$

(13.15)

with initial values satisfying

$$\frac{x(0)}{y(0)} = 1 - \frac{a}{1-\xi}.$$

(13.16)

The algorithm converges to

$$X = \lim_{n\to\infty} x(n) = \lim_{n\to\infty} y(n) = \frac{y(0)}{x(0)}P(q),$$

$$P(q) := \prod_{n=0}^{\infty}(1 - aq^n).$$

Note that the convergence is linear.

ξ ($\neq 1$) is arbitrary in (16) and a, q are given; naturally, one wonders what value of ξ will produce the most efficient algorithm. To explore this note that

$$\frac{x(n+1) - x(n)}{y(n) - y(n+1)} = \frac{x(n)}{y(n)}(\xi - 1)\left[\frac{qx(n) + (1-q)y(n)}{(1-\xi)x(n) + (1-q)y(n)}\right],$$

so, as before, for $q > 0$,

$$\lim_{n \to \infty} \frac{X - x(n)}{y(n) - X} = \frac{\xi - 1}{2 - q - \xi}.$$

This suggests that by taking the right-hand side to be 1, so that in the limit $x(n)$ and $y(n)$ are equally close to X, then the average value of $x(n)$ and $y(n)$ will be a better approximation to X than either one alone. This requires choosing

$$\xi = \frac{3 - q}{2}.$$

The resulting algorithm is

$$x(n+1) = x(n) \frac{(q-1)x(n) + (3-q)y(n)}{2y(n)},$$

$$y(n+1) = x(n+1) \frac{y(n)}{qx(n) + (1-q)y(n)}, \qquad n \geq 0. \tag{13.17}$$

If

$$\frac{x(0)}{y(0)} = \frac{1 + 2a - q}{1 - q}, \qquad 1 + 2a - q \neq 0, \quad |q| < 1, \tag{13.18}$$

then

$$X := \lim_{n \to \infty} x(n) = x(0)P(q),$$

and an improved value of $x(n)$ is

$$t(n) := \frac{x(n) + y(n)}{2}.$$

In most cases of interest, $|a| < 1$, $0 < q < 1$.

Slater (1966, p. 249ff.) has tabulated P to seven significant figures for selected values of a and q using the formula (*ibid*, p. 92 (3.2.2.15))

$$P^{-1}(q) = \sum_{n=0}^{\infty} a^n / [q]_n, \tag{13.19}$$

$$[q]_n := (1 - q)(1 - q^2) \cdots (1 - q^n), \qquad [q]_0 = 1.$$

Obviously these computations will run into trouble for q near 1. In fact, Slater mentions that the critical range seems to be $0.89 < q < 1$. For these ranges of values of q, it is best to use neither (19) nor the algorithm (17) but to use the infinite product directly by means of the algorithm

$$u(n+1) = qu(n), \qquad\qquad u(0) = a,$$

$$v(n+1) = v(n)(1 + u(n)), \qquad v(0) = 1.$$

Although convergence is slow the computations are stable.

Using a combination of approaches, Allasia and Bonardo (1980) have tabulated P extensively for $|a|<1$, $0<q<1$. Gatteschi (1969/70) has discussed the computation of P for other ranges of a and q.

Certain related infinite products which arise in the theory of theta functions can be computed with the algorithm (17); these are

$$\prod_{n=1}^{\infty}(1+q^n), \qquad \prod_{n=1}^{\infty}(1-q^{2n}):=Q_0,$$

$$\prod_{n=1}^{\infty}(1+q^{2n}):=Q_1, \qquad \prod_{n=1}^{\infty}(1+q^{2n-1}):=Q_2,$$

$$\prod_{n=1}^{\infty}(1-q^{2n-1}):=Q_3.$$

The first is obtained from P by setting $a=-q$; and Q_0 and Q_1 may be reduced to P or the first product by putting $q^2=p$, and similarly for Q_2 and Q_3. Further, it is known that

$$Q_0Q_3=\prod_{n=1}^{\infty}(1-q^n), \qquad Q_1Q_2=\prod_{n=1}^{\infty}(1+q^n), \qquad Q_1Q_2Q_3=1$$

(Hancock, 1958, Ch. 18).

13.3 Carlson's results

Carlson (1971) has determined that in many important cases, the limits of two-dimensional algorithms can be expressed in terms of the function

$$R(a;b,b';x^2,y^2):=\frac{1}{B(a,a')}\int_0^{\infty} t^{a'-1}(t+x^2)^{-b}(t+y^2)^{-b'}\,dt, \qquad \textbf{(13.20)}$$

where B denotes the beta function, Re $a>0$, Re $a'>0$, and

$$a+a'=b+b'.$$

For certain values of the parameters, as I shall show, this function may act as an invariant. Among the cases considered are well-known algorithms attributed to Gauss and Borchardt.

It is easy to show that

$$R(a;b,b';x^2,y^2)=R(a;b',b;y^2,x^2)$$

$$=y^{-2a}F\left(\begin{matrix}a,b\\b+b'\end{matrix}\middle|\,1-\frac{x^2}{y^2}\right). \qquad \textbf{(13.21)}$$

In what follows, define the functions $f_i(x, y)$ by

$$f_1(x, y):=\frac{x+y}{2}, \qquad f_2(x, y):=\sqrt{(xy)},$$

$$f_3(x, y):=\left(x\frac{x+y}{2}\right)^{1/2}, \qquad f_4(x, y):=\left(\frac{x+y}{2}y\right)^{1/2}. \qquad \textbf{(13.22)}$$

We start with a

Lemma *Let $x(0), y(0) > 0$ and define*

$$x(n+1) = f_i(x(n), y(n)), \quad y(n+1) = f_j(x(n), y(n)), \qquad n \geq 0.$$

$$\textbf{(13.23)}$$

Then $x(n), y(n)$ have a common limit, denoted by X_{ij}.

Proof It is easy to show that exactly one of the following statements is true according to whether $x < y$, $x = y$, or $x > y$:

$$\left.\begin{array}{l} x < f_3 < f_2 < f_1 < f_4 < y, \\ x = f_1 = f_2 = f_3 = f_4 = y, \\ y < f_4 < f_2 < f_1 < f_3 < x. \end{array}\right\} \qquad \textbf{(13.24)}$$

Define $I(n) = [x(n), y(n)]$. Then

$$|x^2(n+1) - y^2(n+1)| \leq |f_3^2(x(n), y(n)) - f_4^2(x(n), y(n))|$$

$$= \frac{|x^2(n) - y^2(n)|}{2}. \qquad \textbf{(13.25)}$$

(23), (24) show $I(n+1) \subset I(n)$. Thus $\{I(n)\}$ comprises a nested set of intervals with lengths tending to 0 which, by Cantor's principle, possess a single common point, the common limit of $x^2(n)$ and $y^2(n)$. ∎

The convergence of all the algorithms is at least linear. For, defining $\varepsilon(n):=|x(n)^2 - y(n)^2|$ and assuming $x(0) < y(0)$, we have

$$x(0) < x(n) < x(n+1) < y(n+1) < y(n) < y(0),$$

and

$$\varepsilon(n+1) \leq \frac{\varepsilon(n)}{2},$$

so $\varepsilon(n) \leq \varepsilon^n(0)/2^n$. Thus

$$X - x(n) + y(n) - X \leq \frac{[y^2(0) - x^2(0)]}{2^n(x(n) + y(n))},$$

or

$$0 < X - x(n) < \frac{[y^2(0) - x^2(0)]}{2^{n+1}x(0)},$$

and a similar inequality holds for $y(n) - X$. However in some cases, as we shall see, the convergence is hyperlinear.

Now in the integral for R, make the substitution

$$t := \frac{s(s + f_2^2)}{s + f_1^2},$$

s a new variable of integration. After a little algebra one finds

$$R(a; b, b'; x^2, y^2) = \frac{1}{B(a, a')} \int_0^\infty s^{a'-1}(s + f_1^2)^{a-1}(s + f_2^2)^{a'-1}$$
$$\times (s + f_3^2)^{1-2b}(s + f_4^2)^{1-2b'} \, ds. \qquad \textbf{(13.26)}$$

Given any i and j, $i \neq j$, we can choose two parameters so that f_k disappears from the integrand if $k \neq i$ or j. This result will be used to prove

Theorem 13.2 *Let* $x(0) > 0$, $y(0) > 0$. *Then* X_{ij} *is as given in the Table 13.2.*

Table 13.2 $X_{ij} = [R(a, b, b'; x(0)^2, y(0)^2)]^{-1/2a}.$

			a, b, b'		
	j	1	2	3	4
i					
1		—	$\frac{1}{2}, \frac{1}{2}, \frac{1}{2}$	$\frac{1}{4}, \frac{3}{4}, \frac{1}{2}$	$\frac{1}{2}, \frac{1}{2}, 1$
2		$\frac{1}{2}, \frac{1}{2}, \frac{1}{2}$	—	$1, \frac{3}{4}, \frac{1}{2}$	$1, \frac{1}{2}, 1$
3		$\frac{1}{2}, 1, \frac{1}{2}$	$1, 1, \frac{1}{2}$	—	$1, 1, 1$
4		$\frac{1}{4}, \frac{1}{2}, \frac{3}{4}$	$1, \frac{1}{2}, \frac{3}{4}$	$1, 1, 1$	—

Proof As an illustration I will do the case $i = 1$, $j = 3$. Letting $a' = 1$ and $b' = \frac{1}{2}$ in (26) gives

$$R(a; a + \tfrac{1}{2}, \tfrac{1}{2}; x^2, y^2) = R(a; 1 - a, 2a; f_1^2, f_3^2).$$

There is one value of a, $a = \frac{1}{4}$, which causes the parameters of the Rs to coincide. Thus

$$F(x, y) := R(\tfrac{1}{4}; \tfrac{3}{4}, \tfrac{1}{2}; x^2, y^2)$$
$$= R(\tfrac{1}{4}; \tfrac{3}{4}, \tfrac{1}{2}, f_1^2, f_3^2).$$

Letting $x = x(n)$, $y = y(n)$, $x(n)$, $y(n)$ defined as in (23) with $i = 1$, $j = 3$, shows that F is an invariant for the algorithm. We have

$$F(X, X) = F(x(0), y(0))$$

$$= X_{13}^{-1/2} F\left(\begin{matrix} \frac{1}{4}, \frac{3}{4} \\ 1 \end{matrix} \middle| 0\right) = X_{13}^{-1/2},$$

so

$$X_{13} = [R(\tfrac{1}{4}; \tfrac{3}{4}, \tfrac{1}{2}; x^2(0), y^2(0))]^{-2},$$

as claimed. ∎

Example 13.3 $(i = 1,\ j = 2,$ Gauss' algorithm) The algorithm is the famous algorithm of arithmetic and geometric means,

$$x(n+1) = \frac{x(n) + y(n)}{2}, \qquad y(n+1) = \sqrt{x(n)y(n)}, \qquad n \geq 0. \qquad \textbf{(13.27)}$$

Assume without loss of generality that $0 < x(0) < y(0)$.

Using Table 13.2, the formula (21) and a result of Erdélyi (1953, vol. 2, p. 318 (5)) shows that

$$X = \lim_{n \to \infty} x(n) = \lim_{n \to \infty} y(n) = \frac{\pi y(0)}{2} \left\{ K \sqrt{\left(1 - \frac{x^2(0)}{y^2(0)}\right)} \right\}^{-1}, \qquad \textbf{(13.28)}$$

where K is the complete elliptic integral of the first kind,

$$K(k) := \int_0^1 \frac{du}{\sqrt{(1 - u^2)}\sqrt{(1 - k^2 u^2)}}.$$

Let's investigate the convergence of this algorithm. Again, assume $x(0) < y(0)$. Following Carlson (1965) we put

$$\varepsilon(n) = \frac{x(n) - y(n)}{x(n) + y(n)}.$$

Then we find $\varepsilon(n)$ satisfies

$$\varepsilon(n+1) = \frac{\varepsilon^2(n)}{(1 + \sqrt{[1 - \varepsilon^2(n)]})^2}.$$

This implies that $\varepsilon(n) \downarrow$, $n > 0$. Thus

$$\frac{\varepsilon^2(n)}{4} < \varepsilon(n+1) < \frac{\varepsilon^2(n)}{\Delta(0)}, \qquad n \geq 0.$$

$$\Delta(m) := (1 + \sqrt{[1 - \varepsilon^2(m)]})^2.$$

Iterating this relationship shows that

$$4 \left(\frac{\varepsilon(0)}{4}\right)^{2^n} < \varepsilon(n) < \Delta(0) \left(\frac{\varepsilon(0)}{\Delta(0)}\right)^{2^n}, \qquad n \geq 1.$$

By the inequality of the arithmetic and geometric means,

$$y(n) < y(n+1) < X < x(n+1) < x(n), \qquad n \geq 1, \tag{13.29}$$

so

$$x(n) - y(n) = (x(n) - X) + (X - y(n)) > x(n) - X > 0,$$

and we find

$$0 < x(n) - X < 2\Delta(0)y(0)[\varepsilon(0)/\Delta(0)]^{2^n}. \tag{13.30}$$

The same holds for $X - y(n)$. (30) establishes that convergence is quadratic.

A more subtle argument shows that convergence is exactly quadratic, of degree

$$K_2(y(0)/x(0)) := \inf_{0 \leq m} \left(\frac{\varepsilon(m)}{\Delta(m)} \right)^{1/2^m}. \quad \blacksquare$$

Example 13.4 ($i = 1$, $j = 4$, Borchardt's algorithm) The algorithm is

$$x(n+1) = \frac{x(n) + y(n)}{2}, \qquad y(n+1) = \sqrt{[x(n+1)y(n)]}, \qquad n \geq 0. \tag{13.31}$$

This might seem to be little more than Gauss' algorithm where $y(n+1)$ is computed using the latest available information about $x(n)$. However, the slight alteration in the algorithm produces a profound change in its numerical properties. Assume $x(0) < y(0)$. We find

$$X = \lim_{n \to \infty} x(n) = \sqrt{[y^2(0) - x^2(0)]} \left[\arccos \frac{x(0)}{y(0)} \right]^{-1}, \tag{13.32}$$

from Erdélyi *et al.* (1953, vol. 1, p. 102 (13)).

In this case $x(n)$ and $y(n)$ can be computed explicitly as follows. Define

$$\cos v(n) = \frac{x(n)}{y(n)}.$$

The recurrence formulas give

$$\frac{x(n+1)}{y(n+1)} = \sqrt{\left(\frac{1 + x(n)/y(n)}{2} \right)},$$

or

$$\cos v(n+1) = \sqrt{\left(\frac{1 + \cos v(n)}{2} \right)} = \cos \frac{v(n)}{2}.$$

Thus $v(n+1) = v(n)/2$, or $v(n) = v(0)2^{-n}$ and

$$\frac{x(n)}{y(n)} = \cos(2^{-n}v(0)).$$

On the other hand

$$\frac{y(n+1)}{y(n)} = \sqrt{\left(\frac{x(n+1)}{y(n)}\right)} = \frac{x(n+1)}{\sqrt{[x(n+1)y(n)]}} = \frac{x(n+1)}{y(n+1)} = \cos\,(2^{-n-1}v(0)),$$

so

$$y(n) = y(0) \prod_{k=1}^{n} \cos\,(2^{-k}v(0)).$$

But from the formula

$$\cos\alpha = \frac{\sin 2\alpha}{2 \sin \alpha},$$

with $\alpha = 2^{-k}v(0)$ we find

$$\prod_{k=1}^{n} \cos\,(2^{-k}v(0)) = \frac{\sin v(0)}{2^n \sin\,(2^{-n}v(0))}.$$

Thus

$$\left.\begin{array}{l} x(n) = \dfrac{y(0) \sin v(0) \operatorname{ctn}\,(v(0)/2^n)}{2^n}, \\[3mm] y(n) = \dfrac{y(0) \sin v(0) \csc\,(v(0)/2^n)}{2^n}, \\[3mm] v(0) = \arccos\,(x(0)/y(0)), \end{array}\right\} \tag{13.33}$$

and it is clear the algorithm converges for all complex $x(0)$, $y(0)$. (However, (33) must be interpreted properly in view of the branch points of the functions involved. For more on this, see Tricomi (1966, p. 24ff.).)

The convergence behavior of the algorithm may be analyzed by means of the explicit expressions (33) but it is more convenient to proceed as follows. Let

$$\varepsilon(n) := y^2(n) - x^2(n).$$

Then

$$\varepsilon(n+1) = \varepsilon(n)/4,$$

and this implies

$$\varepsilon(n) = \varepsilon(0)/4^n.$$

If $x(0) < y(0)$, then $X - x(n) < y(n) - x(n)$ and an analysis similar to that following the lemma shows

$$\frac{y^2(0) - x^2(0)}{4^n(y(0) + X)} < X - x(n) < \frac{y^2(0) - x^2(0)}{4^n(x(0) + X)}.$$

Similar estimates may be inferred for $x(0)$, $y(0)$ complex. Thus the convergence is linear only. ∎

Example 13.5 $(i = 1, j = 3;$ lemniscate constants) The algorithm is

$$x(n+1) = \frac{x(n) + y(n)}{2}, \qquad y(n+1) = \left(x(n)\frac{x(n) + y(n)}{2}\right)^{1/2}, \qquad n \geqslant 0.$$

$$\text{(13.34)}$$

Define the *lemniscate function* by

$$\operatorname{arcsl} x := \int_0^x (1 - t^4)^{-1/2} \, \mathrm{d}t, \qquad x^2 \leqslant 1.$$

This function, studied by Gauss, rectifies the lemniscate; see Carlson (1965).

One finds

$$X = (x^2(0) - y^2(0))^{1/2} \left[\operatorname{arcsl}\left(1 - \frac{y^2(0)}{x^2(0)}\right)\right]^{-1/2}, \qquad y(0) < x(0).$$

A formula with a related function holds if $y(0) > x(0)$. It is easy to show that convergence of (34) is linear, $O(4^{-n})$. ∎

Example 13.6 (Carlson (1971)) Consider

$$x(n+1) = \left[\frac{x(n)(x^p(n) - y^p(n))}{p(x(n) - y(n))}\right]^{1/p},$$

$$y(n+1) = \left[\frac{y(n)(x^p(n) - y^p(n))}{p(x(n) - y(n))}\right]^{1/p}, \qquad n \geqslant 0, \qquad p > 1. \quad \text{(13.35)}$$

This algorithm is not covered by the previous theory but it is easily verified that

$$F(x, y) = (x^p - y^p)\Big/\ln\frac{x}{y}, \qquad \text{(13.36)}$$

is an invariant for (35). If $\varepsilon(n) := y^p(n) - x^p(n)$, then

$$\varepsilon(n+1) = \varepsilon(n)/p,$$

and we have linear convergence for positive $x(0)$, $y(0)$, $x(0) \neq y(0)$. In fact $x(n)$, $y(n)$ are readily found to have the same limit

$$X = \left[\frac{x^p(0) - y^p(0)}{p \ln (x(0)/y(0))}\right]^{1/p}. \quad ∎$$

Example 13.7 (Tricomi (1966)) This is one of a large number of algorithms which can be used for the evaluation of various elliptic integrals and functions. It is given here because its use depends on Gauss' algorithm.

Define

$$F(\phi, k) := \int_0^\phi \frac{\mathrm{d}t}{\sqrt{(1 - k^2 \sin^2 t)}}, \qquad 0 \leqslant \phi \leqslant \pi/2.$$

It is a fairly simple matter to show that

$$(1+k')F(\phi, k) = F\left(\psi, \frac{1-k'}{1+k'}\right),$$

$$\psi := \phi + \arctan[k' \tan \phi], \qquad k' := \sqrt{(1-k^2)}.$$

Let

$$\sqrt{[1-z^2(n)]} := k, \qquad \phi := \phi(n), \qquad \psi := \phi(n+1),$$
$$(1-k')/(1+k') := \sqrt{[1-z^2(n+1)]}.$$

We get

$$F(\phi(n), \sqrt{[1-z^2(n)]}) = \frac{1}{1+z(n)} F(\phi(n+1), \sqrt{[1-z^2(n+1)]}),$$

where

$$z(n+1) = \frac{2\sqrt{z(n)}}{1+z(n)}, \qquad \phi(n+1) = \phi(n) + \arctan[z(n) \tan \phi(n)],$$

$$n \geq 0. \quad \textbf{(13.37)}$$

Now recall Gauss' algorithm. Let $y(n)/x(n) = z(n)$. Then $z(n)$ satisfies the first equation above and, from (27),

$$\frac{x(n+1)}{x(n)} = \frac{1+z(n)}{2},$$

that is,

$$\lim_{n\to\infty} x(n) = x(0) \prod_{n=0}^{\infty} \frac{(1+z(n))}{2} = \frac{\pi}{2} \Big/ K\sqrt{[1-z^2(0)]}.$$

F is an invariant for the algorithm defined by (37) and we have succeeded in evaluating the associated infinite product. Obviously $z(n) \to 1$. Furthermore,

$$F(\phi, k) = \phi + O(k^2 \phi), \qquad k \to 0.$$

Thus

$$2^{-n}F(\phi_n, \sqrt{[1-z^2(n)]}) = 2^{-n}\phi(n) + O(2^{-n}\phi(n)(1-z^2(n))).$$

Taking limits in a straightforward manner gives

$$\lim_{n\to\infty} 2^{-n}\phi(n) = \frac{\pi F(\phi(0), \sqrt{[1-z^2(0)]})}{2K(\sqrt{[1-z^2(0)]})}. \qquad \textbf{(13.38)}$$

This formula can be used, along with Gauss' algorithm, to compute $F(\phi, k)$.

In the second formula in (37) the arctan is to be determined so that $|\phi(n+1) - 2\phi(n)| < \pi/2$; see Tricomi (1966, p. 20) for details. ∎

A number of important two-dimensional algorithms similar in structure to the ones in this section have been studied by Allasia (1969/70), (1970/71), (1971/72).

13.4 An algorithm of Gatteschi

Gatteschi (1966), (1966/67) has given a generalization of Borchardt's algorithm which is interesting because it involves an unusual generalization of the elementary trigonometric functions and illustrates how multi-dimensional recurrences arise naturally from certain functional equations. (Functional equations are discussed in more generality in Section 13.6.)

Consider the functional equation

$$\phi(\sqrt{(2k)}t) = k\phi^2(t) + 1 - k, \qquad k > 1, \quad t \in \mathcal{R}. \tag{13.39}$$

Given any value $t_0 > 0$, if $\phi(t)$ is defined in the interval $t_0 \le t < t_0\sqrt{(2k)}$, $\phi(t) = h(t)$, say, the equation has a unique solution defined for all $t > 0$. (This can be seen from the fact that the substitution $t = e^{\xi}$, $\phi(e^{\xi}) = \Phi(\xi)$ sends the equation into a difference equation.) A similar statement holds if $t_0 < 0$; just reverse all the inequalities above. I shall, however, be interested only in *analytic* solutions corresponding to initial conditions given at the point $t_0 = 0$. Here the situation is quite different: arbitrary functions $h(t)$ cannot arise.

A solution of (39) $\phi(t) \equiv 1$ or $\phi(t) = (1/k) - 1$ will be called a *trivial solution*.

Theorem 13.3 *The equation (39) possesses a unique nontrivial solution $\Phi(k, t) \equiv \Phi(t)$ analytic at $t = 0$ with the property*

$$\Phi''(t)|_{t=0} = -2.$$

This solution has the following properties:

(i) $\Phi(0) = 1$;

(ii) Φ *is entire*;

(iii) *any other nontrivial solution $\phi(t)$ analytic at 0 can be obtained from Φ by taking $\phi(t) = \Phi(\lambda t)$ for some $\lambda \ne 0$*;

(iv) Φ *is even and real on the real axis*;

(v)

$$\Psi(\sqrt{(2k)}t) = \sqrt{(2k)}\Phi(t)\Psi(t), \qquad \Psi(t) := \Phi'(t); \tag{13.40}$$

(vi) Φ *is strictly decreasing in some minimal interval $[0, \beta]$, where $\Phi(\beta) = 1 - k$, and so the equation $c = \Phi(t_0)$ has a unique solution t_0 in $[0, \beta]$ for $1 - k \le c \le 1$.*

Proof The first part of the proof proceeds by a power series argument. That Φ is even follows from the equation. Assume

$$\Phi(t) = \sum_{n=0}^{\infty} a(2n)t^{2n}.$$

Then we find $a(0) = 1$ or $a(0) = (1/k) - 1$, and

$$a(2n)(2k)^n = k \sum_{i=0}^{n} a(2i)a(2n-2i), \qquad n \geqslant 1,$$

or

$$a(2) - a(0)a(2) = 0,$$

$$2(2^{n-1}k^{n-1} - a(0))a(2n) = \sum_{i=1}^{n-1} a(2i)a(2n-2i), \qquad n \geqslant 2.$$

It is easily seen that taking $a(0) = (1/k) - 1$ gives a trivial solution. Thus $a(0) = 1$. Now put $a(2) = -1$, and assume

$$|a(2n)| \leqslant \frac{1}{n!}, \qquad 0 \leqslant n \leqslant N-1. \tag{13.41}$$

Then

$$|a(2N)| \leqslant \frac{1}{2(2^{N-1}k^{N-1}-1)}\left[\frac{1}{1!\,(N-1)!} + \frac{1}{2!\,(N-2)!} + \cdots + \frac{1}{(N-1)!\,1!}\right]$$

$$\leqslant \frac{1}{(2^N-2)N!}\left[\binom{N}{1} + \binom{N}{2} + \cdots + \binom{N}{N-1}\right]$$

$$= \frac{(2^N-2)}{(2^N-2)N!} = \frac{1}{N!},$$

and since $|a(2)| = 1$ the formula (41) is established for all n by induction.

The above arguments establish (i), (ii) and (iv), while (iii) and (v) are trivial.

Since

$$\Phi'(\sqrt{(2k)}\,t) = \sqrt{(2k)}\,\Phi(t)\Phi'(t),$$

and $a(2) = -1$, Φ' is negative in some right neighborhood of 0, $(0, \delta)$, and hence Φ is decreasing in $(0, \delta)$. Now assume $\Phi > 0$ for all $t > 0$. Then Φ must approach asymptotically some constant $\eta \geqslant 0$, since (40) shows that Φ' is always negative, hence Φ is monotone decreasing. We have

$$\eta = k\eta^2 + 1 - k,$$

and either $\eta = 1$, an impossibility, or $\eta = (1/k) - 1 < 0$, an impossibility. Thus for some minimum value $\alpha > 0$, $\Phi(\alpha) = 0$. This means $\Phi'(\sqrt{(2k)}\,\alpha) = 0$ and $\Phi(\sqrt{(2k)}\,\alpha) = 1 - k$. But the point $\beta := \sqrt{(2k)}\,\alpha$ cannot be a point of inflection, since $\Phi''(\sqrt{(2k)}\,\alpha) = \Phi'(\alpha)^2 > 0$. Thus it is a local minimum, and (vi) is established. ∎

Now let $x(0)$, $y(0)$ be real numbers such that

$$1-k \leqslant \frac{x(0)}{y(0)} < 1,$$

and let τ be defined by

$$\frac{x(0)}{y(0)} := \Phi(\tau), \qquad 0 < \tau \leqslant \beta.$$

Put

$$x(n) = \frac{y(0)\Psi(\tau)\Phi(\tau\sqrt{(2k)^{-n}})}{\sqrt{(2k)^n}\Psi(\tau\sqrt{(2k)^{-n}})},$$

$$y(n) = \frac{y(0)\Psi(\tau)}{\sqrt{(2k)^n}\Psi(\tau\sqrt{(2k)^{-n}})}, \qquad n \geqslant 0.$$

(13.42)

Using (39) and (40) shows that

$$x(n+1) = \frac{x(n)+(k-1)y(n)}{k}, \quad y(n+1) = \sqrt{[x(n+1)y(n)]}, \qquad n \geqslant 0,$$

(13.43)

and this is Gatteschi's generalization of Borchardt's algorithm. It is easy to show that the algorithm converges. In fact

$$\Phi(\tau\sqrt{(2k)^{-n}}) = 1 + O(2k)^{-n},$$
$$\Psi(\tau\sqrt{(2k)^{-n}}) = -2\tau\sqrt{(2k)^{-n}}(1 + O(2k)^{-n}).$$

Thus

$$\lim_{n \to \infty} x(n) = \lim_{n \to \infty} y(n) = \frac{-y(0)\Psi(\tau)}{2\tau}.$$

(13.44)

For $k = 2$ (43) reduces to Borchardt's algorithm, Example 4. Then

$$\Phi(t) = \cos(\sqrt{2})t, \qquad \Psi(t) = -\sqrt{2}\sin(\sqrt{2})t,$$

and (44) yields the known limit (32).

Let's examine the error of the algorithm. Let

$$\lambda := \min\left[\frac{1}{k^2}, \left(\frac{1}{k} - \frac{1}{k^2}\right)\right], \qquad \Lambda := \max\left[\frac{1}{k^2}, \left(\frac{1}{k} - \frac{1}{k^2}\right)\right].$$

Then it is clear that

$$k\lambda(x+y) \leqslant \frac{x+(k-1)y}{k} \leqslant k\Lambda(x+y), \qquad x, y \geqslant 0.$$

(13.45)

Defining $\varepsilon(n) := y^2(n) - x^2(n)$ and using (45) with $x = x(n)$, $y = y(n)$ gives the inequality

$$\lambda\varepsilon(n) \leqslant \varepsilon(n+1) \leqslant \Lambda\varepsilon(n), \qquad n \geqslant 0.$$

Thus

$$\Lambda^n \varepsilon(0) \leqslant \varepsilon(n) \leqslant \Lambda^n \varepsilon(0), \qquad n \geqslant 0.$$

We can obtain an explicit error estimate from this inequality by the procedures used in Section 13.3:

$$0 < X - x(n) < \frac{(y^2(0) - x^2(0))\Lambda^n}{2x(0)}, \qquad 0 < x(0) < y(0).$$

Note that for no value of k can convergence be quadratic.

The following interesting properties of Φ are shown in Gatteschi (1966/67):

(i) If

$$1 < k < k_0 := \frac{1 + \sqrt{5}}{2},$$

there is only one positive zero of Φ. Thereafter the function oscillates close to its limit $(1/k) - 1$ as $t \to \infty$.

(ii) For $k = k_0$ there is an infinity $\{\alpha(n)\}$ of positive zeros which, for $n \geqslant 2$, are also relative maxima. There is also an infinity of relative minima, $\{\beta(n)\}$, and

$$\alpha(1) < \beta(1) < \alpha(2) < \beta(2) < \cdots.$$

When t is large Φ jumps from values close to zero to values close to $1 - k$; it becomes approximately a piecewise constant function. In the variable ξ, $e^\xi = t$, and it is approximately periodic with period $\ln \sqrt{(2k)}$.

(iii) For $k > k_0$ there is an infinity of zeros, an infinity of minima equal to $1 - k$, an infinity of maxima equal to $k(k-1)^2 + 1 - k > 0$, and other extrema also.

(iv) Φ is bounded on the real axis iff $1 < k \leqslant 2$.

13.5 Tricomi's algorithm

Let k be a real number, $0 \leqslant k \leqslant 1$, and let $k' = \sqrt{(1-k^2)}$. Tricomi (1965) has introduced the algorithm

$$x(n+1) = \frac{x(n) + \sqrt{[k'^2 y^2(n) + k^2 x^2(n)]}}{y(n) + \sqrt{[k'^2 y^2(n) + k^2 x^2(n)]}} y(n),$$

$$y(n+1) = \sqrt{[x(n+1)y(n)]}, \qquad n \geqslant 0.$$

(13.46)

A function appropriate to the study of this equation is the elliptic cosine, $\operatorname{cn} v \equiv \operatorname{cn}(v, k)$, which satisfies the half-angle formula,

$$\operatorname{cn} \frac{v}{2} = \left(\frac{\operatorname{cn} v + \sqrt{(k'^2 + k^2 \operatorname{cn}^2 v)}}{1 + \sqrt{(k'^2 + k^2 \operatorname{cn}^2 v)}} \right)^{1/2}.$$

Assume $|x(0)| < y(0)$ and put

$$\frac{x(n)}{y(n)} = \text{cn } v(n), \qquad 0 < v(n) < 2K.$$

(Recall that $4K$ is the real primitive period of the function cn; hence the restriction on $v(n)$.)

The recurrence shows

$$\frac{x(n+1)}{y(n+1)} = \sqrt{\frac{x(n+1)}{y(n)}} = \left[\frac{(x(n)/y(n)) + \sqrt{[k'^2 + k^2(x^2(n)/y^2(n))]}}{1 + \sqrt{[k'^2 + k^2(x^2(n)/y^2(n))]}}\right]^{1/2},$$

and this means

$$v(n+1) = \frac{v(n)}{2} \qquad \text{or} \qquad v(n) = 2^{-n}v(0).$$

Thus

$$\frac{x(n)}{y(n)} = \text{cn }(2^{-n}v(0)), \qquad \frac{y(n+1)}{y(n)} = \frac{x(n+1)}{y(n+1)} = \text{cn }(2^{-n-1}v(0)). \qquad \textbf{(13.47)}$$

Put

$$W(t) = \prod_{n=1}^{\infty} \text{cn }(2^{-n}t).$$

Then from (47) we find

$$X := \lim_{n \to \infty} x(n) = \lim_{y \to \infty} y(n) = y(0)W(v(0)).$$

In contrast to the special case $k = 0$ (Borchardt's algorithm) it is not known in general how to evaluate in terms of known functions the infinite product for $W(t)$.

Allasia (1972) has studied a generalization of this algorithm.

13.6 Solutions of linear functional equations

A kind of functional equation that arises fairly often in applied mathematics is of the form

$$\phi[f(x)] = g(x)\phi(x), \qquad \textbf{(13.48)}$$

f, g being given and ϕ being the unknown function. For an extensive theoretical discussion of this equation, see Kuczma (1968). Multidimensional iteration sequences may often be used to compute solutions of this and related equations.

We begin our discussion with an existence theorem.

Let $a > 0$, $I := [0, a]$, $I^* := (0, a]$.

Lemma 1 *Let*

(i) $f : I \to I$ *be continuous and strictly increasing,* $0 < f(x) < \theta x$, $x \in I^*$, *for some* $0 < \theta < 1$;

(ii) $g : I \to \mathscr{C}$ *be continuous,* $g(x) \neq 0$, $x \in I^*$, *and for some positive* M, μ, δ *let* $|g(x) - 1| \leqslant M x^\mu$, $x \in [0, \delta)$.

Then for every $\eta \in \mathscr{C}$ *there exists a unique continuous solution* $\phi : I \to \mathscr{C}$ *of* (48) *having the property* $\phi(0) = \eta$.

Proof See Kuczma (1968). ∎

In what follows I will assume hypotheses (i), (ii) of the lemma are satisfied. Let $x(0) \in I$ be given and consider the algorithm

$$x(n+1) = f(x(n)), \qquad n \geqslant 0. \tag{13.49}$$

The equation (48) produces

$$\frac{\phi(x(n+1))}{\phi(x(n))} = g(x(n)), \qquad n \geqslant 0,$$

or

$$\phi(x(0))/\phi(x(n+1)) = \prod_{j=0}^{n} g(x(j))^{-1},$$

and since, by (i), $x(n) \to 0$ and ϕ is continuous,

$$\phi(x(0)) = \eta \prod_{j=0}^{\infty} g(x(j))^{-1}.$$

Letting

$$y(n) := \eta \prod_{j=0}^{n} g(x(j))^{-1},$$

produces the following two-dimensional recurrence:

$$\begin{aligned}
x(n+1) &= f(x(n)), & x(0) &= \alpha \in I; \\
y(n+1) &= y(n)/g(x(n+1)), & y(0) &= \eta/g(x(0)), & n \geqslant 0.
\end{aligned} \tag{13.50}$$

Theorem 13.4 *If the conditions* (i), (ii) *of Lemma 1 are satisfied, the algorithm* (50) *converges and*

$$\lim_{n \to \infty} x(n) = 0, \qquad \lim_{n \to \infty} y(n) = \phi(\alpha). \quad \blacksquare$$

The theory concerning the related nonhomogeneous equation

$$\phi(f(x)) = g(x)\phi(x) + F(x) \tag{13.51}$$

is slightly more complicated, but Kuczma (1968) gives several existence and uniqueness results. The following is both simple and useful.

Lemma 2 *Let*

(i) $f:I\to I$ *be continuous and strictly increasing,* $f(x)<x$, $x>0$;
(ii) $g:I\to\mathcal{C}$ *be continuous,* $g(x)\ne 0$, $x\in I^*$, $|g(0)|>1$;
(iii) $F:I\to\mathcal{C}$ *be continuous.*

Then the nonhomogeneous equation (51) *has a unique continuous solution* $\phi:I\to\mathcal{C}$. ∎

Now let (i), (ii), (iii) be satisfied and define

$$f^{(i)}(x):=f(f^{(i-1)}(x)),\qquad i\geq 1,\quad f^{(0)}(x):=x,\quad x\in I.$$

Write the equation (51) as

$$\phi(x)=\frac{\phi(f(x))-F(x)}{g(x)},\qquad x\in I.$$

Using this formula, it is easily established by induction that

$$\phi(x)=\phi(f^{(n+1)}(x))G_n(x)-\sum_{i=0}^{n}F(f^{(i)}(x))G_{i+1}(x),\qquad n\geq 0,\quad x\in I,$$

where

$$G_n(x):=\prod_{i=0}^{n-1}g(f^{(i)}(x))^{-1}.$$

Since $G_n(x)\to 0$ and $\phi(f^{(n+1)}(x))\to\phi(0)$, we obtain

$$\phi(x)=-\sum_{n=0}^{\infty}F(f^{(n)}(x))G_{n+1}(x).$$

Now let $\alpha\in I$, $x\to x(0)$, and define $x(n)$ as before. The result is the following three-dimensional recurrence,

$$
\begin{aligned}
x(n+1)&=f(x(n)), & x(0)&=\alpha\in I,\\
y(n+1)&=y(n)/g(x(n+1)), & y(0)&=g(x(0))^{-1},\\
z(n+1)&=z(n)-F(x(n+1))y(n+1), & n&\geq 0,\\
z(0)&=-F(x(0))y(0).
\end{aligned}
\tag{13.52}
$$

Theorem 13.5 *If the conditions* (i), (ii), (iii) *of Lemma 2 are satisfied, then the algorithm* (52) *converges in the sense that*

$$\lim_{n\to\infty}z(n)=\phi(\alpha).\qquad\blacksquare$$

Example 13.8 The equation

$$\phi(2x) = \tfrac{1}{2}[\phi(x) + x],$$

arises in statics. I write it in the form

$$\phi\left(\frac{x}{2}\right) = 2\phi(x) - \frac{x}{2},$$

so $f(x) = x/2$, $g(x) \equiv 2$, $F(x) = -x/2$. The algorithm is: $x(n+1) = x(n)/2$, $x(0) = \alpha$, $0 \le \alpha < 1$, $y(n+1) = y(n)/2$, $y(0) = \tfrac{1}{2}$, $z(n+1) = z(n) + \tfrac{1}{2}x(n+1)y(n+1)$, or

$$x(n) = \frac{\alpha}{2^n}, \qquad y(n) = \frac{1}{2^{n+1}},$$

and in this case an explicit computation gives the solution $\phi(\alpha) = \alpha/3$. ∎

The special homogeneous equation

$$\phi(f(x)) = c\phi(x), \qquad c \ne 0, 1, \tag{13.53}$$

is one of the most important special cases of (48) and is called *Schröder's equation*.

Existence and uniqueness results given in Kuczma (1968) which we state without proof lead to very simple iterative algorithms for computing the solution of (53).

There are a number of different cases to consider corresponding to different assumptions about f.

Theorem 13.6 *Let*

(i) $f : I \to I$ *have an rth continuous derivative, be strictly increasing and satisfy* $0 < f(x) < x$, $x \in I^*$;
(ii) $f'(0) = c$, $0 < f'(0) < 1$.

Then for $r \ge 2$ *the equation* (53) *for every* $\eta \in \mathcal{R}$ *has exactly one* C^r *solution* $\phi : I \to \mathcal{R}$ *satisfying* $\phi'(0) = \eta$.

The solution is given by the algorithm:

$$x(n+1) = f(x(n)), \qquad n \ge 0, \quad x(0) = \alpha \in I,$$
$$\phi(\alpha) = \eta \lim_{n \to \infty} c^{-n}x(n), \qquad n \ge 0, \tag{13.54}$$

and for $\eta \ne 0$ ϕ *is strictly monotonic in* I, *increasing for* $\eta > 0$, *decreasing for* $\eta < 0$. ∎

Theorem 13.7 *Let hypotheses* (i) (*with* $r = 1$) *and* (ii) *of Theorem 6 hold and suppose*

(iii) $f'(x) \ne 0$, $x \in I$;

(iv) *there exist positive constants K, μ such that*

$$|f'(x) - c| \leqslant Kx^{\mu},$$

holds for x sufficiently small positive.

Then the equation (53) for every $\eta \in \mathcal{R}$ has exactly one C^1 solution $\phi : I \to \mathcal{R}$ satisfying $\phi'(0) = \eta$.

The solution is given by the algorithm (54) and for $\eta \neq 0$ is strictly monotonic in I, increasing for $\eta > 0$, decreasing for $\eta < 0$. ∎

Theorem 13.8 *Let hypotheses* (i) *(with r = 0) and* (ii) *of Theorem 6 hold. Further, let*

(iii) $f(x)/x$ *be increasing in I.*

Consider the algorithm

$$x(n+1) = f(x(n)), \qquad n \geqslant 0, \quad x(0) = \alpha \in I,$$
$$y(n+1) = f(y(n)), \qquad n \geqslant 0, \quad y(0) = \beta \in I, \quad n \geqslant 0, \tag{13.55}$$

β *fixed, arbitrary. Then the limit*

$$\phi(\alpha) = \eta \lim_{n \to \infty} \frac{x(n)}{y(n)},$$

defines a one-parameter (in η) family of solutions of equation (53) *which are continuous and for $\eta \neq 0$ strictly monotonic in I, increasing for $\eta > 0$, decreasing for $\eta < 0$.* ∎

The case $f'(0) = 0$ is also of importance.

Theorem 13.9 *Let hypothesis* (i) *(with r = 1) of Theorem 6 hold and $f'(0) = 0$. Further, suppose*

(ii) *there exist positive numbers b, δ, M and $\mu > 1$ such that*

$$|f'(x) - \mu bx^{\mu-1}| \leqslant Mx^{\mu-1+\delta},$$

holds for x sufficiently small.

Then the algorithm

$$x(n+1) = f(x(n)), \qquad\qquad n \geqslant 0, \quad x(0) = \alpha \in I, \quad n \geqslant 0,$$

$$\phi(\alpha) = \eta \lim_{n \to \infty} \mu^{-n} \ln \frac{1}{x(n)}, \qquad \eta \neq 0,$$

defines a one-parameter family of continuous and strictly decreasing solutions of the equation

$$\phi(f(x)) = \mu\phi(x), \qquad x \in I^*. \quad ∎$$

14 Higher-dimensional algorithms

14.1 The computation of a class of trigonometric integrals

Multidimensional algorithms for computing a very general class of trigonometric integrals can be derived from a result due to Bartky (1938) for the transformation of such integrals. In general these algorithms converge with extraordinary rapidity.

Let H be a function of the variable

$$R := \sqrt{(b^2 \cos^2 \phi + c^2 \sin^2 \phi)}, \tag{14.1}$$

where $b, c > 0$, $0 \le \phi \le \pi/2$, as well as the parameter $\mathbf{a}$, $\mathbf{a} \in \mathscr{C}^p$. Further, let H be continuous in the interval $b \wedge c \le R \le b \vee c$ and in all its parameters. Define

$$F(\mathbf{a}, b, c) := \int_0^{\pi/2} H(R) \frac{d\phi}{R}. \tag{14.2}$$

Lemma (Bartky)

$$F(\mathbf{a}, b, c) = \int_0^{\pi/2} \bar{H}(\bar{R}) \frac{d\phi}{\bar{R}}, \tag{14.3}$$

where

$$\bar{R} := \sqrt{(\bar{b}^2 \cos^2 \phi + \bar{c}^2 \sin^2 \phi)},$$
$$\bar{b} := \frac{b+c}{2}, \qquad \bar{c} := \sqrt{(bc)}, \tag{14.4}$$

$$\bar{H}(\bar{R}) := \tfrac{1}{2}[H(\bar{R} + \sqrt{(\bar{R}^2 - \bar{c}^2)}) + H(\bar{R} - \sqrt{(\bar{R}^2 - \bar{c}^2)})]. \tag{14.5}$$

Proof The proof is rather elegant. The integral F when written in terms of R alone is

$$F = \int_c^b \frac{H(R) \, dR}{\Delta},$$

where

$$\Delta^2 := (R^2 - c^2)(b^2 - R^2).$$

Pick a new variable $\bar{R}$ by

$$\bar{R} := \tfrac{1}{2}(R + bcR^{-1}) := \sqrt{(\bar{b}^2 \cos^2 \bar{\phi} + \bar{c}^2 \sin^2 \bar{\phi})}.$$

As R varies from c to b, $\bar{R}$ diminishes from the arithmetical mean $\bar{b} := (b + c)/2$ to the geometric mean $\bar{c} := \sqrt{(bc)}$, and then increases to $\bar{b}$. Also

$$R = \bar{R} \pm (\bar{R}^2 - \bar{c}^2)^{1/2},$$
$$\Delta^2 = 4R^2(\bar{b}^2 - \bar{R}^2),$$
$$d\bar{R} = \tfrac{1}{2}(1 - \bar{c}^2 R^{-2}) \, dR$$
$$= \pm R^{-1}(\bar{R}^2 - \bar{c}^2)^{1/2} \, dR,$$

and so

$$\frac{dR}{\Delta} = \pm \frac{d\bar{R}}{2\bar{\Delta}},$$
$$\bar{\Delta}^2 = (\bar{R}^2 - \bar{c}^2)(\bar{b}^2 - \bar{R}^2).$$

Thus

$$F = \int_{\bar{c}}^{\bar{b}} \bar{H}(\bar{R}) \frac{d\bar{R}}{\bar{\Delta}} = \int_0^{\pi/2} \bar{H}(\bar{R}) \frac{d\bar{\phi}}{\bar{R}},$$

$\bar{H}$ defined as in (5); and this is the required result. ∎

Now write $H(R) = H(\mathbf{a}, R)$ to illustrate the dependence on $\mathbf{a}$ explicitly and suppose we can write

$$\Phi(\mathbf{a}, b, c)H(\mathbf{a}, R) = H(\bar{\mathbf{a}}, \bar{R}), \tag{14.6}$$

for some Φ. (Note that $\bar{\mathbf{a}}$ may depend on $\bar{b}$, $\bar{c}$.) Of course this may not always be possible but in many cases it is. It is possible for instance when H is a general polynomial in R or a general rational function.

We will then have

$$F(\bar{\mathbf{a}}, \bar{b}, \bar{c}) = \Phi(\mathbf{a}, b, c)F(\mathbf{a}, b, c).$$

Define

$$\mathbf{a}(n+1) = \bar{\mathbf{a}}(n), \qquad b(n+1) = \bar{b}(n),$$
$$c(n+1) = \bar{c}(n), \tag{14.7}$$

with

$$\mathbf{a}(0) = \mathbf{a}, \qquad b(0) = b, \qquad c(0) = c. \tag{14.8}$$

The integral operator F is an *invariant* for the following algorithm:

$$\mathbf{a}(n+1) = \bar{\mathbf{a}}(n); \qquad b(n+1) = \frac{b(n) + c(n)}{2}, \qquad c(n+1) = \sqrt{[b(n)c(n)]},$$
$$\tag{14.9}$$

$$F(n+1) = \Phi(n)F(n), \qquad n \geq 0,$$

where

$$F(n) := F(\mathbf{a}(n), b(n), c(n)),$$
$$\Phi(n) := \Phi(\mathbf{a}(n), b(n), c(n)).$$

Applying the theory of Section 12.2 gives

$$F(\mathbf{a}, b, c) = K(n) \int_0^{\pi/2} \frac{H(\mathbf{a}(n), R(n)) \, d\phi}{R(n)},$$

$$K(n) := \prod_{j=0}^{n-1} \Phi(j)^{-1}, \qquad\qquad (\mathbf{14.10})$$

$$R(n) := \sqrt{[b^2(n) \cos^2 \phi + c^2(n) \sin^2 \phi]}.$$

Now

$$\lim_{n \to \infty} b(n) = \lim_{n \to \infty} c(n) = X,$$

X being the limit of Gauss' algorithm, (13.28). Consequently

$$H(\mathbf{a}(n), R(n)) \sim H(\mathbf{a}(n), b(n)),$$

the latter being independent of ϕ. If taking the limit under the integral sign is permissible (this is usually quite easy to ascertain) and we define

$$G(n) := \frac{\pi}{2} \frac{K(n)}{b(n)} H(\mathbf{a}(n), b(n)), \qquad\qquad (\mathbf{14.11})$$

then we have

$$F(\mathbf{a}, b, c) = \lim_{n \to \infty} G(n). \qquad\qquad (\mathbf{14.12})$$

Equations (8)–(11) constitute the desired algorithm for computing F.

Example 14.1 Compute the integral

$$F := \int_0^{\pi/2} \frac{\cos 2\phi \, d\phi}{\sqrt{(\cos^2 \phi + \tfrac{1}{4} \sin^2 \phi)}}.$$

Write the integrand as

$$H(R) := a_1 R^2 + a_2,$$

with $a_1 = \tfrac{8}{3}$, $a_2 = -\tfrac{5}{3}$, $b = 1$, $c = \tfrac{1}{2}$.

$$\bar{H}(\bar{R}) = \tfrac{1}{2}[a_1(\bar{R} - \sqrt{(\bar{R} - \bar{c}^2)})^2 + a_2 + a_1(\bar{R} + \sqrt{(\bar{R} - \bar{c}^2)})^2 + a_2]$$
$$= 2a_1 \bar{R}^2 + a_2 - a_1 \bar{c}^2,$$

Table 14.1

n	$G(n)$
1	0
2	$-0.359\,341\,389\,635\,099$
3	$-0.364\,709\,428\,094\,467$
4	$-0.364\,710\,005\,650\,180$
5	$-0.364\,710\,005\,650\,203$
6	$-0.364\,710\,005\,650\,211$
7	$-0.364\,710\,005\,650\,211$

so if $\mathbf{a}:=(a_1, a_2)$, then $\bar{\mathbf{a}}:=(2a_1, a_2 - a_1\bar{c}^2)$. The format becomes

$$b(n+1) = \frac{b(n)+c(n)}{2}, \qquad c(n+1) = \sqrt{[b(n)c(n)]},$$

$$a_1(n+1) = 2a_1(n), \qquad a_2(n+1) = a_2(n) - a_1(n)c^2(n+1),$$

$$\left. G(n) = \frac{\pi(a_1(n)b^2(n) + a_2(n))}{2b(n)}, \qquad n \geqslant 0, \right\}$$

$$a_1(0) = \tfrac{8}{3}, \qquad a_2(0) = -\tfrac{5}{3}, \qquad b(0) = 1, \qquad c(0) = \tfrac{1}{2}.$$

The algorithm converges with stunning rapidity, as Table 14.1 shows. Such enthusiastic convergence seems to be typical of these algorithms. ■

Example 14.2   Consider the integral

$$F := \int_0^{\pi/2} \ln(A + BR)\,\frac{\mathrm{d}\phi}{R}.$$

Then

$$H(R) := \ln(A + BR),$$
$$\bar{H}(\bar{R}) = \tfrac{1}{2}\ln(2AB\bar{R} + A^2 + B^2\bar{c}^2).$$

Letting $\bar{A} = A^2 + B^2\bar{c}^2$, $\bar{B} = 2AB$ gives the algorithm

$$b(n+1) = \frac{b(n)+c(n)}{2}, \qquad c(n+1) = \sqrt{[b(n)c(n)]},$$

$$\left. A(n+1) = A^2(n) + B^2(n)c^2(n+1), \right.$$

$$\left. B(n+1) = 2A(n)B(n), \qquad n \geqslant 0, \right\} \qquad \textbf{(14.13)}$$

$$\left. c(0) = c, \qquad b(0) = b, \qquad A(0) = A, \qquad B(0) = B. \right\}$$

$\Phi \equiv 2$ and

$$F = \frac{\pi}{2}\lim_{n\to\infty}\left\{\frac{\ln[A(n) + B(n)b(n)]}{2^n b(n)}\right\}. \quad ■$$

I will now discuss the convergence of the algorithm (8)–(11) for a wide class of functions, namely, rational functions.

Let

$$H(x) := \sum_{j=0}^{p} a_j x^j + \sum_{j=1}^{q} \frac{\alpha_j}{x + \beta_j}. \tag{14.14}$$

Recall that the Chebyshev polynomial of degree n, $T_n(x)$, has the explicit representation

$$T_n(x) = \frac{(x + \sqrt{(x^2 - 1)})^n + (x - \sqrt{(x^2 - 1)})^n}{2}, \qquad n \geq 0.$$

Let

$$\sigma(j, r) := \partial_j T_r(x), \qquad 0 \leq j \leq r.$$

A little algebra shows that

$$\tfrac{1}{2}[H(\bar{R} + \sqrt{(\bar{R}^2 - \bar{c}^2)}) + H(\bar{R} - \sqrt{(\bar{R}^2 - \bar{c}^2)})] = \sum_{j=0}^{p} a_j \bar{c}^j T_j\left(\frac{\bar{R}}{\bar{c}}\right) + \sum_{j=1}^{q} \frac{\alpha_j}{2\beta_j}$$

$$+ \sum_{j=1}^{q} \frac{(\alpha_j/4)}{\bar{R} + (\beta_j^2 + \bar{c}^2)/2\beta_j} \left(1 - \frac{\bar{c}^2}{\beta_j^2}\right). \tag{14.15}$$

Inspection shows h is a function having the property (6) and further

$$\left.\begin{array}{l} \bar{a}_j = \displaystyle\sum_{r=j}^{p} a_r \bar{c}^{r-j} \sigma(j, r), \qquad 1 \leq j \leq p, \\[4mm] \bar{a}_0 = \displaystyle\sum_{r=0}^{p} a_r \bar{c}^r \sigma(0, r) + \sum_{j=1}^{q} \frac{\alpha_j}{2\beta_j}, \\[4mm] \bar{\alpha}_j = \dfrac{\alpha_j}{4}\left(1 - \dfrac{\bar{c}^2}{\beta_j^2}\right), \qquad 1 \leq j \leq q, \\[4mm] \bar{\beta}_j = \dfrac{\beta_j^2 + \bar{c}^2}{2\beta_j}, \qquad 1 \leq j \leq q. \end{array}\right\} \tag{14.16}$$

Equations (14)–(16) lead to the algorithm

$$b(n+1) = \frac{b(n) + c(n)}{2}, \qquad c(n+1) = \sqrt{[b(n)c(n)]},$$

$$b(0) = b, \qquad c(0) = c; \tag{14.17}$$

$$a_j(n+1) = \sum_{r=j}^{p} a_r(n) c^{r-j}(n+1)\sigma(j, r), \qquad 1 \leq j \leq p,$$

$$a_j(0) = a_j; \tag{14.18}$$

$$a_0(n+1) = \sum_{r=0}^{p} a_r(n) c^r(n+1)\sigma(0, r) + \sum_{j=1}^{q} \frac{\alpha_j(n)}{2\beta_j(n)},$$

$$a_0(0) = a_0; \tag{14.19}$$

$$\alpha_j(n+1) = \frac{\alpha_j(n)}{4}\left(1 - \frac{c^2(n+1)}{\beta_j^2(n)}\right), \qquad 1 \leqslant j \leqslant q, \quad \alpha_j(0) = \alpha_j; \qquad \textbf{(14.20)}$$

$$\beta_j(n+1) = \frac{\beta_j^2(n) + c^2(n+1)}{2\beta_j(n)}, \qquad 1 \leqslant j \leqslant q, \quad \beta_j(0) = \beta_j. \qquad \textbf{(14.21)}$$

I will now establish a convergence result.

Theorem 14.1 *Let* $b, c > 0$. *The convergence of each of the quantities* $b(n)$, $c(n)$, $\alpha_j(n)$, $\beta_j(n)$ *above is quadratic. In particular,* $b(n)$, $c(n)$, $\beta_j(n)$ *converge quadratically to* X (*the limit of the Gauss algorithm* (13.28)) *and* $\alpha_j(n)$ *converges quadratically to* 0. *Further*

$$a_p(n) = a_p(0)2^{n(p-1)}, \qquad p > 0, \qquad \textbf{(14.22)}$$

$$a_j(n) = K_j 2^{n(p-1)}[1 + o(1)], \qquad 0 \leqslant j \leqslant p - 1, \qquad \textbf{(14.23)}$$

and

$$P(n) := \frac{\pi}{2} \sum_{j=0}^{p} a_j(n) b^{j-1}(n), \qquad \textbf{(14.24)}$$

converges quadratically to F.

Proof We start with equation (21) for fixed j. Obviously $\beta_j(n)$ is either always <0 or >0. Assume without loss of generality the latter. By the inequality of the arithmetic and geometric means $\beta_j(n) > c(n)$, $n \geqslant 1$, and the sequence $\{f(n)\}$ defined by

$$f(n+1) := \frac{f^2(n) + X^2}{2f(n)}, \qquad f(0) := \beta_j,$$

majorizes $\{\beta_j(n)\}$ since $c(n) < X$ (see (13.29)). Thus

$$c(n) < \beta_j(n) < f(n), \qquad n \geqslant 1.$$

But $c(n)$, $f(n)$ converge quadratically to X and so must $\beta_j(n)$.
  We can write

$$\beta_j(n) - c(n+1) = O(\Lambda_j^{2^n}), \qquad 1 \leqslant j \leqslant q.$$

Then

$$\frac{\alpha_j(n+1)}{\alpha_j(n)} = \left[\frac{\beta_j(n) + c(n+1)}{4\beta_j^2(n)}\right] O(\Lambda_j^{2^n}),$$

or

$$\left|\frac{\alpha_j(n+1)}{\alpha_j(n)}\right| < M\Lambda_j^{2^n}, \qquad n \geqslant 0,$$

and taking products shows

$$|\alpha_j(n)| < |\alpha_j(0)| \, M^n \Lambda_j^{1+2+4+\cdots+2^{n-1}}, \qquad n \geq 0$$
$$= |\alpha_j(0)| \, M^n \Lambda_j^{2^n-1} = O(\hat{\Lambda}_j^{2^n}),$$

for some $\hat{\Lambda}_j$, $\Lambda_j < \hat{\Lambda}_j < 1$. This gives the quadratic convergence to 0 of $\alpha_j(n)$.

Letting $j = p$ in (18) shows

$$a_p(n+1) = a_p(n) 2^{p-1}, \qquad p \geq 1,$$

and taking products gives (22). Equation (18) can now be used in a straightforward way to establish (23) inductively. That (24) converges quadratically to F then follows easily by using the above estimates in (14), (11) and (12). ∎

For a large number of applications of Bartky's lemma to the evaluation of particular complete integrals, as well as a thorough discussion of the error and a compendium of Algol procedures, see Bulirsch (1965a,b, 1969b). The transformation may also be used to compute certain incomplete integrals providing certain auxiliary functions, such as the Jacobi zeta function, are utilized; see Bulirsch (1969a).

14.2 Incomplete elliptic integrals

Incomplete elliptic integrals are often the invariants for multidimensional recurrence relations, and thus those recurrences can be used to establish computational algorithms for the integrals. I shall give a number of examples. The first two involve the computation of $F(\phi, k)$ and provide simpler algorithms than that studied in Example 13.7.

Example 14.3 (The descending Gauss transformation for $F(\phi, k)$) Let

$$F(\phi, k) := \int_0^\phi \frac{d\theta}{\sqrt{(1 - k^2 \sin^2 \theta)}}, \qquad 0 \leq k \leq 1,$$
$$k' := \sqrt{(1 - k^2)}.$$

$$(14.25)$$

The substitution

$$\sin \bar{\theta} := \frac{(1 + k') \sin \theta}{1 + \sqrt{(1 - k^2 \sin^2 \theta)}},$$

produces after a considerable amount of algebra the relationship

$$\int_0^{\bar{\phi}} \frac{d\theta}{\sqrt{(1 - \bar{k}^2 \sin^2 \theta)}} = \frac{2}{(1 + k')} \int_0^\phi \frac{d\theta}{\sqrt{(1 - k^2 \sin^2 \theta)}},$$

where

$$\bar{k} := \frac{1 - \sqrt{(1 - k^2)}}{1 + \sqrt{(1 + k^2)}},$$

$$\sin \bar{\phi} := \frac{(1 + k') \sin \phi}{1 + \sqrt{(1 - k^2 \sin^2 \phi)}}.$$

Now let $\phi \to \phi(n)$, $\bar{\phi} \to \phi(n+1)$, etc., and let

$$\sin \phi(n) := \frac{b(n)}{t(n)}, \qquad k'(n) := \frac{c(n)}{b(n)}. \qquad (14.26)$$

The integral is then seen to be an invariant for the algorithm

$$\left.\begin{array}{l} t(n+1) = \frac{1}{2}[t(n) + \sqrt{(t^2(n) - b^2(n) + c^2(n))}], \\[2mm] b(n+1) = \dfrac{b(n) + c(n)}{2}, \\[2mm] c(n+1) = \sqrt{[b(n)c(n)]}, \qquad n \geqslant 0, \\[2mm] \qquad t(0) = \csc \phi, \qquad b(0) = 1, \qquad c(0) = k'. \end{array}\right\} \qquad (14.27)$$

Thus

$$\frac{1}{b(n)} F[\phi(n), k(n)] = \frac{1}{b(n+1)} F[\phi(n+1), k(n+1)], \qquad (14.28)$$

and so, since $k'(n) \to 1$, taking products and the limit gives

$$F[\phi, k] = \frac{1}{X} \arcsin \frac{X}{T} \approx \frac{1}{b(n)} \arcsin \frac{b(n)}{t(n)},$$

where

$$X = \lim_{n \to \infty} b(n), \qquad T = \lim_{n \to \infty} t(n).$$

Since $c(n) < b(n)$, all n, we see $t(n+1) < t(n)$. Assume $t(j) > b(j)$, $0 \leqslant j \leqslant n$. Then

$$t(n+1) - b(n+1) = \frac{1}{2}[t(n) - b(n) - c(n) + \sqrt{(t^2(n) + c^2(n) - b^2(n))}],$$
$$> \frac{1}{2}[-c(n) + c(n)] = 0,$$

so $t(n) > b(n)$ for all n.

We also have

$$t(n) - t(n+1) = \frac{t(n)}{2}\left[1 - \sqrt{\left(1 - \frac{(b^2(n) - c^2(n))}{t^2(n)}\right)}\right].$$

Since $1 - \sqrt{(1 - x)} < x$, $0 < x < 1$, it follows that

$$t(n) - t(n+1) < \frac{b^2(n) - c^2(n)}{2T} = \frac{2c(n+1)}{T} \varepsilon(n),$$

in the notation of Example 13.3.

Using the fact that $c(n) < X$ and summing give

$$t(n) - T < \frac{2\Delta(0)X}{T} \sum_{k=n}^{\infty} \left[\frac{\varepsilon(0)}{\Delta(0)}\right]^{2^k} = O\left(\frac{\varepsilon(0)}{\Delta(0)}\right)^{2^n}.$$

Thus $t(n)$ converges quadratically to T and its convergence is monitored by that of Gauss' algorithm,

$$\frac{|\varepsilon(0)|}{\Delta(0)} := \frac{|\varepsilon(0)|}{(1 + \sqrt{(1 - \varepsilon^2(0))})^2} = \frac{(1 - \sqrt{(1 - \varepsilon^2(0))})^2}{|\varepsilon(0)|^3} < |\varepsilon(0)|$$

$$= \frac{1 - k'}{1 + k'} = \frac{(1 - k')^2}{k^2} < k^2.$$

Thus the descending Gauss transformation is a good algorithm to use when k is small. ∎

Example 14.4 (The ascending Landen transformation) In $F(\phi, k)$ let

$$\tan \bar{\theta} = \frac{(1 + k)\tan \theta}{1 + \sqrt{(1 + k'^2 \tan^2 \theta)}}. \tag{14.29}$$

Copious amounts of algebra then show

$$\tfrac{1}{2}(1 + k)F(\phi, k) = F(\bar{\phi}, \bar{k}),$$

$\bar{k}$ as in Example 3, $\bar{\phi}$ as determined by (29). The resulting algorithm is

$$\left. \begin{array}{l} s(n + 1) = \tfrac{1}{2}[s(n) + \sqrt{(s^2(n) + b^2(n) - c^2(n))}], \\[2mm] b(n + 1) = \dfrac{b(n) + c(n)}{2}, \\[2mm] c(n + 1) = \sqrt{[b(n)c(n)]}, \qquad n \geqslant 0, \\[2mm] s(0) = \cot \phi, \qquad b(0) = 1, \qquad c(0) = k. \end{array} \right\} \tag{14.30}$$

By a now familiar argument,

$$F(\phi, k) = \frac{1}{X}\sinh^{-1}\frac{X}{S} \approx \frac{1}{b(n)}\sinh^{-1}\frac{b(n)}{s(n)},$$

where $\lim s(n) = S$.

It can be shown that $s(n)\uparrow$ while $a(n)/s(n)\downarrow$. As before we see that convergence is quadratic,

$$S - s(n) = O\left(\frac{|\varepsilon(0)|}{\Delta(0)}\right)^{2^n},$$

where

$$\frac{|\varepsilon(0)|}{\Delta(0)} < |\varepsilon(0)| = \frac{1-k}{1+k} < k'^2,$$

and thus the algorithm is the appropriate one to use when k is near 1. ■

Example 14.5 (Jacobian elliptic functions) The two previous algorithms may be reversed to yield algorithms for computing the Jacobian elliptic functions. Recall the definition

$$\text{sn}\,(u, k) := \sin \phi, \qquad \text{where} \qquad u := F(\phi, k), \quad 0 \leqslant u \leqslant K(k).$$

In Example 3 note

$$\text{sn}\,u = 1/t(0),$$

$$T = \lim_{n \to \infty} t(n) = X/\sin(Xu). \tag{14.31}$$

First Gauss' algorithm is used to compute X, and (27) to compute T. Then the recursion for $t(n)$ is reversed to give $t(0)$. Thus we write the algorithm in two stages:

$$\left.\begin{aligned}
b(n+1) &= \frac{b(n)+c(n)}{2}, \\[2mm]
c(n+1) &= \sqrt{[b(n)c(n)]}, \qquad n = 0, 1, 2, \ldots, \nu-1, \\[2mm]
b(0) &= 1, \qquad c(0) = k',
\end{aligned}\right\} \tag{14.32}$$

$$t(n-1) = t(n) + \frac{c^2(n-1) - b^2(n-1)}{4t(n)},$$

$$n = \nu, \nu-1, \nu-2, \ldots, 1, \tag{14.33}$$

$$t(\nu) = b(\nu)/\sin[b(\nu)u].$$

Then approximately

$$\text{sn}\,u = 1/t(0), \qquad \text{cn}\,u = \sqrt{(t^2(0)-1)}\,\text{sn}\,u,$$

$$\text{dn}\,u = [2t(1)-t(0)]\,\text{sn}\,u.$$

This is the algorithm to use if $k^2 < \tfrac{1}{2}$.

The algorithm corresponding to the second example is useful when $k^2 \geqslant \tfrac{1}{2}$. We have

$$\left.\begin{aligned}
b(n+1) &= \frac{b(n)+c(n)}{2}, \\[2mm]
c(n+1) &= \sqrt{[b(n)c(n)]}, \qquad n = 0, 1, 2, \ldots, \nu-1, \\[2mm]
b(0) &= 1, \qquad c(0) = k,
\end{aligned}\right\} \tag{14.34}$$

$$s(n-1) = s(n) - \frac{b(n)\sqrt{[b^2(n) - c^2(n)]}}{s(n)},$$

$$n = v, v-1, v-2, \ldots, 1, \quad \textbf{(14.35)}$$

$$s(v) = b(v)/\sinh[b(v)u],$$

and approximately,

$$\text{sn}(u) = (1 + s^2(0))^{-1/2}, \qquad \text{cn } u = s(0) \text{ sn } u,$$

$$\text{dn } u = (2s(1) - s(0)) \text{ sn } u.$$

For more details about these algorithms, see Gautschi (1975), Carlson (1965), Hofsummer and van de Riet (1963), Salzer (1962). ■

Appendix A The general theory of linear difference equations

Let $\{y^{(h)}(n)\}$, $1 \leqslant h \leqslant \sigma$, be a set of (complex-valued) functions defined on Z^0. The functions are said to be *linearly dependent* if and only if a relation

$$c_1 y^{(1)}(n) + c_2 y^{(2)}(n) + \cdots + c_\sigma y^{(\sigma)}(n) = 0, \qquad n \geqslant 0, \tag{A.1}$$

holds for some constants c_j not all zero. Otherwise the functions are said to be *linearly independent*.

Casorati's determinant is defined as

$$D(n) := \begin{vmatrix} y^{(1)}(n) & y^{(2)}(n) & \cdots & y^{(\sigma)}(n) \\ y^{(1)}(n+1) & y^{(2)}(n+1) & \cdots & y^{(\sigma)}(n+1) \\ \vdots & \vdots & & \vdots \\ y^{(1)}(n+\sigma-1) & y^{(2)}(n+\sigma-1) & \cdots & y^{(\sigma)}(n+\sigma-1) \end{vmatrix},$$

$$n \geqslant 0. \tag{A.2}$$

In what follows I will give a brief survey of properties of the linear difference operator of order σ,

$$\mathscr{L}[y(n)] := \sum_{\nu=0}^{\sigma} A_\nu(n) y(n+\nu), \tag{A.3}$$

where the $A_\nu(n)$ are defined for $n \geqslant 0$ and $A_0(n) A_\sigma(n) \neq 0$.

Good sources for these theorems are Milne-Thomson (1960) and Nörlund (1954).

It is convenient to use the notation

$$\overset{n}{\underset{k=0}{S}} \phi(k), \tag{A.4}$$

to stand for either of the quantities

$$\sum_{k=0}^{n-1} \phi(k), \qquad -\sum_{k=n}^{\infty} \phi(k), \tag{A.5}$$

where an obvious convergence condition is required for the latter interpretation. (If the sum does not converge a meaning may often be assigned to it by use of the Nörlund sum, see Milne-Thomson (1960).)

Differences are defined by

$$\Delta\phi(n) := \phi(n+1) - \phi(n),$$
$$\Delta^{(k)}\phi(n) := \Delta^{(k-1)}[\phi(n+1) - \phi(n)], \qquad k = 2, 3, \ldots . \tag{A.6}$$

Note

$$\Delta \mathop{S}_{k=0}^{n} \phi(k) = \phi(n), \qquad n \geqslant 0.$$

A-I *The functions $\{y^{(h)}(n)\}$ are linearly dependent only if $D(n) \equiv 0$, $n \geqslant 0$.*

A-II *The nonhomogeneous linear difference equation of order $\sigma(\geqslant 1)$*

$$\mathscr{L}[y(n)] := \sum_{\nu=0}^{\sigma} A_\nu(n)y(n+\nu) = f(n), \qquad n \geqslant 0, \tag{A.7}$$

possesses a unique solution satisfying the conditions

$$y(k+\nu) := \alpha_\nu, \qquad 0 \leqslant \nu \leqslant \sigma - 1, \tag{A.8}$$

for any constants $\alpha_0, \alpha_1, \ldots, \alpha_{\sigma-1}$ and any $k \geqslant 0$.

A-III *The homogeneous equation*

$$\mathscr{L}[y(n)] := \sum_{\nu=0}^{\sigma} A_\nu(n)y(n+\nu) = 0, \qquad n \geqslant 0, \tag{A.9}$$

possesses a linearly independent set of solutions $\{y^{(h)}(n)\}$, $1 \leqslant h \leqslant \sigma$, called a fundamental set, and any solution of (9) can be expressed as a linear combination of these functions.

To construct such a set define

$$y^{(h)}(n) := \delta_{h-1,n}, \qquad 0 \leqslant n \leqslant \sigma - 1, \quad 1 \leqslant h \leqslant \sigma, \tag{A.10}$$

and compute $y^{(h)}(n)$, $n \geqslant \sigma$, from (9).

A set of σ functions $\{y^{(h)}(n)\}$ satisfying (9) is a fundamental set iff $D(n) \neq 0$ for any value of n.

If the set of σ functions $\{y^{(h)}(n)\}$ satisfying (9) has the property

$$\lim_{n\to\infty} y^{(h)}(n)/y^{(h+1)}(n) = 0, \qquad 1 \leqslant h \leqslant \sigma - 1, \tag{A.11}$$

then the set is a fundamental set.

A-IV *There exists a solution of $\mathscr{L}[y(n)] = 0$ which vanishes for no value of n.*

Proof Let $\{y^{(h)}(n)\}$ be a fundamental set. Pick any number c_1 different from all the numbers

$$-y^{(1)}(n)/y^{(2)}(n), \qquad n \geqslant 0, \quad y^{(2)}(n) \neq 0.$$

Then the function

$$y_1(n) := y^{(1)}(n) + c_1 y^{(2)}(n)$$

vanishes iff $y^{(1)}(n) = y^{(2)}(n) = 0$.

Next pick a number c_2 different from all the numbers

$$-y_1(n)/y^{(3)}(n), \qquad n \geqslant 0, \quad y^{(3)}(n) \neq 0.$$

The function

$$y_2(n) := y^{(1)}(n) + c_1 y^{(2)}(n) + c_2 y^{(3)}(n)$$

vanishes iff $y^{(1)}(n) = y^{(2)}(n) = y^{(3)}(n) = 0$. Continuing this construction I arrive at the function

$$y_\sigma(n) := y^{(1)}(n) + c_1 y^{(2)}(n) + \cdots + c_{\sigma-1} y^{(\sigma)}(n),$$

which vanishes iff

$$y^{(1)}(n) = y^{(2)}(n) = \cdots = y^{(\sigma)}(n) = 0.$$

But there is no value of n for which this happens since for that value $D(n) = 0$, contradicting A-III. ■

A-V *Let $\{y^{(h)}(n)\}$ be a fundamental set for (9) and $D(n)$ be defined as in (2). Then $D(n)$ satisfies*

$$D(n+1)/D(n) = (-1)^\sigma A_0(n)/A_\sigma(n), \qquad n \geqslant 0, \tag{A.12}$$

that is,

$$D(n) = D(0)(-1)^{n\sigma} \prod_{j=0}^{n-1} A_0(j)/A_\sigma(j), \qquad n \geqslant 0. \tag{A.13}$$

A-VI *The equation*

$$\sum_{v=0}^{\sigma} A_{\sigma-v}(n+v)y(n+v) = 0, \qquad n \geqslant 0, \tag{A.14}$$

is called the equation adjoint to (9). The functions

$$\hat{y}^{(h)}(n) := T^{(h)}(n)/A_0(n)D(n), \tag{A.15}$$

$$T^{(h)}(n) := (-1)^{h-1}$$

$$\times \begin{vmatrix} y^{(1)}(n+1) & \cdots & y^{(h-1)}(n+1) \\ \vdots & & \vdots \\ y^{(1)}(n+\sigma-1) & \cdots & y^{(h-1)}(n+\sigma-1) \\ & y^{(h+1)}(n+1) & \cdots & y^{(\sigma)}(n+1) \\ & \vdots & & \vdots \\ & y^{(h+1)}(n+\sigma-1) & \cdots & y^{(\sigma)}(n+\sigma-1) \end{vmatrix}, \tag{A.16}$$

are linearly independent and satisfy (14), and hence are a fundamental set.

Furthermore,

$$\sum_{j=1}^{\sigma} \hat{y}^{(j)}(n) y^{(j)}(n+k) = \begin{cases} 1/A_0(n), & k=0; \\ 0, & 1 \leqslant k \leqslant \sigma-1; \\ -1/A_\sigma(n), & k=\sigma. \end{cases} \qquad \text{(A.17)}$$

The proof for this statement follows by expanding the determinant

$$\begin{vmatrix} y^{(1)}(n+k) & y^{(2)}(n+k) & \cdots & y^{(\sigma)}(n+k) \\ y^{(1)}(n+1) & y^{(2)}(n+1) & \cdots & y^{(\sigma)}(n+1) \\ \vdots & \vdots & & \vdots \\ y^{(1)}(n+\sigma-1) & y^{(2)}(n+\sigma-1) & \cdots & y^{(\sigma)}(n+\sigma-1) \end{vmatrix} \qquad \text{(A.18)}$$

by minors of the first row.

The adjoint of the adjoint

$$\sum_{\nu=0}^{\sigma} A_\nu(n+\sigma) y(n+\nu) = 0 \qquad \text{(A.19)}$$

has solutions $\{y^{(h)}(n+\sigma)\}$, $1 \leqslant h \leqslant \sigma$. These are precisely the solutions obtained by using $\hat{y}^{(h)}(n)$ for $y^{(h)}(n)$ in the definition of $T^{(h)}(n)$ above.

A-VII *Let $\{y^{(h)}(n)\}$ be a fundamental set for the homogeneous equation (9). Any solution $y(n)$ of the nonhomogeneous equation (7) may be written*

$$y(n) = c_1 y^{(1)}(n) + c_2 y^{(2)}(n) + \cdots + c_\sigma y^{(\sigma)}(n) + p(n), \qquad \text{(A.20)}$$

where $p(n)$ is a particular solution of (7).

A-VIII *A particular solution $p(n)$ of equation (7) is given by*

$$p(n) := -\sum_{j=1}^{\sigma} y^{(j)}(n) \overset{n}{\underset{k=0}{S}} f(k) \hat{y}^{(j)}(k). \qquad \text{(A.21)}$$

Proof Choose the first interpretation in (5). I have

$$\mathcal{L}[p(n)] = -\sum_{\nu=0}^{\sigma} A_\nu(n) \sum_{j=1}^{\sigma} y^{(j)}(n+\nu) \left(\sum_{k=0}^{n-1} + \sum_{k=n}^{n+\nu-1} f(k) \hat{y}^{(j)}(k) \right)$$

$$= -\sum_{\nu=1}^{\sigma} A_\nu(n) \sum_{k=0}^{\nu-1} f(n+k) \sum_{j=1}^{\sigma} y^{(j)}(n+\nu) \hat{y}^{(j)}(n+k). \qquad \text{(A.22)}$$

By (17) the inner sum vanishes for $\nu+1-\sigma \leqslant k \leqslant \nu-1$, so only those terms corresponding to $\nu=\sigma$, $k=0$, remain. The right-hand side of (22) becomes

$$-A_\sigma(n) f(n) \sum_{j=1}^{\sigma} y^{(j)}(n+\sigma) \hat{y}^{(j)}(n) = f(n),$$

by (17). The proof for the second interpretation in (5) is done similarly. ∎

A-IX *A nonvanishing solution of the first-order homogeneous equation*

$$y(n) + A_1(n)y(n+1) = 0,$$ (A.23)

is

$$y^{(1)}(n) := \prod_{j=0}^{n-1} [-A_1(j)]^{-1}.$$ (A.24)

The complete solution of the nonhomogeneous equation

$$y(n) + A_1(n)y(n+1) = f(n),$$ (A.25)

is

$$y(n) := cy^{(1)}(n) + p(n),$$ (A.26)

$$p(n) := -y^{(1)}(n) \sum_{k=0}^{n} \frac{f(k)}{y^{(1)}(k)}.$$ (A.27)

A-X *If $y^{(1)}(n)$ is one member of a fundamental set for the second-order equation*

$$\mathscr{L}[y(n)] := y(n) + A_1(n)y(n+1) + A_2(n)y(n+2) = 0,$$ (A.28)

another can be found by reduction of order:

$$y^{(2)}(n) := y^{(1)}(n) \sum_{k=0}^{n} \frac{\prod_{j=0}^{k-1} A_2(j)^{-1}}{y^{(1)}(k)y^{(1)}(k+1)}.$$ (A.29)

The complete solution of the nonhomogeneous equation $\mathscr{L}[y(n)] = f(n)$ is then

$$y(n) := c_1 y^{(1)}(n) + c_2 y^{(2)}(n) + p(n),$$

$$p(n) := \sum_{k=0}^{n} [y^{(1)}(k+1)y^{(2)}(n) - y^{(2)}(k+1)y^{(1)}(n)] \frac{f(k)}{D(k)}.$$ (A.30)

Appendix B The asymptotic theory of linear difference equations

B.1 General theory

The most important results here are the theorems of Poincaré and Perron. In what follows I will assume the coefficients in the equation $\mathcal{L}[y(n)] = 0$ (see (A.3)) satisfy $A_\sigma(n) = 1$, $A_0(n) \neq 0$,

$$\lim_{n \to \infty} A_\nu(n) = \mu_\nu, \qquad \mu_0 \neq 0. \tag{B.1}$$

The equation

$$\sum_{\nu=0}^{\sigma} \mu_\nu t^\nu = 0 \tag{B.2}$$

is called the *characteristic equation* for the difference equation (A.9). I shall in this section assume the roots of this equation have distinct moduli. Denote them by t_i, $1 \leq i \leq \sigma$.

Theorem B.1 (Poincaré's theorem) *For any solution y(n) of the difference equation $\mathcal{L}[y(n)] = 0$*

$$\lim_{n \to \infty} \frac{y(n+1)}{y(n)} = t_i, \tag{B.3}$$

for some i.

Proof A proof can be found in Milne-Thomson ((1960), p. 527) or Nörlund ((1954), p. 300ff.). ■

A refinement of this result was given by Perron who showed that $\mathcal{L}[y(n)]$ possesses a fundamental set having the property that $\lim |y^{(h)}(n)|^{1/n}$ approaches a root of the characteristic equation. Perron's theorem has been made even more precise by Evgrafov (1953), whose proof is very elegant.

Theorem B.2 (Perron's theorem) *There is a fundamental set for the* *equation* $\{y^{(h)}(n)\}$ *with the property*

$$\lim_{n\to\infty}\frac{y^{(h)}(n+1)}{y^{(h)}(n)}=t_h, \qquad 1\leqslant h\leqslant\sigma. \tag{B.4}$$

Proof   See Meschkowski ((1959), p. 110). ■

Example B.1   Consider the equation

$$(n+1)(n+2)(1-\lambda)y(n)$$
$$+(n+2)[(n+a+1)\lambda-2n-b_1-b_2-1]y(n+1)$$
$$+(n+b_1+1)(n+b_2+1)y(n+2)=0. \quad \textbf{(B.5)}$$

Dividing by $(n+b_1+1)(n+b_2+1)$ gives an equation for which

$$\mu_0=1-\lambda, \qquad \mu_1=\lambda-2, \qquad \mu_2=1,$$

cf. (1). Thus the characteristic equation has roots $t_1=1-\lambda$, $t_2=1$. For $\lambda\neq0,1$, there is a fundamental set $\{y^{(h)}(n)\}$ having the property

$$y^{(1)}(n)=(1-\lambda)^n e^{n\nu_n}, \qquad y^{(2)}(n)=e^{n\mu_n},$$

where $\{\nu_n\}$, $\{\mu_n\}$ are null sequences (see Wimp (1981, Section 1.4)). This information is about as much as Perron's theorem will furnish. This equation in fact has solutions

$$\mathscr{F}_h(n):=(n+b_h)_{1-b_h}F\!\left(\begin{matrix}1-b_h-n, & 1+a-b_h\\ & 1+b_s-b_h\end{matrix}\,\middle|\,\lambda\right),$$

$$s\neq h, \qquad h=1,2, \quad \textbf{(B.6)}$$

and

$$\mathscr{G}(n):=(n+1+a)_{-a}F\!\left(\begin{matrix}1+a-b_1, & 1+a-b_2\\ & 1+a+n\end{matrix}\,\middle|\,\frac{1}{\lambda}\right), \tag{B.7}$$

providing of course that the parameters b_h, a, λ are such that these functions are defined. $y^{(1)}(n)$ may be identified with either $\mathscr{F}_1$ or $\mathscr{F}_2$ and $y^{(2)}(n)$ with $\mathscr{G}$. Further, as we will see in Section B.2, there exist complete Poincaré-type asymptotic developments for these functions, e.g.,

$$n^a\mathscr{G}(n)\sim1+\frac{c_1}{n}+\frac{c_2}{n^2}+\cdots,$$

and a similar series for $\mathscr{F}_1$ multiplied by $(1-\lambda)^n$. ■

A generalization of Perron's theorem will be given later (Theorem 10).

B.2 The construction of formal series solutions

In this section, the standard form of the difference equation will be

$$\sum_{\nu=0}^{\sigma} A_\nu(n)y(n+\nu) = 0, \qquad A_0 = 1, \quad A_\sigma \neq 0, \quad \sigma \geqslant 2, \qquad n \geqslant 0, \quad \textbf{(B.8)}$$

where the $A_\nu(n)$ can be represented as asymptotic series,

$$A_\nu(n) \sim n^{K_\nu/\omega}[a_{0\nu} + a_{1\nu}n^{-1/\omega} + a_{2\nu}n^{-2/\omega} + \cdots], \qquad n \to \infty, \qquad \textbf{(B.9)}$$

K_ν an integer, ω an integer $\geqslant 1$, and $a_{0\nu} \neq 0$ unless $A_\nu \equiv 0$. The theory to follow is quite general since virtually every difference equation encountered in practical applications has this form; in particular the theory includes all difference equations with polynomial or rational (in n) coefficients. An equation in such a form will be called an equation of *Birkhoff type*.

Note that the equation adjoint to the above after division by $A_\sigma(n)$ will also be in this general form.

As we shall see, there exists a fundamental set for the above equation whose members share an unusual property: each has an asymptotic expansion valid as $n \to \infty$ which consists of an exponential leading term multiplied by a descending series of the kind (9) (where, however, ω may be replaced by an integral multiple of ω and where logarithmic terms may appear). These series play the same role as the so-called subnormal series play in the theory of ordinary linear differential equations with polynomial coefficients, see Ince (1926). The existence of such solutions is quite important since the very form of the series often enables one to analyze the behavior, for a particular difference equation, of the computational algorithms given in this book.

Although pieces of the theory are very old, some of its features having been described by J. Horn (1910), P. M. Batchelder (1927) and C. R. Adams (1928), it was given in the form to be presented here in two very long papers by Birkhoff (1930) and Birkhoff and Trjitzinsky (1932).

I begin with several definitions.

Consider the series

$$e^{Q(\rho;n)}s(\rho; n), \qquad \textbf{(B.10)}$$

where

$$Q(n) \equiv Q(\rho; n) := \mu_0 n \ln n + \sum_{j=1}^{\rho} \mu_j n^{(\rho+1-j)/\rho}, \qquad \textbf{(B.11)}$$

$$s(n) = s(\rho; n) := n^\theta \sum_{j=0}^{t} (\ln n)^j n^{r_j - j/\rho} q_j(\rho; n), \qquad \textbf{(B.12)}$$

$$q_j(\rho; n) \equiv q_j(n) := \sum_{s=0}^{\infty} b_{sj} n^{-s/\rho}, \qquad \textbf{(B.13)}$$

and ρ, r_j, $\mu_0\rho$ are integers, $\rho \geqslant 1$, μ_j, θ, b_{ij} are complex, $b_{0j} \neq 0$ unless $b_{sj} = 0$ for all s, $r_0 = 0$, $-\pi \leqslant \text{Im } \mu_1 < \pi$.

Definition B.1   *The series* (10), *called a formal series* (FS), *will be called a formal series solution* (FSS) *of* (8), *if, when it is substituted in* (8), *the equation is divided by* $e^{Q(n)}$ *and the obvious algebraic manipulations* (*see below*) *are performed, then the coefficient of each quantity*

$$n^{\theta+(r/\rho)+(s/\omega)}(\ln n)^j, \qquad 0 \leqslant j \leqslant t, \quad r, s = 0, \pm 1, \pm 2, \ldots,$$

is equal to zero. ∎

A concept of formal equality of two FS can be defined by requiring that, when the series are written with the same value of ρ (as is always possible), the parameters t, θ, b_{sj}, r_j, μ_j for both series are the same. Formal equality of FSS also arises in the theory of ordinary linear differential equations; see Ince (1926).

The construction of FSS may be carried out by using the identities,

$$e^{\mu(n+\nu)^\delta} = e^{\mu n^\delta}\left\{1+\sum_{k,j} a_{kj} n^{k(\delta-1)+1-j}\right\}, \tag{B.14}$$

$$(n+\nu)^{\mu_0(n+\nu)} = n^{\mu_0 n+\mu_0\nu} e^{\mu_0\nu}\left\{1+\frac{\mu_0\nu^2}{2n}+\cdots\right\}, \tag{B.15}$$

$$(n+\nu)^\alpha = n^\alpha\left\{1+\frac{\alpha\nu}{n}+\cdots\right\}, \tag{B.16}$$

$$[\ln(n+\nu)]^r = \left[\ln n+\frac{\nu}{n}-\frac{\nu^2}{2n^2}+\cdots\right]^r, \tag{B.17}$$

although in practice it is difficult to obtain by hand other than the first few terms this way, particularly in those (rare) instances when logarithmic terms enter the solutions. For some additional details about this procedure, see Birkhoff (1930).

It is easily seen that FS have the following properties:

$$\left.\begin{aligned}
e^{Q_1(n+\nu)}s_1(n+\nu) &= e^{Q_2(n)}s_2(n), \qquad \nu = 1, 2, \ldots, \\
Q_1 \text{ given}, \quad &\nu \text{ fixed}, \quad \text{for some } Q_2, s_2, \\
e^{Q_1(\rho;\,n)}s_1(\rho;\,n) \times e^{Q_2(\rho';\,n)}s_2(\rho';\,n) &= e^{Q_3(\rho^*;\,n)}s_3(\rho^*;\,n), \\
\rho^* :&= \text{least common multiple of } \rho, \rho'.
\end{aligned}\right\} \tag{B.18}$$

A sum of two FS is not in general a FS but if Q, θ are the same for both series one has

$$e^{Q(n)}n^\theta q_1(\rho;\,n) + e^{Q(n)}n^\theta q_2(\rho';\,n) = e^{Q(n)}n^\theta q_3(\rho^*;\,n).$$

Definition B.2 *Let $\{f(n)\}$ be a complex sequence; then*

$$f(n) \sim e^{Q(n)} s(n), \qquad n \to \infty, \tag{B.19}$$

means that for every $k \geqslant 1$ we can determine functions $A_{kj}(n)$, $0 \leqslant j \leqslant t$, such that

$$e^{-Q(n)} n^{-\theta} f(n) = \sum_{j=0}^{t} (\ln n)^j n^{r_t - j/\rho} \sum_{s=0}^{k-1} b_{sj} n^{-s/\rho}$$

$$+ n^{-k/\rho} \sum_{j=0}^{t} (\ln n)^j n^{r_t - j/\rho} A_{kj}(n), \tag{B.20}$$

and $|A_{kj}|$ is bounded as $n \to \infty$. ■

Note that the representation in (19) is unique. (It is readily verified that 0 has no nontrivial representation of the form (10).)

Definition B.3 *Let W_k denote the determinant with entries*

$$e^{Q_i(n+j)} s_i(n+j), \qquad 1 \leqslant i,\ j \leqslant k,$$

and write

$$W_k := \exp \left\{ \sum_{j=1}^{k} Q_j(n) \right\} \bar{s}(n) := e^{\bar{Q}(n)} \bar{s}(n) \tag{B.21}$$

(as is obviously possible by using (18)). We say the k FS $\{e^{Q_h(n)} s_h(n)\}$ are formally linearly independent if $\bar{s}(n) \not\equiv 0$. Otherwise they are formally linearly dependent. ■

Theorem B.3 (Birkhoff–Trjitzinsky) *There exist σ formally linearly independent FSS of the difference equation (8), where $\rho = \upsilon \omega$ for some integer $\upsilon \geqslant 1$, and each FSS represents asymptotically in the sense of (20) some solution of the equation. The σ solutions so represented constitute a fundamental set for the equation.* ■

Definition B.4 *The fundamental sets mentioned in the preceding theorem will be called* Birkhoff sets *and each member and its corresponding series will be called a* Birkhoff solution *and a* Birkhoff series *respectively.* ■

Sometimes short cuts can be taken with the construction of FSS. Since (8) and (9) are (formally) unchanged by the substitution $n \to n e^{2k\pi i \omega}$ other FSS corresponding to a $\rho > \omega$ can be generated by this substitution. Also no series containing a term $(\ln n)^k$, $k \geqslant 1$, occurs without associated series containing terms $(\ln n)^j$, $0 \leqslant j \leqslant k-1$, in accordance with the following theorem of Wimp (1969).

Theorem B.4 *Let* (8), (9) *hold and*

$$\phi^{(k)}(n) := e^{Q(\rho;n)} n^{\theta} \sum_{l=0}^{k} (\ln n)^{l} q_{t+l-k}(\rho; n) n^{r_{k}-l/\rho}(t+1-k)_{l}/l!,$$

$$0 \le k \le t. \quad \textbf{(B.22)}$$

Then if $\phi^{(t)}(n)$ *is a FSS so is* $\phi^{(h)}(n)$, $0 \le h \le t-1$. ∎

In trying to apply Theorem 3, one should first attempt the construction of FSS without the use of logarithmic terms, i.e., assume

$$\left.\begin{array}{l} y(n) := E(n) e K(n), \\[4pt] E(n) := e^{\mu_0 n \ln n + \mu_1 n} n^{\theta}, \\[4pt] K(n) := \exp\{\alpha_1 n^{\beta} + \alpha_2 n^{\beta-(1/\rho)} + \cdots\}, \qquad \alpha_1 \ne 0, \\[8pt] \hspace{5cm} \beta = j/\rho, \quad 0 \le j < \rho. \end{array}\right\} \quad \textbf{(B.23)}$$

Then

$$\left.\begin{array}{l} \dfrac{y(n+k)}{y(n)} = n^{\mu_0 k} \lambda^{k} \left\{ 1 + \dfrac{(k\theta + (k^2 \mu_0)/2)}{n} + \cdots \right\} \\[14pt] \hspace{2cm} \times \exp\left\{ \alpha_1 \beta k n^{\beta-1} + \alpha_2 \left(\beta - \dfrac{1}{\rho}\right) k n^{\beta-(1/\rho)-1} + \cdots \right\}, \\[14pt] \lambda := e^{\mu_0 + \mu_1}. \end{array}\right\} \quad \textbf{(B.24)}$$

Example B.2 Construct Birkhoff series for

$$(n+b)(n+c)y(n) - (n+1)[(2n+b+c+1)+z]y(n+1)$$
$$+ (n+1)(n+2)y(n+2) = 0, \quad \textbf{(B.25)}$$

(see Wimp (1974)).

Making the substitution (23), (24) we find the equation to be formally satisfied is

$$1 - 2n^{\mu_0} \lambda \left[1 + \frac{1}{n}\left(\theta + \frac{\mu_0 + 3 + z - b - c}{2}\right) + O(n^{-2}) \right]$$

$$\times \exp\left\{ \alpha_1 \beta n^{\beta-1} + \alpha_2 \left(\beta - \frac{1}{\rho}\right) n^{\beta-(1/\rho)-1} + \cdots \right\}$$

$$+ n^{2\mu_0} \lambda^2 \left[1 + \frac{1}{n}(2\theta + 2\mu_0 + 3 - b - c) + O(n^{-2}) \right]$$

$$\times \exp\left[2\alpha_1 \beta n^{\beta-1} + 2\alpha_2 \left(\beta - \frac{1}{\rho}\right) n^{\beta-(1/\rho)-1} + \cdots \right] = 0.$$

Obviously $\mu_0 = 0$ and this implies $\lambda = 1$. This further requires $\rho \ne 1$ or else $z = 0$. Expanding the exponentials gives

$$\frac{-z}{n} + \alpha_1^2 \beta^2 n^{2\beta-2} + (\text{lower order terms}) = 0.$$

Unless $\beta = \frac{1}{2}$ either $\alpha_1 = 0$ or $z = 0$. Thus $\beta = \frac{1}{2}$, $\alpha_1 = 2z^{1/2}$. Equating coefficients of $n^{-3/2}$ then yields

$$\theta = \frac{b+c}{2} - \frac{5}{4}.$$

Another series is obtained from this one by letting $n \to ne^{2\pi i}$.
The Birkhoff series are thus

$$y^{(1)}(n) \sim n^{(b+c)/2 - 5/4} e^{-2\sqrt{(nz)}} \left\{ 1 + \frac{c_1}{n^{1/2}} + \frac{c_2}{n} + \cdots \right\}, \qquad \textbf{(B.26)}$$

$$y^{(2)}(n) \sim n^{(b+c)/2 - 5/4} e^{2\sqrt{(nz)}} \left\{ 1 - \frac{c_1}{n^{1/2}} + \frac{c_2}{n} + \cdots \right\}. \qquad \textbf{(B.27)}$$

As shown in the cited reference the functions

$$\frac{(b)_n (c)_n}{n!} \Psi(n+b, b+1-c; z), \qquad \frac{(b)_n}{n!} \Phi(n+b, b+1-c; z), \qquad \textbf{(B.28)}$$

are a fundamental set for the equation.

Without some additional information, there is no way to relate these functions to the Birkhoff series. However a result given in Slater ((1960), p. 80) allows us to identify $y^{(1)}$ with a constant multiple of the first function in (28) for $|\arg z| < \pi$. ∎

Matching the Birkhoff series to known solutions can be a very difficult problem. Sometimes the Olver growth theorems, see Section B.3, can be useful in accomplishing this.

Example B.3 Construct Birkhoff series for

$$xy(n) + (n+2)y(n+1) - 2y(n+3) = 0. \qquad \textbf{(B.29)}$$

Proceeding as before I find

$$x + n^{\mu_0 + 1} \lambda \left[1 + \frac{1}{n} \left(\theta + 2 + \frac{\mu_0}{2} \right) + \cdots \right]$$

$$\times \exp \left(\alpha_1 \beta n^{\beta - 1} + \alpha_2 \left(\beta - \frac{1}{\rho} \right) n^{\beta - (1/\rho) - 1} + \cdots \right)$$

$$- 2n^{3\mu_0} \lambda^3 \left[1 + \frac{(3\theta + (a/2)\mu_0)}{n} + \cdots \right]$$

$$\times \exp \left(3\alpha_1 \beta n^{\beta - 1} + 3\alpha_2 \left(\beta - \frac{1}{\rho} \right) n^{\beta - (1/\rho) - 1} + \cdots \right) = 0.$$

One value of μ_0 is $\mu_0 = \frac{1}{2}$. Corresponding to this I have $\lambda = 2^{-1/2}$ so

$$-\frac{2\theta}{n} - 2\alpha_1 \beta n^{\beta - 1} - 4\alpha_1^2 \beta^2 n^{2\beta - 2} + (\text{lower order terms}) = 0.$$

This requires $\beta = 0$, $\theta = 0$. (Note ρ is still undetermined.) Next we consult the terms of order $n^{-(1/\rho)-1}$:

$$x + \frac{n^{3/2}}{\sqrt{2}}\left[\frac{2\alpha_2}{\rho} n^{-(1/\rho)-1} + (\text{lower order terms})\right] = 0.$$

We thus find that either $x = 0$ or $\rho = 2$. We take $\rho = 2$ to get the series

$$y^{(1)}(n) \sim n^{n/2}(2e)^{-n/2}\left(1 + \frac{c_1}{n^{1/2}} + \frac{c_2}{n} + \cdots\right). \tag{B.30}$$

Letting $n \to ne^{2\pi i}$ gives another series,

$$y^{(2)}(n) \sim n^{n/2}(-1)^n(2e)^{-n/2}\left(1 - \frac{c_1}{n^{1/2}} + \frac{c_2}{n} + \cdots\right). \tag{B.31}$$

There is one more series, namely that corresponding to $\mu_0 = -1$. Obviously for this series $\rho = 1$ (otherwise one could construct *two* Birkhoff series, giving a set containing one too many FSS) and a rapid computation shows $\lambda = -x$, $\theta = -\frac{3}{2}$, so

$$y^{(3)}(n) \sim n^{-n}(-xe)^n n^{-3/2}\left[1 + \frac{d_1}{n} + \frac{d_2}{n^2} + \cdots\right] \tag{B.32}$$

is the required series.

The preceding difference equation is satisfied by

$$T(n; x) = \int_0^\infty t^n e^{-t^2 - (x/t)}\, dt, \tag{B.33}$$

see Cole and Pescatore (1979). Laplace's method allows us to identify T with $y^{(1)}(n)$. In fact,

$$T(n; x) \sim \sqrt{\left(\frac{\pi}{2}\right)}\, y^{(1)}(n), \qquad n \to \infty. \quad\blacksquare \tag{B.34}$$

The following are useful results concerning sums of functions which have Birkhoff series representations.

Theorem B.5   (Exponential sums)   *Let*

$$s(n) := \sum_{k=0}^{n} a(k), \tag{B.35}$$

be convergent, $\lim_{n\to\infty} s(n) = s$, *and*

$$a(n) \sim e^{Q(\rho; n)} n^\theta q(\rho; n), \tag{B.36}$$

where Q, q are as in (10)–(13). *Then*

$$s(n) - s \sim e^{Q(\rho; n)} n^{\theta^*} q^*(\rho; n), \tag{B.37}$$

where θ^* and the leading coefficient b_0^* of q^* are as follows:

CASE I $\quad Q \not\equiv 0$; *denote the first nonzero μ_j in the sequence $\{\mu_0, \mu_1, \ldots, \mu_\rho\}$ by μ_τ. Then*

$$\theta^* = \begin{cases} \theta + \mu_0, & \tau = 0; \\ \theta + \dfrac{\tau - 1}{\rho}, & 1 \leqslant \tau \leqslant \rho; \end{cases} \tag{B.38}$$

$$b_0^* = \begin{cases} -b_0 e^{\mu_0 + \mu_1}, & \tau = 0; \\ b_0 (1 - e^{-\mu_1})^{-1}, & \tau = 1; \\ b_0 \rho / \mu_\tau (\rho + 1 - \tau), & 2 \leqslant \tau \leqslant \rho. \end{cases} \tag{B.39}$$

CASE II $\quad Q \equiv 0$;

$$\theta^* = \theta + 1; \qquad b_0^* = b_0 / (\theta + 1). \tag{B.40}$$

In the above, b_0 is the leading coefficient of q. ∎

Theorem B.6 (Exponential sums) *Let $s(n)$ be divergent and $a(n)$, $Q(\rho; n)$, θ, $q(\rho; n)$ be as in the previous theorem.*
  Then for some constant c,

$$s(n) - c \sim e^{Q(\rho; n)} n^{\theta^*} q^*(\rho; n), \tag{B.41}$$

where θ^*, b_0^* are as in (39), (40), except in the two following cases.

CASE I $\quad \mu \neq 0$; *then $b_0^* = b_0$, $\theta^* = \theta$.*

CASE II $\quad Q \equiv 0$ *and q contains a term n^{-1}. Then for some c, d,*

$$s(n) - c + d \ln n \sim n^{\theta + 1} q^*(\rho; n). \quad ∎ \tag{B.42}$$

For the proofs of these two results see Wimp (1981).
  Often it is important to decide whether a homogeneous difference equation has a fundamental set of solutions which are free from zeros for n sufficiently large. This need not be the case, obviously. But when the difference equation has coefficients which are rational in n, which is the most common situation, the very appearance of the Birkhoff series corresponding to some basis of solutions makes it clear that this is always true.

Theorem B.7 *Let the coefficients $A_\nu(n)$ in (8) be rational in n. Then the equation has a basis of solutions $\{y^{(h)}(n)\}$ such that*

  (i) $\quad y^{(h)}(n)$ *is free from zeros for n sufficiently large;*
  (ii) $\quad \lim_{n \to \infty} |y^{(h)}(n)| / |y^{(i)}(n)|$ *exists (being possibly infinite),*
  $\quad\quad 1 \leqslant h, j \leqslant \sigma.$ ∎

Certain types of Birkhoff sets will be important in our discussion of the generalized Miller algorithm.

Definition B.5 *A Birkhoff set $\{y^{(h)}(n)\}$ is a canonical set if*

$$y^{(h)}(n) = M^{(h)}(n)[1 + o(1)], \qquad n \to \infty,$$

where $M^{(h)}(n)$ is a constant multiple of $M^{(j)}(n)$ if and only if $h = j$. ∎

Every equation of the type (8)–(9) has a canonical set as does its adjoint. No two members of a canonical set display the same asymptotic behavior as $n \to \infty$ so not every Birkhoff set is canonical, e.g., $\{1, n^2, n^2 + n\}$ is a Birkhoff set for

$$y(n) - 3y(n+1) + 3y(n+2) - y(n+3) = 0,$$

but not a canonical set, as is $\{1, n, n^2\}$ or $\{1, n+1, n^2+n\}$.

It is important to know how the solutions of the adjoint behave given a Birkhoff set for the equation. The following result (Wimp (1969)) contains the necessary information.

Theorem B.8 *Let $\{y^{(h)}(n)\}$ be a Birkhoff set for (8), (9) with*

$$y^{(h)}(n) \sim e^{Q^{(h)}(n)} s^{(h)}(\rho; n), \qquad 1 \le h \le \sigma. \tag{B.43}$$

Then

  (i)

$$D(n) \sim e^{Q(\omega;n)} s(\omega; n), \tag{B.44}$$

where, in $Q(\omega; n)$,

$$\left.\begin{array}{ll} \mu_0 = -K_\sigma/\omega, & e^{\mu_1} = (-1)^\sigma e^{K_\sigma/\omega}/a_{0\sigma}, \\[2mm] \mu_2 = -\omega a_{1\sigma}/(\omega-1)a_{0\sigma}, & \mu_3 = \dfrac{\omega}{(\omega-2)}\left[\dfrac{a_{1\sigma}^2}{2a_{0\sigma}^2} - \dfrac{a_{2\sigma}}{a_{0\sigma}}\right], \dots, \end{array}\right\} \tag{B.45}$$

$a_{j\sigma}$ being as in (9) and $s(\omega; n)$ is free from logarithms.

  (ii)

$$Q(\omega; n) = \sum_{h=1}^{\sigma} Q^{(h)}(\rho; n), \tag{B.46}$$

  (iii)

$$\hat{y}^{(h)}(n) = \frac{T^{(h)}(n)}{D(n)} \sim e^{-Q^{(h)}(n)} \hat{s}^{(h)}(\rho; n), \qquad 1 \le h \le \sigma, \tag{B.47}$$

form a Birkhoff set for the adjoint equation.

Proof The proofs are purely computational; see Wimp (1969). ∎

The results in this and the previous section can sometimes be used to determine the existence of minimal solutions for linear difference equations (cf. Chapter 2).

In what follows $[(x_1, y_1), (x_2, y_2)]$ will denote the closed line segment in R^2 joining the points (x_1, y_1), (x_2, y_2).

Theorem B.9 *Let the equation* $\mathscr{L}[y(n)] = 0$ *have rational coefficients with*

$$A_\nu(n) \sim a_{0\nu} n^{K_\nu}, \qquad n \to \infty, \tag{B.48}$$

where $K_\nu \in Z$ *and* $a_{0\nu} \neq 0$ *unless* $A_\nu \equiv 0$ *and let the segment* $P := [(0, K_0), (1, K_1)]$ *but not* $[(0, K_0), (2, K_2)]$ *belong to the boundary of the convex hull S of those points* $(0, K_0), (1, K_1), \ldots, (\sigma, K_\sigma)$ *which lie on or above* $[(0, K_0), (\sigma, K_\sigma)]$. *Then the equation has a minimal solution* $w(n)$ *and*

$$\lim_{n \to \infty} |w(n)|^{1/(n \ln n)} = s, \tag{B.49}$$

where s is the negative of the slope of P,

$$s = K_0 - K_1. \tag{B.50}$$

Remark If $A_\nu \equiv 0$ interpret (ν, K_ν) as $(\nu, -\infty)$.

Proof The result is a consequence of the Birkhoff–Trjitzinsky theory. Since the coefficients A_ν are rational they possess complete (in fact, convergent) asymptotic expansions of Poincaré type. After substituting a FS of the kind and performing the reduction indicated in (24) we are left with $\sigma + 1$ FS with leading terms

$$a_{0\nu} n^{K_\nu + \nu\mu_0}.$$

Neither K_0 nor $K_\sigma = -\infty$ since neither $A_0(n)$ nor $A_\sigma(n)$ can be zero. If the desired formal identity is to be possible, then two of the terms must involve the same power of n, so

$$\mu_0 = \frac{-(K_l - K_m)}{(l - m)}, \tag{B.51}$$

for some l, m, and all other terms must be of less or equal degree, i.e.,

$$K_j + j\mu_0 \leqslant K_m + m\mu_0, \qquad 0 \leqslant j \leqslant \sigma,$$

or

$$K_j - K_m \leqslant -\mu_0(j - m), \qquad 0 \leqslant j \leqslant \sigma.$$

These relationships have a simple geometric interpretation; see Birkhoff ((1930), p. 211). The negatives of the slopes of those segments constituting S correspond to the permissible values of μ_0. The conditions given in the theorem guarantee that there is a minimum value of μ_0 and to it there corresponds exactly one FSS with $\mu_0 = K_0 - K_1$ (see Birkhoff (1930)); thus the result follows. ∎

The situation for Example 3 is illustrated in Fig. 1.

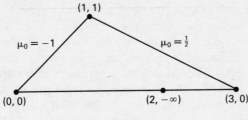

Fig. 1

For the case of a three-term recurrence relation this result, which becomes the Perron–Kreuser Theorem, can be strengthened considerably, the coefficients being required only to be $\sim cn^{\alpha}$.

Theorem B.10   (Perron–Kreuser)   *Let $\sigma = 2$,*

$$A_\nu(n) \sim a_{0\nu} n^{k_\nu}, \qquad a_{0\nu} \neq 0,$$

K_ν *real. Let $\alpha := (K_0 - K_2)/2$.*

(i) *If $(1, K_1)$ lies above $[(0, K_0), (2, K_2)]$, then $\mathcal{L}[y(n)] = 0$ has a fundamental set $y^{(h)}(n)$ with the property*

$$\frac{y^{(h)}(n+1)}{y^{(h)}(n)} \sim c_h n^{\theta_h}, \qquad h = 1, 2, \tag{B.52}$$

$$\theta_1 = K_0 - K_1; \qquad \theta_2 = K_1 - K_2; \qquad c_1 = -\frac{a_{00}}{a_{01}}, \qquad c_2 = -\frac{a_{01}}{a_{02}}.$$

(ii) *If $(1, K_1)$ lies on $[(0, K_0), (2, K_2)]$, then $\mathcal{L}[y(n)] = 0$ has a fundamental set with the property*

$$\frac{y^{(h)}(n+1)}{y^{(h)}(n)} \sim t_h n^{\alpha}, \qquad \alpha := \frac{K_0 - K_2}{2}, \qquad h = 1, 2, \tag{B.53}$$

where t_h are the roots of

$$a_{00} + a_{01} t + a_{02} t^2,$$

provided $|t_1| \neq |t_2|$. If this is not the case, then

$$\varlimsup_{n \to \infty} \left| \frac{y(n)}{n!^{\alpha}} \right|^{1/n} = |t_1|, \tag{B.54}$$

for all nontrivial solutions of the equation.

(iii) *If* $(1, K_1)$ *lies below* $[(0, K_0), (2, K_2)]$, *then*

$$\varlimsup_{n \to \infty} \left| \frac{y(n)}{n!^{\alpha}} \right|^{1/n} = \sqrt{\left| \frac{a_{00}}{a_{02}} \right|} , . \tag{B.55}$$

for all nontrivial solutions of the equation.

Proof The proof follows by a straightforward application of the theorems of Poincaré and Perron which are given in Section B.1. For instance to show the relationship (52) for $h = 1$ make the substitution in the difference equation $z(n) = y(n)/(n!)^{\theta_1}$. Perron's theorem shows that the new equation has a solution with the property $z(n+1)/z(n) \sim c$ which implies (52) with $h = 1$. Another solution may be constructed by the method of reduction of order (see (A.29)) and elementary asymptotic estimates for this solution based on the formulas in A-IX give (52) for $h = 2$. Details of the remainder of the proof are left to the reader. ∎

B.3   The Olver growth theorems

Olver (1967b) has given several theorems which are useful in assessing the growth of the solutions of the difference equation when the coefficients are real and satisfy certain positivity conditions.

Theorem B.11   (Olver's comparison theorem)   *Let*

$$y(n) + a(n)y(n+1) + b(n)y(n+2) = 0, \tag{B.56}$$
$$Y(n) + A(n)Y(n+1) + B(n)Y(n+2) = 0, \tag{B.57}$$

where

$$b(n) \geqslant B(n) > 0; \qquad \frac{A(n)+1}{B(n)} \leqslant \frac{a(n)+1}{b(n)} \leqslant -1. \tag{B.58}$$

If

$$Y(1) - y(1) \geqslant Y(0) - y(0) \geqslant 0; \qquad y(1) \geqslant \max(y(0), 0); \tag{B.59}$$

then $y(n)$, $Y(n)$ *are nondecreasing and* $y(n) \leqslant Y(n)$.

Proof We start with the identity

$$y(n+2) - y(n+1) = -y(n+1)\left(\frac{a(n)+1}{b(n)} + 1\right) + \frac{(y(n+1) - y(n))}{b(n)},$$

and observe that both $n \geqslant 0$,   **(B.60)**

$$-\left(\frac{a(n)+1}{b(n)} + 1\right), \qquad \frac{1}{b(n)},$$

are nonnegative. From (59) it follows that $y(1)\geqslant 0$, $y(1)-y(0)\geqslant 0$. Thus from (60) with $n=0$ we have $y(2)-y(1)\geqslant 0$. By induction we establish that $y(n+1)\geqslant y(n)$, $n\geqslant 0$. Similarly from the identity obtained from (60) by capitalizing all letters we obtain the fact that $Y(n)$ is nondecreasing. Subtraction of these two yields

$$(Y(n+2)-y(n+2))-(Y(n+1)-y(n+1))$$

$$=-\left(\frac{a(n)+1}{b(n)}+1\right)(Y(n+1)-y(n+1))$$

$$-\left[\left(\frac{A(n)+1}{B(n)}\right)-\left(\frac{a(n)+1}{b(n)}\right)\right]Y(n+1)$$

$$+\frac{1}{b(n)}[(Y(n+1)-y(n+1))-(Y(n)-y(n))]$$

$$+\left(\frac{1}{B(n)}-\frac{1}{b(n)}\right)(Y(n+1)-Y(n)).$$

Assume that $n\geqslant 0$ and $Y(n+1)-y(n+1)\geqslant Y(n)-y(n)\geqslant 0$ as is the case when $n=0$. Using the facts that $Y(n)\geqslant 0$ and $Y(n+1)-Y(n)\geqslant 0$ we deduce that $(Y(n+2)-y(n+2))-(Y(n+1)-y(n+1))\geqslant 0$ and so, by induction, $Y(n)\geqslant y(n)$. ■

Theorem B.12 *Let* (56) *hold and*

$$-a(n)\geqslant b(n)+1; \qquad b(n)>0, \tag{B.61}$$

and let the solution in question of the equation, $w(n)$, satisfy

$$w(0)>0, \qquad \kappa:=\frac{w(1)}{w(0)}\geqslant 1.$$

Then $w(n)$ is nondecreasing and

$$w(0)[\min(\kappa,\lambda)]^n\leqslant w(n)\leqslant w(0)[\max(\kappa,\Lambda)]^n, \tag{B.62}$$

where λ, Λ are the largest roots of the equations

$$1+\alpha\lambda+\beta\lambda^2=0, \qquad 1+A\Lambda+B\Lambda^2=0, \tag{B.63}$$

respectively, and

$$\beta:=\sup b(n), \qquad \alpha:=-1+\beta\sup\frac{(a(n)+1)}{b(n)},$$

$$B:=\inf b(n), \qquad A:=-1+B\inf\frac{(a(n)+1)}{b(n)}.$$

Remark B and α always exist. When $\beta=\infty$ the left-hand inequality is to be omitted and when $A=-\infty$ the right-hand inequality is to be omitted.

When A, β exist we have

$$-\alpha \geqslant 1+\beta, \qquad A \geqslant -1+B, \tag{B.64}$$

$$\beta \geqslant B, \qquad \frac{\alpha}{\beta} \leqslant \frac{A}{B}, \qquad \Lambda \geqslant \frac{1}{B}, \qquad \Lambda \geqslant \lambda; \tag{B.65}$$

$$\lambda = \frac{-\alpha}{2\beta} + \sqrt{\left(\frac{\alpha^2}{4\beta^2} - \frac{1}{\beta}\right)} \geqslant \frac{\beta+1+|\beta-1|}{2\beta} = \max\left(\frac{1}{\beta}, 1\right). \tag{B.66}$$

Proof   By Theorem 12, $w(n)$ is a nondecreasing function of n. First assume $\kappa \geqslant \lambda$. Define

$$h(n) := w(0)\lambda^n.$$

Then $h(0) = w(0)$, $h(1) = w(0)\lambda$ and by (63) we get

$$h(n) + \alpha h(n+1) + \beta h(n+2) = 0. \tag{B.67}$$

We now apply Theorem 11 with (67) taking the place of (56). Because of the definitions of α and β conditions (58) are fulfilled. Further, (59) is satisfied since $\kappa \geqslant \lambda \geqslant 1$. Therefore

$$w(n) \geqslant h(n) = w(0)\lambda^n,$$

verifying the first part of the inequality (62).

Now consider the case $\kappa < \lambda$. Let

$$h(n) := w(0)\kappa^n.$$

Then $h(0) = w(0)$, $h(1) = w(1)$ and $h(n)$ satisfies

$$h(n) + \hat{\alpha} h(n+1) + \hat{\beta} h(n+2) = 0,$$

where

$$\hat{\alpha} := -\kappa\hat{\beta} - \frac{1}{\kappa}, \qquad \hat{\beta} := \max\left(\beta, \frac{1}{\kappa}\right).$$

Theorem 11 will be used in the same way as before. (59) is clearly satisfied. The remaining conditions are

$$\hat{\beta} \geqslant b(n) > 0, \qquad \frac{a(n)+1}{b(n)} \leqslant \frac{\hat{\alpha}+1}{\hat{\beta}} \leqslant -1.$$

The first inequality is clearly satisfied. Since

$$\frac{\hat{\alpha}+1}{\hat{\beta}} + 1 = (1-\kappa)\left(\hat{\beta} - \frac{1}{\kappa}\right)\Big/ \hat{\beta} \geqslant 0,$$

$$\frac{(\alpha+1)}{\beta} \geqslant \frac{(a(n)+b(n))}{b(n)},$$

it is sufficient to show

$$\frac{\hat{\alpha}+1}{\hat{\beta}} \ge \frac{\alpha+1}{\beta}.$$
(**B.68**)

If $\hat{\beta} = \kappa^{-1}$, then $\hat{\alpha} = -1 - 1/\kappa$, so (68) follows immediately from (64).

On the other hand if $\beta = \hat{\beta} > \kappa^{-1}$, then

$$\frac{(\hat{\alpha}+1)}{\hat{\beta}} - \frac{(\alpha+1)}{\beta} = \frac{-\alpha}{\beta} - \kappa - \frac{1}{\kappa\beta}$$

$$= \lambda + \frac{1}{\beta\lambda} - \left(\kappa + \frac{1}{\kappa\beta}\right)$$

$$= (\lambda - \kappa)\left(1 - \frac{1}{\beta\kappa\lambda}\right) > 0.$$

This completes the proof of the inequality on the left.

To prove the right-hand inequality, first suppose $\kappa \le \Lambda$. Defining $H(n) := w(0)\Lambda^n$ we get $H(0) = w(0)$, $H(1) = w(0)\Lambda$ and

$$H(n) + AH(n+1) + BH(n+2) = 0.$$
(**B.69**)

Theorem 11 is applicable to (56) and (69) with an obvious identification and we find $w(n) \le w(0)\Lambda^n$.

When $\kappa > \Lambda$ define $H(n) := w(0)\kappa^n$, so $H(0) = w(0)$, $H(1) = w(1)$ and we have

$$H(n) + \hat{A}H(n+1) + \hat{B}H(n+2) = 0,$$

where $\hat{B} = B$ and $\hat{A} = -B\kappa - \kappa^{-1}$. Theorem 11 can be applied to the preceding equation and (56). Conditions (59) are satisfied. The first of conditions (58) becomes $b(n) \ge B > 0$ which is true and the second

$$-\kappa - \frac{1}{\kappa B} + \frac{1}{B} \le \frac{a(n)+1}{b(n)} \le -1.$$

This will follow if we can show

$$\frac{A+1}{B} \ge -\kappa - \frac{1}{\kappa B} + \frac{1}{B},$$

or

$$B(\Lambda - \kappa) - \frac{1}{\Lambda\kappa}(\Lambda - \kappa) = (\Lambda - \kappa)\left(B - \frac{1}{\Lambda\kappa}\right) \le 0.$$

But the first term is negative and the second, by (65) and the fact that $\kappa \ge 1$, is positive, so the statement is true. This concludes the proof of the theorem. ∎

When $a(n)$, $b(n)$ have opposite signs the analysis is somewhat simpler and the comparison theorem is not needed.

Theorem B.13 *Let* (56) *hold with*

$$a(n) \geqslant 0, \qquad b(n) < 0, \qquad w(0) > 0, \qquad \kappa := \frac{w(1)}{w(0)} \geqslant 1,$$

where $w(n)$ is the solution in question.
  Then

$$w(0)[\min(\kappa, \lambda)]^n \leqslant w(n) \leqslant w(0)[\max(\kappa, \lambda)]^n, \tag{B.70}$$

where λ, Λ are the largest roots of the equations

$$1 + \alpha\lambda + \beta\lambda^2 = 0, \qquad 1 + A\Lambda + B\Lambda^2 = 0,$$

respectively, and

$$\beta := \inf b(n), \qquad \alpha := \beta \sup \frac{a(n)}{b(n)},$$

$$B := \sup b(n), \qquad A := B \inf \frac{a(n)}{b(n)}.$$

Proof To prove the left-hand inequality in (70) when $\kappa \geqslant \lambda$ assume that $w(n) \geqslant w(0)\lambda^n$ and $w(n-1) \geqslant w(0)\lambda^{n-1}$, which is true when $n = 1$. The equation provides

$$w(n+2) = \frac{a(n)}{[-b(n)]} w(n+1) + \frac{1}{[-b(n)]} w(n)$$

$$\geqslant \frac{-w(0)\lambda^n}{\beta} [\alpha\lambda + 1] = w(0)\lambda^{n+2},$$

which is the required inequality.

Now let $\kappa < \lambda$. The roots of the equation $1 + \alpha x + \beta x^2$ are λ and $1/\beta\lambda$. Since $(1/\beta\lambda) < \kappa < \lambda$ (recall, β is negative) we have

$$1 + \alpha\kappa + \beta\kappa^2 > 0.$$

Assume that $w(n) \geqslant w(0)\kappa^n$, $w(n-1) \geqslant w(0)\kappa^{n-1}$, which is true when $n = 1$. Then

$$w(n+2) \geqslant \frac{-w(0)\kappa^n}{\beta} (\alpha\kappa + 1) > w(0)\kappa^{n+2},$$

as was to be shown.
  The right-hand inequality is demonstrated in a similar way. ∎

The above theorems when used in conjunction with the asymptotic results in Section 2 can be very helpful in establishing whether or not a given solution of a difference equation is a minimal solution. In fact the theorems can sometimes be used to match given solutions to their Birkhoff series.

Example B.4 Let δ denote the differential operator

$$\delta = x^{-1}D,$$

and consider the functions

$$\mathcal{K}_n(x):\equiv y(1)(n):=(-x)^n\delta^n\left(\frac{e^{-x}}{x}\right), \qquad x>0$$

$$\text{(B.71)}$$

$$= \sqrt{\left(\frac{2}{\pi x}\right)}K_{n+(1/2)}(x),$$

so

$$\mathcal{K}_0(x)=\frac{e^{-x}}{x}, \qquad \mathcal{K}_1(x)=e^{-x}\left(\frac{1}{x}+\frac{1}{x^2}\right), \qquad \text{(B.72)}$$

etc. These functions arise for example in the computation of Coulomb and kinetic energy integrals and have been discussed by many writers (Todd *et al.* (1970) and the references given there). To maintain the drama of the analysis, assume the representation of $\mathcal{K}_n$ as a Bessel function is not known. How can we deduce its asymptotic character?
$\mathcal{K}_n(x)$ satisfies

$$y(n)+\frac{(2n+3)}{x}y(n+1)-y(n+2)=0.$$

Another linearly independent solution is

$$y^{(2)}(n):=(-1)^{n+1}\mathcal{K}_n(-x)=x^n\delta^n\left(\frac{e^x}{x}\right).$$

The equation is of the kind required for an application of the Birkhoff–Trjitzinsky theory. Using the computational techniques of Section B.2 we determine that there are Birkhoff series,

$$z^{(1)}(n)=n^n\left(\frac{2}{ex}\right)^n\left[1+\frac{c_1}{n}+\cdots\right],$$

$$z^{(2)}(n)=n^{-n-1}\left(\frac{-ex}{2}\right)^n\left[1+\frac{d_1}{n}+\cdots\right].$$

We have no information on how the series are to be connected with the functions $\{y^{(h)}(n)\}$. Theorem 13, however, will make the connection clear.
Let $y^{(1)}(n):=\mathcal{K}_n(x):=w(n)$. Then

$$\kappa := y^{(1)}(1)/y^{(1)}(0)=1+\frac{1}{x}\geqslant 1.$$

Since $a(n)=(2n+3)/x$, $b(n)=-1$, we have

$$B=\beta=-1, \qquad \alpha=3/x, \qquad A=\infty,$$

$$\lambda=\frac{3}{2x}+\sqrt{\left(\frac{a}{4x^2}+1\right)}>\kappa.$$

The theorem shows that

$$\mathcal{K}_n(x) \geqslant \frac{e^{-x}}{x}\left(1+\frac{1}{x}\right)^n.$$

This is a very weak inequality but it does the job. $\mathcal{K}_n(x)$ must have exponential growth at least so we conclude

$$\mathcal{K}_n(x) \sim C z^{(1)}(n), \qquad n \rightarrow \infty.$$

Obviously $y^{(2)}(n)$ and $y^{(1)}(n)$ have the same asymptotic expansion, a surprising fact. A solution $\mu(n)$ having the asymptotic expansion $z^{(2)}(n)$ is the minimal solution. By the way, Theorem 4.2 guarantees that the equation has a solution to which the simplified Miller algorithm converges. (To get the equation in the required form, let $y(n) = (-1)^n Y(n)$.) Any multiple of $\mu(n)$ is such a solution. ■

Appendix C Recursion formulas for hypergeometric functions

Let $\{u_j\}$ be a sequence with elements in $\mathscr{C}$ and z an indeterminate. The expression

$$u(z) := u_0 + u_1 z + u_2 z^2 + \cdots, \tag{C.1}$$

is called a *formal power series*. We write $u(z) \doteq v(z)$ to mean $u_j = v_j$, $j \geq 0$. Clearly ' $\doteq$ ' is an equivalence relation and the set $\mathscr{V}$ of all such series is a linear algebra over $\mathscr{C}$. The concepts of product (Cauchy product), addition, scalar multiplication, linear independence, linear combination, and bases of subspaces of $\mathscr{V}$ are defined in the obvious way. (In fact, if one requires $u_0 \neq 0$, $u_j \in \mathscr{R}$, $\mathscr{V}$ is a nonarchimedian ordered field, see Lightstone and Robinson (1975).)

In addition, we will consider the set $\mathscr{V}^{-1}$ of formal power series in powers of $1/z$,

$$u(z) := u_0 + u_1 z^{-1} + u_2 z^{-2} + \cdots.$$

The algebra for $\mathscr{V}^{-1}$ is constructed analogously to that for $\mathscr{V}$.

Let a_j, $1 \leq j \leq p$, b_j, $1 \leq j \leq q$, be complex parameters, $b_j \neq 0, -1, -2, \ldots$. The formal power series

$$_pF_q\!\left(\begin{matrix} a_p \\ b_q \end{matrix} \;\middle|\; z\right) := \sum_{k=0}^{\infty} \frac{(a_1)_k (a_2)_k \cdots (a_p)_k z^k}{(b_1)_k (b_2)_k \cdots (b_q)_k k!}, \tag{C.2}$$

$$(\alpha)_k := \alpha(\alpha+1) \cdots (\alpha + k - 1), \qquad (\alpha)_0 := 1,$$

is called a *generalized hypergeometric series*. Note that if any a_j is zero or a negative integer the series terminates. For the moment we shall assume that if z is assigned a value in $\mathscr{C}$, the series in (2) must either converge or terminate.

Now let c_j, f_j, $1 \leq j \leq m$, be formal power series representing functions analytic at $z = 0$ and let the functions represented by the series possess analytic continuations into a common region $\mathscr{D}$ of the complex plane.

Theorem C.1 (The permanence principle) *Let*

$$c_1 f_1 + c_2 f_2 + \cdots + c_m f_m \doteq 0. \tag{C.3}$$

Then

$$c_1 f_1 + c_2 f_2 + \cdots + c_m f_m = 0, \qquad z \in \mathcal{D}. \tag{C.4}$$

Proof See Nehari ((1952), p. 107). ∎

The main result of this section is that recurrence relations obtain among generalized hypergeometric functions some of whose parameters differ by integers. The above theorem then allows us to replace the symbol '$\doteq$' in the recurrence by '$=$' for complex values of z in a common domain (containing zero) of analyticity of the functions.

It will save much space to use an abbreviated notation concerning products involving the parameter groups a_i and b_j. I will write

$$f(a_p) := \prod_{j=1}^{p} f(a_j). \tag{C.5}$$

Whenever a parameter is subscripted by the maximum index in its parameter group it should be understood this convention is being used.

Now let p, q be integers ≥ 0 and consider the formal series

$$E(n, z) := {}_{p+3}F_q \left(\begin{matrix} -n, & n+\lambda, & a_p, & 1 \\ & b_q & \end{matrix} \middle| z \right), \tag{C.6}$$

$$K(n, z) := \frac{(b_q - 1)z^{-1}}{(n+1)(n+\lambda-1)(a_p-1)}$$
$$\times {}_{q+1}F_{p+2} \left(\begin{matrix} & 2-b_q, & 1 \\ n+2, & 2-n-\lambda, & 2-a_p \end{matrix} \middle| \frac{(-1)^{q-p}}{z} \right), \tag{C.7}$$

$$F_r(n, z) := (n+b_r)_{1-b_r}(n+\lambda)_{1-b,p+2}F_{q-1}$$
$$\times \left(\begin{matrix} 1-b_r-n, & 1-b_r+n+\lambda, & 1-b_r+a_p \\ & (1-b_r+b_q)^* \end{matrix} \middle| z \right), \qquad 1 \leq r \leq q, \tag{C.8}$$

where the * means the term corresponding to $b_j = b_r$ is to be omitted, and

$$G_r(n, z) := (n+1+a_r)_{-a_r}(n+\lambda)_{-a_r} {}_q F_{p+1}$$
$$\times \left(\begin{matrix} & 1+a_r-b_q \\ 1+a_r+n, & 1+a_r-n-\lambda, & (1+a_r-a_p)^* \end{matrix} \middle| \frac{(-1)^{q-p}}{z} \right), \qquad 1 \leq r \leq p, \tag{C.9}$$

$$G_{p+1}(n, z) := \frac{n! \, \Gamma(n+\lambda+1-b_q)(-1)^{n(q-p-1)}z^{-n}}{\Gamma(n+\lambda)\Gamma(2n+\lambda+1)\Gamma(n+\lambda+1-a_p)}$$
$$\times {}_q F_{p+1} \left(\begin{matrix} n+\lambda+1-b_q \\ 2n+\lambda+1, & n+\lambda+1-a_p \end{matrix} \middle| \frac{(-1)^{q-p}}{z} \right). \tag{C.10}$$

$$G_{p+2}(n, z) := \frac{n!\,\Gamma(n+\lambda+1-b_q)(-1)^{n(q-p-1)}z^{-n}}{\Gamma(n+\lambda)\Gamma(2n+\lambda+1)\Gamma(n+\lambda+1-a_p)}$$

$$\times {}_qF_{p+1}\left(\begin{matrix} n+\lambda+1-b_q \\ 2n+\lambda+1, \quad n+\lambda+1-a_p \end{matrix} \,\middle|\, \frac{(-1)^{q-p}}{z} \right). \qquad \text{(C.11)}$$

For these series to exist, obvious conditions must be imposed on the parameters. I will assume that in any given situation these conditions are satisfied.

Theorem C.2 *Consider the difference equation*

$$\sum_{v=0}^{t} [A_v(n) + zB_v(n)]\Phi_{n+v} \doteq \frac{(b_q-1)(n+t+\lambda)_{n+t}}{(n+t+b_q-1)(n+\lambda)_{n+t}}, \qquad \text{(C.12)}$$

$$t := \max(p+2, q), \qquad A_t := 1, \qquad B_0 = B_t := 0,$$

with

$$A_v(n) := \frac{(n+v+1)_{t-v}(2n+\lambda+t)_t(n+v+b_q-1)}{(t-v)!\,(n+\lambda+v)_{t-v}(2n+\lambda+v)_v(n+t+b_q-1)}$$

$$\times {}_{q+2}F_{q+1}\left(\begin{matrix} v-t, \quad 2n+\lambda+v, \quad n+v+b_q \\ 2n+2v+\lambda+1, \quad n+v+b_q-1 \end{matrix} \,\middle|\, 1 \right)$$

$$= \frac{(-1)^q(n+v+1)_{t-v}(2n+2v+\lambda)(2n+\lambda+t)_t(n+\lambda+1-b_q)}{v!\,(n+\lambda+v)_{t-v}(2n+\lambda+v)_{t+1-v}(n+t+b_q-1)}$$

$$\times {}_{q+2}F_{q+1}\left(\begin{matrix} -v, \quad 2n+\lambda+v, \quad n+\lambda+2-b_q \\ 2n+\lambda+t+1, \quad n+\lambda+1-b_q \end{matrix} \,\middle|\, 1 \right), \qquad \text{(C.13)}$$

and

$$B_v(n) := \frac{(n+v+1)_{t-v}(2n+\lambda+t)_t(n+v+a_p)}{(t-v-1)!\,(n+v+\lambda)_{t-v}(2n+v+\lambda+1)_{v-1}(n+t+b_q-1)}$$

$$\times {}_{p+2}F_{p+1}\left(\begin{matrix} v+1-t, \quad 2n+\lambda+v+1, \quad n+v+a_p+1 \\ 2n+2v+\lambda+1, \quad n+v+a_p \end{matrix} \,\middle|\, 1 \right)$$

$$= \frac{(-1)^p(n+v+1)_{t-v}(2n+2v+\lambda)(2n+\lambda+t)_t(n+v+\lambda+1-a_p)}{(v-1)!\,(n+\lambda+v)_{t-v}(2n+\lambda+v+1)_{t-v}(n+t+b_q-1)}$$

$$\times {}_{p+2}F_{p+1}\left(\begin{matrix} 1-v, \quad 2n+\lambda+v+1, \quad n+\lambda+2-a_p \\ 2n+\lambda+t+1, \quad n+\lambda+1-a_p \end{matrix} \,\middle|\, 1 \right). \qquad \text{(C.14)}$$

This equation is satisfied by the functions $E(n, z)$ *and, when divided by* z, *by* $K(n, z)$.

The homogeneous equation

$$\sum_{v=0}^{t} [A_v(n) + zB_v(n)]\Phi_{n+v} \doteq 0, \qquad \text{(C.15)}$$

is satisfied by the functions $F_r(n, z)$ and, when divided by z, by the functions $G_r(n, z)$.

If no a_i equals any b_j none of the above-mentioned functions satisfies a difference equation of the form specified of order smaller than t.

Proof A complete proof together with many examples is given in Luke ((1969), v. 2). ■

The following two recursion relationships for the Gaussian hypergeometric function are important cases of the above result.

The recursion relation,

$$(2n+\lambda+3)(n+\lambda-b_p+1)y(n)$$
$$+\{(2n+\lambda+2)\langle(2n+\lambda+3)(n+b_p)-(2n+\lambda+1)(n+b_p+1)\rangle$$
$$+z(2n+\lambda+1)_3\}y(n+1)+(2n+\lambda+1)(n+b_p+1)y(n+2)=0,$$

(C.16)

$p=2$, has as solutions,

$$\phi^{(h)}(n)=\frac{\Gamma(n+\lambda+1-b_h)}{\Gamma(n+b_h)}F\left(\begin{matrix}1-b_h-n, & n+\lambda+1-b_h\\ & 1+b_j-b_h\end{matrix}\middle|z\right),$$

$$j\neq h, \quad h=1,2, \quad \text{(C.17)}$$

$$\phi^{(3)}(n)=\frac{(-z)^n\Gamma(2n+\lambda)}{\Gamma(n+b_s)}F\left(\begin{matrix}1-b_1-n & 1-b_2-n\\ & 1-2n-\lambda\end{matrix}\middle|\frac{1}{z}\right),$$

(C.18)

$$\phi^{(4)}(n)=\frac{(-z)^{-n}\Gamma(n+\lambda+1-b_s)}{\Gamma(2n+\lambda+1)}F\left(\begin{matrix}n+\lambda+1-b_1, & n+\lambda+1-b_2\\ & 2n+\lambda+1\end{matrix}\middle|\frac{1}{z}\right).$$

(C.19)

(The implicit assumption is that the parameters are such that the above quantities are defined.)

Letting $p=1$, $q=2$, $z\to z/\lambda$ and $\lambda\to\infty$ gives a recursion relation

$$(1-z)(n+1)(n+2)y(n)+(n+2)$$
$$\times[(n+a+1)z-2n-b_1-b_2-1]y(n+1)$$
$$+(n+b_q+1)y(n+2)=0,$$

(C.20)

which has as solutions

$$\psi^{(h)}(n)=(n+b_h)_{1-b_h}F\left(\begin{matrix}1-b_h-n, & 1+a-b_h\\ & 1+b_j-b_h\end{matrix}\middle|z\right),$$

$$j\neq h, \quad h=1,2, \quad \text{(C.21)}$$

$$\psi^{(3)}(n)=(n+1+a)_{-a}F\left(\begin{matrix}1+a-b_1, & 1+a-b_2\\ & 1+a+n\end{matrix}\middle|\frac{1}{z}\right),$$

(C.22)

$$\psi^{(4)}(n)=\frac{n!\,\Gamma(n+a)(-z)^n}{\Gamma(n+b_s)}F\left(\begin{matrix}1-b_1-n, & 1-b_2-n\\ & 1-a-n\end{matrix}\middle|\frac{1}{z}\right).$$

(C.23)

Index of higher mathematical functions discussed

Bibliography

Abramowitz, M. and Stegun, I. A. (ed.) (1964) *Handbook of Mathematical Functions*. National Bureau of Standards Applied Mathematics Series #55, Washington, D.C.

Aczél, J. (1969) *On Applications and Theory of Functional Equations*. Academic Press, N.Y.

Adams, C. R. (1928) On the irregular cases of the linear ordinary difference equation, *Trans. Amer. Math. Soc.* **30**, 507–541.

Allasia, G. (1969/70) Su una classe di algoritmi iterativi bidimensionali, *Rend. Sem. Mat. Univ. Torino* **29**, 269–296.

Allasia, G. (1970/71) Relazioni tra una classe di algoritmi iterativi bidimensionali ed una di equazioni differenziali, *Rend. Sem. Mat. Univ. Torino* **30**, 187–207.

Allasia, G. (1971/72) Alcune generalizzazioni dell'algoritmo della media aritmetico-armonica, *Rend. Sem. Mat. Univ. Torino* **31**, 197–221.

Allasia, G. (1972) Su alcuni algoritmi iterativi bidimensionali, *Ist. Cal. Num. Univ. Torino*, Quaderno #1.

Allasia, G. and Bonardo, F. (1980) On the numerical evaluation of two infinite products, *Math. Comp.* **35**, 917–931.

Amos, D. E. (1973) Bounds on iterated coerror functions and their derivatives, *Math. Comp.* **27**, 413–427.

Amos, D. E. (1974) Computation of modified Bessel functions and their ratios, *Math. Comp.* **28**, 239–251.

Amos, D. E. and Burgmeier, J. W. (1973) Computation with three term, linear nonhomogeneous recursion relations, *SIAM Review* **15**, 335–351.

Arscott, F. M. (1975) The connection between some differential-equation eigenvalue problems and related difference-equation problems, *Proc. Fifth Manitoba Conf. on Numer. Math. and Comp.*, 211–212.

Arscott, F. M., Lacroix, R. and Shymanski, W. T. (1978) A three-term recursion and the computation of Mathieu functions, *Proc. Eighth Manitoba Conf. on Numer. Math. and Comp.*, 107–115.

Askey, R. and Gasper, G. (1971) Jacobi polynomial expansions of Jacobi polynomials with non-negative coefficients, *Proc. Camb. Phil. Soc.* **70**, 243–255.

Bardo, R. D. and Ruedenberg, K. (1971) Numerical analysis and evaluation of normalized repeated integrals of the error function and related functions, *J. Comp. Phys.* **8,** 167–174.

Bartky, W. (1938) Numerical calculation of a generalized complete elliptic integral, *Rev. Mod. Phys.* **10,** 264–269.

Batchelder, P. M. (1927) *An Introduction to Linear Difference Equations.* Cambridge, Mass.

Birkhoff, G. D. (1930) Formal theory of irregular difference equations, *Acta Math.* **54,** 205–246.

Birkhoff, G. D. and Trjitzinsky, W. J. (1932) Analytic theory of singular difference equations, *Acta Math.* **60,** 1–89.

Blair, J. M., Edwards, C. A. and Johnson, J. H. (1978) Rational Chebyshev approximations for the Bickley functions $Ki_n(x)$, *Math. Comp.* **32,** 876–886.

Blanch, G. (1946) On the computation of Mathieu functions, *J. Math. Phys.* **26,** 1–20.

Blanch, G. (1964) Numerical evaluation of continued fractions, *SIAM Rev.* **6,** 383–421.

Blanch, G. and Clemm, D. S. (1962) *Tables Relating to the Radial Mathieu Functions*, U.S. Government Printing Office, Washington, D.C.

Blanch, G. and Rhodes, I. (1955) Tables of characteristic values of Mathieu's equation for large values of the parameter, *J. Wash. Acad. Sci.* **45,** 166–196.

Boas, R. P., Jr. (1954) *Entire Functions.* Academic Press, N.Y.

Boggs, R. A. C. and Smith, Francis, J. (1971) A note on the integration of ordinary differential equations using Chebyshev series, *Comput. J.* **14,** 270–271.

Borchardt, C. W. (1888) *Gesammelte Werke.* Berlin.

Branders, M. (1976) Toepassingen van Chebyshev-veeltermen in de numerieke integratie, Thesis, Louvain.

Brezinski, C. (1977) *Accélération de la convergence en analyse numérique.* Springer Verlag, Berlin.

Bruijn, N. G. de (1961) *Asymptotic Methods in Analysis.* North-Holland Publishing Co., Amsterdam.

Bulirsch, R. (1965a) Numerical calculation of elliptic integrals and elliptic functions, *Numer. Math.* **7,** 78–90.

Bulirsch, R. (1965b) Numerical calculation of elliptic integrals and elliptic functions II, *Numer. Math.* **7,** 353–354.

Bulirsch, R. (1969a) An extension of the Bartky transformation to incomplete elliptic integrals of the third kind, *Numer. Math.* **13,** 266–284.

Bulirsch, R. (1969b) Numerical calculation of elliptic integrals and elliptic functions III, *Numer. Math.* **13,** 305–315.

Carlson, B. C. (1965) On computing elliptic integrals and functions, *J. Math. Phys.* **44,** 36–51.

Carlson, B. C. (1971) Algorithms involving arithmetic and geometric means, *Amer. Math. Monthly* **78**, 496–504.

Carlson, B. C. (1972) The logarithmic mean, *Amer. Math. Monthly* **79**, 615–618.

Carlson, B. C. (1975) Invariance of an integral average of a logarithm, *Amer. Math. Monthly* **82**, 379–382.

Carlson, B. C. (1978) Short proofs of three theorems on elliptic integrals, *SIAM J. Math. Anal.* **9**, 524–528.

Cash, J. R. (1977a) A class of iterative algorithms for the integration of stiff systems of ordinary differential equations, *J. Inst. Math. Appl.* **19**, 325–335.

Cash, J. R. (1977b) A note on the iterative solution of recurrence relations, *Numer. Math.* **27**, 165–170.

Cash, J. R. (1978) An extension of Olver's method for the numerical solution of linear recurrence relations, *Math. Comp.* **32**, 497–510.

Cash, J. R. (1980a) A note on Olver's algorithm for the solution of second-order linear difference equations, *Math. Comp.* **35**, 767–772.

Cash, J. R. (1980b) A note on the numerical solution of linear recurrence relations, *Numer. Math.* **34**, 371–386.

Cash, J. R. (1981) *Stable Recursions with Applications to Numerical Solutions of Stiff Systems.* Academic Press, N.Y.

Cash, J. R. and Miller, J. C. P. (1978) On an iterative approach to the numerical solution of difference schemes, *Comput. J.* **22**, 184–187.

Clenshaw, C. W. (1955) A note on the summation of Chebyshev series, *M.T.A.C.* **9**, 118–120.

Clenshaw, C. W. (1957) The numerical solution of linear differential equations in Chebyshev series, *Proc. Camb. Phil. Soc.* **53**, 134–149.

Clenshaw, C. W. (1962) Chebyshev series for mathematical functions, *Nat'l. Phy. Lab. Math Tables* **5**, London, H.M.S.O.

Clenshaw, C. W. and Norton, H. J. (1963) The solution of nonlinear ordinary differential equations in Chebyshev series, *Comp. J.* **6**, 88–92.

Cole, R. J. and Pescatore, C. (1979) Evaluation of $\int_0^\infty t^n \exp\left(-t^2 - t/x\right) \mathrm{d}t$, *J. Comp. Phys.* **32**, 280–287.

Corbató, F. J. and Uretsky, J. L. (1959) Generation of spherical Bessel functions in digital computers, *J. ACM* **6**, 336–375.

Curry, J. H. (1979) On the Hénon transformation, *Commun. Math. Phys.* **68**, 129–140.

Delft University group on numerical analysis (1973) On the computation of Mathieu functions, *J. Eng. Math.* **7**, 39–61.

Deprit, André (1979) Note on the summation of Legendre series, *Celest. Mech.* **20**, 319–323.

Drane, C. J., Jr. (1964) Dolph–Chebyshev excitation coefficient approximation, *IEEE Trans. Antennas Propag.* (Commun.) **AP-12**, 781–782.

Elliott, D. (1960) On the expansion of functions in ultraspherical polynomials, *J. Austral. Math. Soc.* **1**(2), 428–438.

Elliott, D. (1968) Error analysis of an algorithm for summing certain finite series, *J. Austral. Math. Soc.* **8**(2), 213–221.

Erdélyi, A., Magnus, W., Oberhettinger, F. and Tricomi, F. G. (1953) *Higher Transcendental Functions*, 3 vols., McGraw-Hill, N.Y.

Erdéyli, A., Magnus, W., Oberhettinger, F. and Tricomi, F. G. (1954) *Tables of Integral Transforms*, 2 vols., McGraw-Hill, N.Y.

Evgrafov, M. A. (1953) A new proof of a theorem of Perron, *Izv. Akad. Nauk SSSR Ser. Mat.* **17**, 77–82.

Feigenbaum, M. J. (1980) Universal behavior in non-linear systems, *Los Alamos Science*, Summer, 4–27.

Feit, S. D. (1978) Characteristic exponents and strange attractors, *Commun. Math. Phys.* **61**, 249–260.

Field, D. A. (1978) Error bounds for continued fractions, *Num. Math.* **29**, 261–267.

Field, D. A. and Jones, W. B. (1972) A priori estimates for truncation error of continued fractions $K(1/b_n)$, *Numer. Math.* **19**, 283–302.

Fields, J. L. and Wimp, J. (1963) Basic series corresponding to a class of hypergeometric polynomials, *Proc. Camb. Phil. Soc.* **59**, 599–605.

Freud, G. (1966) *Orthogonal Polynomials*. Pergamon Press, N.Y.

Gatteschi, L. (1966) Su una generalizzazione dell'algoritmo iterativo del Borchardt, *Mem. Accad. Sci. Torino*, 4a, #4.

Gatteschi, L. (1966/67) Su di una equazione funzionale generalizzante quella del coseno, *Rend. Sem. Mat. Univ. Torino* **26**, 65–86.

Gatteschi, L. (1969/70) Procedimenti iterativi per il calcolo numerico di due prodotti infiniti, *Rend. Sem. Mat. Torino* **29**, 187–201.

Gautschi, W. (1961a) Recursive computation of certain integrals, *J. ACM* **8**, 21–40.

Gautschi, W. (1961b) Recursive computation of repeated integrals of the error function, *Math. Comp.* **15**, 227–232.

Gautschi, W. (1966) Computation of successive derivatives of $f(z)/z$, *Math. Comp.* **20**, 209–214.

Gautschi, W. (1967) Computational aspects of three-term recurrence relations, *SIAM Rev.* **9**, 24–82.

Gautschi, W. (1969) An application of minimal solutions of three-term recurrences to Coulomb wave functions, *Aequationes Math.* **2**, 171–176.

Gautschi, W. (1970) Efficient computation of the complex error function, *SIAM J. Numer. Anal.* **7**, 187–198.

Gautschi, W. (1972a) Zur Numerik rekurrenter Relationen, *Computing* **9**, 107–126.

Gautschi, W. (1972b) The condition of orthogonal polynomials, *Math. Comp.* **26**, 923–924.

Gautschi, W. (1973) Numerical aspects of recurrence relations, *ARL Report* 73-0005, USAF (translation of Gautschi (1972a)).

Gautschi, W. (1975) Computational methods in special functions, in *Theory and Application of Special Functions*, ed. R. Askey. Academic Press, N.Y.

Gautschi, W. (1977) Evaluation of the repeated integrals of the coerror function, *ACM Trans. Math. Software* **3**, 240–252.

Gautschi, W. (1978) Questions of numerical condition related to polynomials, in *Recent Advances in Numerical Analysis*, eds. C. de Boer and G. H. Golub. Academic Press, N.Y.

Gautschi, W. (1979a) On generating Gaussian quadrature rules, in *Numerische Integration*, ed. G. Hammerlin. Birkhäuser Verlag, Berlin.

Gautschi, W. (1979b) Un procedimento di calcolo per le funzioni gamma incomplete, *Rend. Sem. Mat. Univers. Politecn. Torino* **37**, 1–9.

Gautschi, W. (1979c) A computational procedure for incomplete gamma functions, *A.C.M. Trans. Math. Software* **5**, 466–489.

Gautschi, W. (1979d) The condition of polynomials in power form, *Math. Comp.* **33**, 343–352.

Gautschi, W. (1981) Minimal solutions of three-term recurrence relations and orthogonal polynomials, *Math. Comp.* **36**, 547–554.

Gautschi, W. (1982) On the convergence behavior of continued fractions with real elements, *Math. Comp.* (to appear).

Gautschi, W. and Klein, B. J. (1970) Recursive computation of certain derivatives—a study of error propagation, *Comm. ACM* **13**, 7–9.

Gautschi, W. and Slavik, J. (1978) On the computation of modified Bessel function ratios, *Math. Comp.* **32**, 865–875.

Gentleman, W. M. (1969). An error analysis of Goertzel's (Watt's) method for computing Fourier coefficients, *Comput. J.* **12**, 160–165.

Goldstein, M. and Thaler, R. M. (1959) Recurrence techniques for the calculation of Bessel functions, *MTAC* **13**, 102–108.

Goldstein, S. (1927) Mathieu functions, *Trans. Camb. Phil. Soc.* **23**, 303–336.

Golub, G. H. and Welsh, J. H. (1969) Calculation of Gaussian quadrature rules, *Math. Comp.* **23**, 221–230.

Guckenheimer, J., Oster, G. and Ipaktchi, A. (1977) The dynamics of density dependent population models, *J. Math. Biology* **4**, 101–147.

Hamming, R. W. (1962) *Numerical Methods for Scientists and Engineers*. McGraw-Hill, N.Y.

Hancock, H. (1958) *Lectures on the Theory of Elliptic Functions*, Dover, N.Y.

Hardy, G. H., Littlewood, J. E. and Pólya, G. (1952) *Inequalities*. Cambridge Univeristy Press.

Hénon, M. (1976) A two-dimensional mapping with a strange attractor, *Commun. Math. Phys.* **50**, 69–77.

Henrici, P. (1977) *Applied and Computational Complex Analysis*, vol. 2, Wiley-Interscience, N.Y.

Henrici, P. (1979) Fast Fourier methods in computational complex analysis, *Siam Rev.* **21,** 481–527.

Henrici, P. and Pfluger, P. (1966) Truncation error estimates for Stieltjes fractions, *Numer. Math.* **9,** 120–138.

Hitotumatu, S. (1963) Note on the computation of Bessel functions through recurrence formula, *J. Math. Soc. Japan* **15,** 353–359.

Hofsommer, D. J. and van de Riet, R. P. (1963) On the numerical calculation of elliptic integrals of the first and second kind and the elliptic functions of Jacobi, *Numer. Math.* **5,** 291–302.

Horn, J. (1910) Über das Verhalten der Integrale linearer Differenzen- und Differentialgleichungen für grosse Werte der Veränderlichen, *J. Reine Ang. Math.* **138,** 159–191.

Horner, T. S. (1980) Recurrence relations for the coefficients in Chebyshev series solutions of ordinary differential equations, *Math. Comp.* **35,** 893–905.

Householder, A. S. (1953) *Principles of Numerical Analysis.* McGraw-Hill, N.Y.

Ince, E. L. (1926) *Ordinary Differential Equations.* Dover, N.Y.

Isaacson, E. and Keller, H. B. (1966) *Analysis of Numerical Methods.* John Wiley and Sons, N.Y.

Jones, W. B. and Snell, R. I. (1969) Truncation error bounds for continued fractions, *SIAM J. Numer. Anal.* **6,** 210–221.

Jones, W. B. and Thron, W. J. (1968) Convergence of continued fractions, *Canad. J. Math.* **20,** 1037–1055.

Jones, W. B. and Thron, W. J. (1971) A posteriori bounds for the truncation error of continued fractions, *SIAM J. Numer. Anal.* **8,** 693–705.

Kaye, J. (1955) A table of the first eleven repeated integrals of the error function, *J. Math. Phys.* **34,** 119–125.

Khovanskii, A. N. (1963) *The Application of Continued Fractions and Their Generalizations to Problems in Approximation Theory.* P. Noordhoff, Groningen.

Kirkpatrick, E. T. (1960) Tables of values of the modified Mathieu functions, *Math. Comp.* **14,** 118–129.

Kress, R. (1972) On the general Hermite cardinal interpolation, *Math. Comp.* **26,** 925–933.

Kuczma, M. (1968) *Functional Equations in a Single Variable.* Polish Scientific Publishers, Warsaw.

Lewanowicz, S. (1976) Construction of a recurrence relation of the lowest order for coefficients of the Gegenbauer series, *Appl. Math.* **15,** 345–396.

Lewanowicz, S. (1979a) Recurrence relations for coefficients of expansions in Bessel functions, Report #N-57, Institute of Computer Science, Wroclaw University.

Lewanowicz, S. (1979b) Construction of a recurrence relation for modified moments, *J. Comp. Appl. Math.* **5**, 193–206.

Lewanowicz, S. (1980) Construction of the lowest-order recurrence relation for the Jacobi coefficients, Report #N-75, Institute of Computer Science, Wroclaw University.

Lightstone, A. H. and Robinson, A. (1975) *Nonarchimedean Fields and Asymptotic Expansions.* North-Holland, Amsterdam.

Lorenz, E. N. (1963) Deterministic nonperiodic flow, *J. Atmos. Sci.* **20**, 130–141.

Lozier, D. W. (1980) Numerical solution of linear difference equations, Report NBSIR 80-1976, Math. Analysis Division, NBS.

Luke, Y. L. (1969) *The Special Functions and Their Approximations*, 2 vols., Academic Press, N.Y.

Luke, Y. L. (1975) *Mathematical Functions and Their Approximations.* Academic Press, N.Y.

Luke, Y. L., and Wimp, Jet (1963) Jacobi polynomial expansions of a generalized hypergeometric function over a semi-infinite ray, *Math. Comp.* **17**, 395–404.

McLachlan, N. W. (1947) *Theory and Application of Mathieu Functions.* Oxford University Press.

Makinouchi, S. (1966) Notes on the recurrence techniques for the calculation of Bessel functions $J_\nu(x)$ and $I_\nu(x)$, *Inform. Process. Japan* **6**, 47–58.

Marzec, C. J. and Spiegel, E. A. (1980) Ordinary differential equations with strange attractors, *SIAM J. Appl. Math.* **38**, 403–421.

Mattheij, R. M. M. (1980) Characterizations of dominant and dominated solutions of linear recursions, *Num. Math.* **35**, 421–442.

Mattheij, R. M. M. and van der Sluis, A. (1976) Error estimates for Miller's algorithm, *Numer. Math.* **26**, 61–78.

Merkes, E. P. (1966) On truncation errors for continued fraction computations, *SIAM J. Num. Anal.* **3**, 486–496.

Meschkowski, H. (1959) *Differenzengleichungen.* Vandenhoeck and Ruprecht, Göttingen.

Mesztenyi, C. and Witzgall, C. (1967) Stable evaluation of polynomials, *J. Res. N.B.S.* **71B**, 11–17.

Metropolis, N. (1965) Algorithms in unnormalized arithmetic. I. Recurrence relations, *Numer. Math.* **7**, 104–112.

Miller, J. C. P. (1952) *Bessel Functions.* Part II, *Math. Tables*, vol. 10, British Assoc. Adv. Sci., Cambridge Univ. Press.

Miller, K. S. (1968) *Linear Difference Equations.* Benjamin, N.Y.

Milne-Thomson, L. M. (1960) *The Calculus of Finite Differences.* Macmillan, London.

Mitrinović, D. S. (1964) *Elementary Inequalities.* Noordhoff, Groningen.

Mori, M. (1980) Analytic representations suitable for numerical computation of some special functions, *Numer. Math.* **35**, 163–174.

Nehari, Z. (1952) *Conformal Mapping.* McGraw-Hill, N.Y.

Newbery, A. C. R. (1974) Error analysis for polynomial evaluation, *Math. Comp.* **28**, 789–793.

Newbery, A. C. R. (1975) Polynomial evaluation schemes, *Math. Comp.* **29**, 1046–1050.

Nörlund, N. E. (1954) *Vorlesungen über Differenzenrechnung*. Chelsea, N.Y.

Norton, H. J. (1964) The iterative solution of nonlinear ordinary differential equations in Chebyshev series, *Comp. J.* **7**, 76–85.

Oliver, J. (1967) Relative error propagation in the recursive solution of linear recurrence relations, *Numer. Math.* **9**, 323–340.

Oliver, J. (1968a) The numerical solution of linear recurrence relations, *Numer. Math.* **11**, 349–360.

Oliver, J. (1968b) An extension of Olver's error estimation technique for linear recurrence relations, *Numer. Math.* **12**, 459–467.

Oliver, J. (1977) An error analysis of the modified Clenshaw method for evaluating Chebyshev and Fourier series, *JIMA* **20**, 379–391.

Oliver, J. (1979) Rounding error propagation in polynomial evaluation schemes, *J. Comp. Appl. Math.* **5**, 85–97.

Olver, F. W. J. (1964) Error analysis of Miller's recurrence algorithm, *Math. Comp.* **18**, 65–74.

Olver, F. W. J. (1967a) Numerical solution of second order linear difference equations, *J. Res. NBS* **71** (B), 111–129.

Olver, F. W. J. (1967b) Bounds for the solutions of second order linear difference equations, *J. Res. NBS* **71** (B), 161–166.

Olver, F. W. J. (1968) An extension of Miller's algorithm, *Apl. Mat.* **13**, 174–176.

Olver, F. W. J. and Sookne, D. J. (1972) Note on backward recurrence algorithms, *Math. Comp.* **26**, 941–947.

Paszkowski, S. (1975) *Zastosowania numeryczne wielomianów i szeregów Czebyszewa*. Warsaw.

Perron, O. (1954) *Vorlesungen über Differenzenrechnung*. Chelsea, N.Y.

Perron, O. (1957) *Die Lehre von dem Kettenbrüchen*, vol. 2, Teubner Verlag, Stuttgart.

Piessens, R. and Branders, M. (1976) Numerical solution of integral equations of mathematical physics using Chebyshev polynomials, *J. Comp. Phys.* **21**, 178–196.

Pincherle, S. (1894). Delle funzioni ipergeometriche e di varie questioni ad esse attinenti, *Giorn. Mat. Battaglini* **32**, 209–291.

Randels, J. B. and Reeves, R. F. (1958) Note on empirical bounds for generating Bessel functions, *Comm. ACM* **1**, 3–5.

Rayleigh, Lord (1910) The incidence of light upon a transparent sphere of dimensions comparable with the wavelength, *Proc. Roy. Soc. London Ser. A* **84**, 25–46.

Rogosinski, W. (1950) *Fourier Series.* Chelsea, N.Y.

Rotenberg, A. (1960) The calculation of toroidal harmonics, *Math. Comp.* **14,** 274–276.

Ruelle, D. (1980) *Strange Attractors.* La Recherche #108.

Ruelle, D. (1981) Small random perturbations of dynamical systems and the definition of attractors, *Commun. Math. Phys.* **82,** 137–151.

Ruelle, D. and Takens, F. (1971) On the nature of turbulence, *Commun. Math. Phys.* **20,** 167–192; **23,** 343–344.

Sadowski, W. L. and Lozier, D. W. (1972) Use of Olver's algorithm to evaluate certain definite integrals of plasma physics involving Chebyshev polynomials, *J. Comp. Phy.* **10,** 607–613.

Salzer, H. E. (1962) Quick calculation of Jacobian elliptic functions, *Comm. ACM* **5,** 399.

Salzer, H. E. (1973) A recurrence scheme for converting from one orthogonal expansion into another, *Comm. ACM* **16,** 705–707.

Salzer, H. E. (1975) Calculating Fourier coefficients for Chebyshev patterns by recurrence formulas, *Proc. IEEE,* Jan., 195–196.

Salzer, H. E. (1976) Converting interpolation series into Chebyshev series by recurrence formulas, *Math. Comp.* **30,** 295–302.

Scraton, R. E. (1972) A modification of Miller's recurrence algorithm, *BIT* **12,** 242–251.

Shapiro, H. N. (1973) A micronote on a functional equation, *Amer. Math. Monthly* **80,** 1041.

Shintani, H. (1965) Note on Miller's recurrence algorithm, *J. Sci. Hiroshima Univ.* **29** (A-I), 121–133.

Shohat, J. A. and Tamarkin, J. D. (1943) *The Problem of Moments.* American Mathematical Society, Providence.

Slater, L. J. (1960) *Confluent Hypergeometric Functions.* Cambridge University Press.

Slater, L. J. (1966) *Generalized Hypergeometric Functions.* Cambridge University Press.

Smith, F. J. (1965) An algorithm for summing orthogonal polynomial series and their derivatives with applications to curve fitting and interpolation, *Math. Comp.* **19,** 33–36.

Snell, J. (1970) The solution in Chebyshev series of systems of linear differential equations with general boundary conditions, *Comput. J.* **13,** 103–106.

Stegun, I. A. and Abramowitz, M. (1957) Generation of Bessel functions on high-speed computers, *MTAC* **11,** 255–257.

Sweezy, W. B. and Thron, W. J. (1967) Estimates of the speed of convergence of certain continued fractions, *SIAM J. Numer. Anal.* **4,** 254–270.

Szegö, G. (1959) *Orthogonal Polynomials.* American Mathematical Society Colloquium Publications, vol. 23, Providence, R.I.

Tait, R. (1967) Error analysis of recurrence equations, *Math. Comp.* **21,** 629–638.

Temme, N. M. (1975) On the numerical evaluation of the modified Bessel function of the third kind, *J. Comp. Phys.* **19.**

Thacher, H. C., Jr. (1972) Series solutions to differential equations by backward recurrence, *Proc. Int. Federation of Information Processing (IFIP) Congress 71.*

Thacher, H. C., Jr. (1979) New backward recurrences for Bessel functions, *Math. Comp.* **33,** 744–764.

Thomas, G. W., Keys, R. G. and Reynolds, A. C., Jr. (1978) The computation of leaky aquifer functions, *J. Hydrology* **36,** 173–178.

Todd, H. D., Kay, K. G. and Silverstone, H. J. (1970) Unified treatment of two-center overlap, Coulomb, and kinetic-energy integrals, *J. Chem. Phys.* **53,** 3951–3956.

Tricomi, F. G. (1965) Sull algoritmo iterativo del Borchardt e su di una sua generalizzazione, *Rend. Circ. Mat. Palermo* (2) **14,** 85–94.

Tricomi, F. G. (1966) *Lectures on the Use of Special Functions by Calculations with Electronic Computers.* University of Maryland Lecture Series #47.

Van der Cruyssen, P. (1979a) Linear difference equations and generalized continued fractions, *Computing* **22,** 269–278.

Van der Cruyssen, P. (1979b) A reformulation of Olver's algorithm for the numerical solution of second-order linear difference equations, *Num. Math.* **32,** 159–166.

Van der Cruyssen, P. (1981) A continued fraction algorithm, *Num. Math.* **37,** 149–156.

Van der Maas, G. J. (1954) A simplified calculation for Dolph–Tchebycheff arrays, *J. Appl. Phys.* **25,** 121–124.

Van der Sluis, A. (1976) Estimating the solutions of slowly varying recursions, *SIAM J. Math. Anal.* **7,** 662–695.

Wall, H. S. (1948) *Analytic Theory of Continued Fractions.* Chelsea, N.Y.

Watson, G. N. (1962) *A Treatise on the Theory of Bessel Functions.* Cambridge University Press.

Wimp, J. (1967) Recursion formulae for hypergeometric functions, *Math. Comp.* **21,** 639–646.

Wimp, J. (1969) On recursive computation, Aerospace Research Laboratories Report ARL 69-1086, Office of Aerospace Research.

Wimp, J. (1974) On the computation of Tricomi's Ψ function, *Computing* **13,** 195–203.

Wimp, J. (1981) *Sequence Transformations.* Academic Press, N.Y.

Wimp, J. (1983) Minimal recurrences and representation theorems for $_3F_2(1)$, to appear.

Wimp, J. and Luke, Y. L. (1969) An algorithm for generating sequences defined by non-homogeneous difference equations, *Rend. Circ. Mat. Palermo* (2) **18,** 251–275.

Zahar, R. V. M. (1977) A mathematical analysis of Miller's algorithm, *Numer. Math.* **27,** 427–447.

Index